Agricultural Engineering: Emerging Trends

Agricultural Engineering: Emerging Trends

Edited by Nathan Hamilton

SYRAWOOD
PUBLISHING HOUSE
New York

Published by Syrawood Publishing House,
750 Third Avenue, 9th Floor,
New York, NY 10017, USA
www.syrawoodpublishinghouse.com

Agricultural Engineering: Emerging Trends
Edited by Nathan Hamilton

International Standard Book Number: 978-1-64740-051-4 (Hardback)

Cataloging-in-Publication Data

Agricultural engineering : emerging trends / edited by Nathan Hamilton.
p. cm.
Includes bibliographical references and index.
ISBN 978-1-64740-051-4
1. Agricultural engineering. 2. Farm equipment. 3. Agricultural mechanics. I. Hamilton Nathan.
S675 .A37 2022
630--dc23

TABLE OF CONTENTS

PREFACE

Agricultural engineering is a field of engineering which studies agricultural production and processing in order to develop solutions for achieving agricultural sustainability. There are diverse aspects to agricultural engineering, chief among which are agricultural resource management, soil management and conservation, livestock production, food engineering, waste management and bioresource engineering. Planning, building, managing and supervising effective systems for irrigation, flood control and drainage is a major focus area of this field. It also involves conducting environmental impact assessments at regular intervals. This book studies, analyzes and upholds the pillars of agricultural engineering and its utmost significance in modern times. Most of the topics introduced herein cover new techniques and the applications of this field. The book is appropriate for students seeking detailed information in this area as well as for experts.

The researches compiled throughout the book are authentic and of high quality, combining several disciplines and from very diverse regions from around the world. Drawing on the contributions of many researchers from diverse countries, the book's objective is to provide the readers with the latest achievements in the area of research. This book will surely be a source of knowledge to all interested and researching the field.

In the end, I would like to express my deep sense of gratitude to all the authors for meeting the set deadlines in completing and submitting their research chapters. I would also like to thank the publisher for the support offered to us throughout the course of the book. Finally, I extend my sincere thanks to my family for being a constant source of inspiration and encouragement.

Editor

1

Soybean productivity in Rhodic Hapludox compacted by the action of furrow openers

Robson Gonçalves Trentin[1], Alcir José Modolo[2*], Thiago de Oliveira Vargas[2], José Ricardo da Rocha Campos[2], Paulo Fernando Adami[3] and Murilo Mesquita Baesso[4]

*[1]Departamento de Engenharia Mecânica, Universidade Tecnológica Federal do Paraná, Via do Conhecimento, km 01, 85501-970, Pato Branco, Paraná, Brasil. [2]Departamento de Agronomia, Universidade Tecnológica Federal do Paraná, Pato Branco, Paraná, Brasil. [3]Departamento de Agronomia, Universidade Tecnológica Federal do Paraná, Dois Vizinhos, Paraná, Brasil. [4]Faculdade de Zootecnia e Engenharia de Alimentos, Universidade de São Paulo, Pirassununga, São Paulo, Brasil. *Author for correspondence. E-mail: alcir@utfpr.edu.br*

ABSTRACT. The heavy traffic of machines in no-tillage systems causes problems as soil compaction and loss of crops productivity. The objective of this paper is to evaluate the productivity of soybeans in reference to furrow openers and the levels of soil compaction in two crops. The experiment was conducted on Rhodic Hapludox by tracing random blocks with subdivided parcels. The soil bulk density levels were laid out in the parcels (1.16, 1.20, 1.22, and 1.26 Mg m^{-3}) and the furrowers in the sub-parcels (double disc and shanks). The resistance to penetration, depth of the furrow, mobilized soil area, final plant stands, height of plants, mean number of beans by pod, 1,000 bean mass, number of pods per plant and productivity of the culture were evaluated. The resistance to penetration increased with the levels of soil compaction regardless of the farming year and up to a depth of 0.20 m. In the first crop, higher productivity with the use of the shank was observed. In the second crop, the use of the shank resulted in an increase in depth of the furrow, mobilized soil, height of the plants and final stand of the plants, but this did not indicate an increase in productivity.

Keywords: no-tillage, seeder-fertilizer, soil density.

Introduction

Over the last 40 years, there have been important technological improvements that have brought about an expressive evolution in agriculture, particularly in the amounts produced by a large segment of the culture. Among these, the culturing of soybean stands out due to its high productive capacity accompanied by good profitability. In addition, soybeans are the chief crop in Brazil.

The no-tillage system played an important role in the establishment of the production levels and became one of the chief soil management systems used in Brazil and one of the best options to reduce soil and water losses. This system consists of stirring up the soil only in the sowing furrow (Palma et al., 2010; Botta, Tolon-Becerra, Lastra-Bravo, & Tourn, 2010). It is also marked by the intense traffic of machines and equipments, which results in serious problems with soil compacting that may cause a loss in production (Moraes et al., 2016; Naderi-Boldaji & Keller, 2016; Botta et al., 2016).

In the Southwest of Paraná State, a region of family agriculture that is based on the production of milk, the land must be intensively used by bringing in animals in the winter to where soybean and corn cultures are grown in the summer in an integrated farming-livestock system.

Intensive exploitation, together with inappropriate conditions and equipment, is an aggravating element in the compaction process that results in the degradation of the soil and a consequent loss in productivity. To solve this problem, a large number of farmers have adopted the use of the shank to replace the disc at the time of sowing (Modolo et al., 2013; Trogello, Modolo, Scarsi, & Dallacort, 2013). However, the use of the bar has become common, even in areas where the levels of soil compaction are not high, which favours soil erosion and increases weeds in the sowing line (Siczek, Horn, Lipiec, Usowicz, & Łukowski, 2015), as well as increases fuel consumption (Santos, Volpato, & Tourino, 2008).

Several studies have examined the effects of soil compaction in the development of plants in several types of soil, but they are not conclusive with regard to the values of soil density and resistance to penetration that are considered limiting factors since good moisture conditions may supplement the lower volume of soil used by the roots to supply the plant with water and nutrients (Oliveira et al., 2012; Gao, Whalley, Tian, Liu, & Ren, 2016; Moraes et al., 2016).

In their studies of the effect of soil density on root growth of several plants in a Rhodic Hapludox, Reinert, Albuquerque, Reichert, Aita, and Andrada (2008) observed that direct planting resulted in an increase in soil density and that restrictions to root development occurred at levels above 1.75 Mg m^{-3}, which resulted in morphological deformities in the roots. Secco, Reinert, Reichert, and Ros (2004) determined that in Rhodic Hapludox with a 42.7% clay content, there was no reduction in the yield of soybeans. Lima, Reinert, Reichert, and Suzuki (2010) studied the productivity of soybeans and bean cultures in Rhodic Hapludox and found that resistance to the penetration of approximately 1.9 MPa is considered critical to the growth and productivity of grains.

In this context, the present paper aims to evaluate the productivity of soybeans in reference to furrow openers and levels of soil compaction over the period of two agricultural crops.

Material and methods

The experiment was conducted in a Rhodic Hapludox (Soil Survey Staff, 2014) with a high clay content (77.40% clay, 20.31% sand and 2.29% silt) whose chemical characteristics are shown in Table 1. The area is defined by the coordinates 26°16'36"S and 52°41'20" W with an average altitude of 750 m and a maximum slope of 3%.

Treatments combined double disc split furrowers and shank in a direct planting sowing-fertilizer, and four levels of soil compaction were obtained using a tractor moving over the entire parcel so that the tires could compress parallel areas. The number of passages of the tractor varied in accordance with the treatment, as follows: Level 0 – no additional compaction, Level 1 – additional compaction using two passages by the tractor, Level 2 – additional compaction using four passages of the tractor, and Level 3 – additional compaction using six passages of the tractor, corresponding to soil densities of 1.16, 1.20, 1.22, and 1.26 Mg m^{-3}, respectively.

A New Holland® tractor model TL75E 4 x 2 TDA (Front Auxiliary Traction) with a maximum allowed weight (4,630 kg) and standard front tires 12.4 x 24, rear tires 18.4 x 30, and a mounted Jacto® sprayer (250 kg) with 600 L of water, which made a total mass of 5,480 kg, was used. The compaction was conducted in November 2013 (2013/2014 crop) immediately after the rainy season with a soil humidity of approximately 38.1%. This procedure was repeated in October 2014 (2014/2015 crop) with the same experimental equipment and a soil moisture of approximately 37.5%.

The climate is a humid subtropical type Cfa according to Köppen's classification. During the culture cycle from 2013/2014, the crop monthly precipitation was over 150 mm in February 2014 and 329 mm in March 2014 with average temperatures between 21 and 27°C. In 2015, the crop precipitation varied from 117 mm in October 2014 to 279 mm in January 2015 with average temperatures between 21 and 25°C (Figure 1).

Table 1. Chemical features of the soil in layers 0.0-0.10 and 0.10-0.20 m deep before the start of the experiment.

Prof.	pH	M.O	P	H+Al	K	Ca	Mg	SB	CTC	V
(m)		$CaCl_2$		g dm^{-3}	mg dm^{-3}		$cmol_c$ dm^{-3}		%	
0.0-0.10	4.85	52.27	13.08	5.30	0.21	4.33	1.52	6.06	11.36	53.26
0.10-0.20	4.85	45.91	8.09	5.28	0.16	4.54	1.27	5.86	11.14	52.55

MO extracted by humid digestion; P and K extracted by Mehlich – 1, pH in $CaCl_2$; Ca, Mg and Al exchangeably extracted in KCL 1 mol L^{-1}.

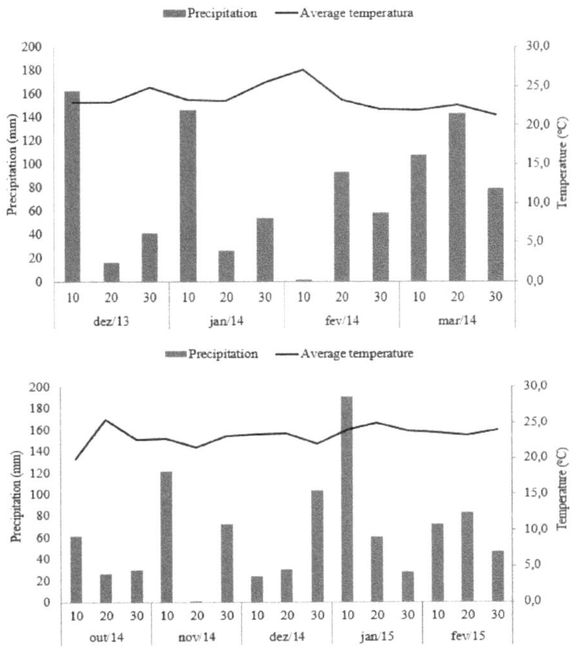

Figure 1. Meteorological data observed during the period of the experiment.
Source: Agronomy Institute of Paraná.

Random blocks encompassing eight treatments and four repetitions were utilized in subdivided parcels with the chief parcels formed by four levels of soil density and the sub-parcels by two furrow openers. The area was divided into four random blocks making up a total of 32 experimental units, each with an area of 75 m^2 (3.75 x 20 m) and with 10-m spacing between the blocks, which was used for the manoeuvring and stabilization of the velocity of the tractor.

The sowing of the soybeans took place on December 12, 2013 (2013/2014 crop), and October 22, 2014 (2014/2015 crop), under the black oats straw (*Avena strigosa*). The Don Mario 5.8i (BMX Apolo) variety was used with a density of 13.5 seeds per linear metre and with a final stand of approximately 265,000 plants ha^{-1}. The average depth of the sowing was 0.05 m. In the moment of sowing, 390 kg ha^{-1} of 02-20-18 compound was used.

A seeder-fertilizer, Vence Tudo brand, was used with a mechanic seed dispenser that contained five sowing rows with inter rows spaced 0.45 m apart. The average sowing speed was 5.0 km h^{-1}.

The furrowers used for laying the fertilizer were shanks with a width point (cone) of 17.76 mm, an angle of attack of 20° and a height of 0.45 m, or a split double disc type furrower with a diameter of 381 mm (15"), according to the treatment.

The mechanical resistance of the soil to penetration was obtained using a Falker penetrometer model PLG1020 with a cone of 1.0 cm^2 and 10 randomly distributed measuring points at each experimental unit. The values of resistance to penetration were obtained for depths between 0 and 0.4 m.

The transversal area of the mobilized soil was determined by a wooden profilometer with vertical rulers graded in centimetres and placed every 0.02 m in the transversal direction with regard to the sowing line. At each experimental unit, the outlines of the natural surface and the final surface of the soil were found. A collection was made along two sowing lines with two repetitions for each line.

The depth of the fertilizer furrow was obtained considering the highest difference between the profiles of the original and internal surface of the soil in the sowing furrow (Araújo, Casão Júnior, Ralisch, & Siqueira, 1999). The evaluation of the final stand of the plants was determined in the physiological maturation of the culture and by counting the plants at 5 m in each of the three central sowing lines in each experimental unit. The result was extrapolated to the number of plants per hectare.

The final height of the plants was measured by the distance between the neck of the plant and the apex of the main stem. The total number of pods per plant was determined by counting all of the pods inserted in the plant including those deemed empty. The mean number of grains per pod was obtained by the relation between the number of pods per plant and the number of grains per plant. The mass of 1,000 grains was defined on the basis of the total mass of the grains per plant and the number of grains per plant.

The harvest was done manually and the threshing by a stationary combine harvester. To calculate the yield, the humidity of the grains was corrected to 13%. The productivity in kg ha^{-1} was calculated through the total mass of grains produced per parcel.

The data obtained were tabulated and underwent a variance analysis. When significant differences occurred ($p \leq 0.05$), the furrower's means were compared to the Tukey test ($p \leq 0.05$), whereas for the soil compaction factor, which was represented by the variable soil density, a polynomial regression analysis was adopted. Models were selected by the criteria of higher R^2 and significance ($p \leq 0.05$) of the parameters of the equation. All of the data were submitted to the Cocharan test ($p \leq 0.05$) to verify the homogeneity of the variances. For non-homogeneous variables, the normality of the data was verified through the Lilliefors test ($p \leq 0.05$), and the variables that did not comply with the homogeneity and normality assumptions were transformed into the square root.

For the analysis of the data, SAEG 9.1 software was used.

Results and discussion

The highest values of resistance to penetration were observed in the treatment with the highest number of passages and up to the depth of approximately 0.20 m, followed by the treatments with four and two passages, respectively. Between the depths of 0.20 and 0.40 m, the values of resistance to penetration were not influenced by the tested density levels, which indicates that the traffic imposed by the tractor increased resistance to penetration in the superficial layers only up to a depth of 0.20 m (Figure 2). This behaviour with higher values of resistance to penetration in the superficial layers is characteristic of soils under no-tillage system, whereas the cumulative effect of the tensions generated by the traffic of machines and equipment is dissipated in the superficial layers (Beutler, Centurion, Centurion, & Silva, 2006; Bonini, Gabriel Filho, Secco, Souza, & Tavares, 2008; Gao et al., 2016; Botta et al., 2016).

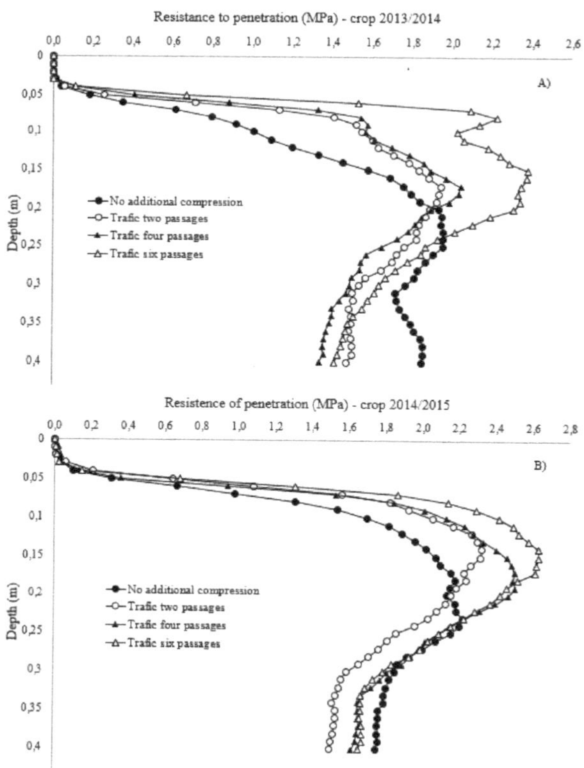

Figure 2. Resistance of the soil to penetration (MPa) in reference to the number of passages of the tractor. A) 2013/2014 crop; B) 2014/2015 crop.

In the 2013/2014 crop, the values of resistance to penetration reached 2.1 MPa when the soil presented 37.7% moisture; in the 2014/2015 crop, resistance reached 2.6 MPa with 36.7% soil moisture. Similar results were obtained by Bonini et al. (2008) in Hapludox under four levels of soil compaction with densities varying between 1.23 and 1.35 Mg m^{-3} and with observed values of resistance of the soil to penetration in the 0.10 to 0.20 m layer of 2.50 and 2.21 MPa, respectively.

Several studies determined the critical values of resistance to penetration with results varying between 1.5 and 4.0 MPa, which are influenced by the type of soil and vegetal species; however, the value of 2.0 MPa is considered to prevent root growth (Botta et al., 2010; Valicheski, Grossklaus, Stürmer, Tramontin, & Baade, 2012). In the present paper, it was observed in the first year of the experiment that only the six passages with the traffic presented values of resistance to penetration higher than 2.0 MPa, unlike the results obtained in the 2014/2015 crop where only areas without additional compaction showed results below that value in the layers up to a depth of 0.2 m. This was probably influenced by the lesser amount of humidity presented at the time of the test and by the cumulative effect of the compacting levels imposed in two consecutive years in spite of the similarity in observed values of soil density in the two years of the experiment.

In the first year of the experiment, the depth of the furrow was not influenced by the density of the soil and by the furrow openers with a mean value of 5.9 cm. However, in the second year, the shank resulted in a furrow depth that was 79% higher than the double disc (Table 2). This result verifies the higher ability of the shank to burst through the compacted layer, thus positioning the fertilizer below the seed. The positioning becomes important since in some cultures, direct contact between the seed and the fertilizer may harm germination and the development of the seedlings as observed by Souza, Farinelli, and Rosolem (2007) when evaluating the depths of fertilizer deposition in the culturing of cotton.

Table 2. Depth of the fertilizer furrow (m) and mobilized soil area (m^2) with reference to the furrow openers.

Furrows	Depth of the fertilizer furrow (m)	Mobilized soil area (m^2)
	Crop 2014/2015	
Disc	0.048 b	0.0031 b
Shank	0.086 a	0.0064 a

Means followed by different lower case letters differ ($p \leq 0.05$) by the Tukey test.

The mobilized soil area in the 2013/2014 crop did not show differences between the treatments with a mean value of 0.0051 m^2. In the 2014/2015 crop, the mobilized area showed a linear relation with soil density (Figure 3).

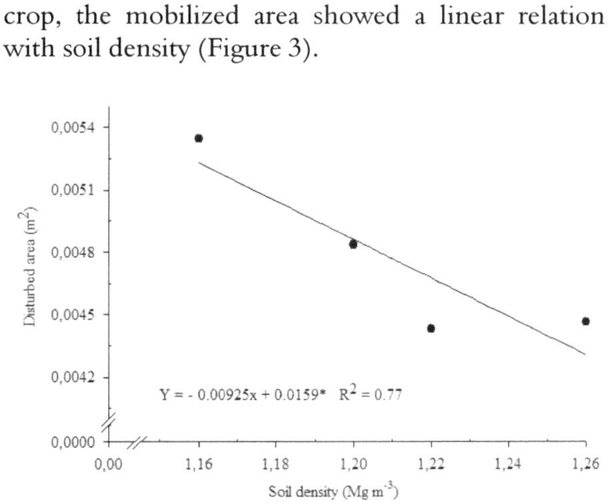

Figure 3. Mobilized soil area (m^2) with reference to soil density in the 2014/2015 crop.

In the second year of the experiment, the shank mobilized twice the soil area in comparison with the double disc (Table 2), which could be explained by the deeper furrow obtained with the shank and the lower soil humidity, which directly influences the cohesive force among the particles and keeps the aggregates more stable and thus brings a greater volume of mobilized material. These results corroborate those obtained by Levien, Furlani, Gamero, Conte, and Cavichioli (2011), who observed a 52% greater soil mobilization when utilizing the shank in comparison with the double disc on a Hapludox in the sowing of a corn culture.

In a study conducted by Mion, Benez, Viliotti, Moreira, and Salvador (2009) on a Hapludox, no difference was observed between the shank and the double disc for the mobilized area on a soil with an average density of 1.42 Mg m^{-3} and in the layer up to 0.15 m and managed through direct planting. The homogeneity in the mobilized soil area observed in the above-mentioned study may be related to the low clay content in the soil (6 g kg^{-1}). In this condition, the lack of a more expressive amount of the colloidal fraction renders the formation of more stable aggregates more difficult as it reduces the cohesion among the soil particles, which facilitates its pulverization and reduces the mobilized volume in the line of planting.

Trogello et al. (2013) did not find a significant difference in the mobilized area of the soil between the same furrowing mechanisms when they studied the speed of operation and cover managements on the same soil and utilized the same sowing as in the present study.

However, Modolo et al. (2013), when studying the action of furrowing mechanisms under different intensities of grazing, obtained a larger area of mobilized soil by utilizing the shank in comparison with the double disc, but did not observe any influence of the grazing intensities on the mobilized soil area. Mion and Benez (2008) evaluated the effects of different furrowing tools on Hapludox and also found a larger area of mobilized soil when utilizing the shank.

In the 2013/2014 crop, a greater stand of plants (260,000 plants ha^{-1}) was observed in the highest soil density (1.26 Mg m^{-3}) when the double disc was utilized (Figure 4). The smallest final stand presented by the smallest soil density may be related to the greater sowing depth obtained at this density (5.1 cm), which makes the still fragile seedling require more vigour to emerge at the surface. However, when the shank was utilized, no influence of the soil density on the final plant stand was noted; the average value was 233,888 plants ha^{-1}.

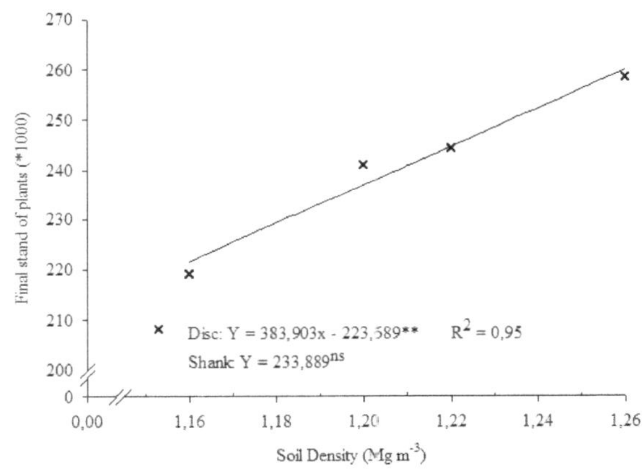

Figure 4. The final stand of soy plants (plants ha^{-1}) in the 2013/2014 crop in reference to the soil density and furrow openers.

In the 2014/2015 crop, the final plant stand was influenced only by the furrowers and there was no effect of soil density on this variable. In Table 3, it is possible to observe that when the shank was utilized, the final stand was 20% greater than that shown by the double disc because the bar presented a greater soil mobilization, thus improving the burying of the seeds and providing better germinating conditions.

Table 3. The final stand of soy plants (plants ha^{-1}) and final height of plants (m) in relation to furrow openers.

Furrows	Final stand of plants (*1000)	Final height of plants (m)
	Corn 2014/2015	
Disc	205.5 b	0.72 b
Shank	247.5 a	0.75 a

Means followed by different lower case letters in a column differ ($p \leq 0.05$) by the Tukey test.

Studies in Rhodic Hapludox comparing the effects of the furrowing systems by bar and disc are conflicting. Trogello et al. (2013) did not find a difference in the final stand for the two mechanisms. However, Arf, Afonso, Romanini Junior, Silva, and Buzetti (2008) and Kaneko et al. (2010) observed two consecutive years where the shank mechanism significantly increased this variable in relation to the double disc. The fact that the above-mentioned authors found conflicting results may not be related to the tools in question, but rather, to the characteristics of the soils utilized in the experiments. In spite of the fact that the soils presented the same classification, they may present considerable differences regarding their physical features and management. Specifically, regarding the Oxisol class, the clay content may range from 20 to 100% without a change in the classification, which renders extrapolation and comparison of the results more difficult.

In the 2013/2014 crop, no influence of the furrow opener mechanisms and soil density on the final height of the plants was observed. The mean height observed was 77.83 cm. In the second year of the experiment, the furrowing bar provided greater aerial development (Table 3), which for its part, may be related to a greater mobilization and rupture of compacted layers caused by the shank in comparison with the double disc (Table 2); this allowed the roots to more efficiently explore the volume of the soil (Giarola, Brachtvogel, Fontaniva, Pereira, & Fioreze, 2009). This result was not observed in the 2013/2014 crop because the mobilized area did not differ between the furrowing tools and thus afforded a similar condition for the growth of the roots.

Corroborating the results observed in the present study, Giarola et al. (2009) obtained the same plant height when evaluating two levels of compaction of a Rhodic Hapludox with densities of 1.38 and 1.45 Mg m^{-3} (at a depth of 0.0 to 0.10 m). However, a study by Oliveira et al. (2012) working on irrigated Rhodic Hapludox and under compacting levels with soil densities ranging between 1.20 and 1.36 Mg m^{-3} in the 0.0 to 0.10 layer, obtained smaller plant heights with an increase in soil compaction. A similar result was also observed by Beutler et al. (2006) in Rhodic Hapludox under compacting levels, and soil densities ranged from 1.31 to 1.71 Mg m^{-3} (at a depth of 0.08 to 0.11 m). It is important to highlight that despite presenting the same classification, the soils used in the three studies mentioned above present differences that may strongly influence vegetal development. Giarola et al. (2009) worked with a soil similar to the present study, which showed a clay content of much higher than 65% (very argillaceous). Oliveira et al. (2012) worked with an argillaceous soil, that is, a clay content between 35 and 65%. However, Beutler et al. (2006) used a medium texture soil in their experiment (a clay content between 15 and 35%). These differences with regard to the content of clay have a strong influence on the hydraulic conductivity of the soil, which may influence the absorption of water and consequently its development (Moraes et al., 2016; Nadei-Boldaji & Keller, 2016).

The mean number of grains per pod was not influenced by the variation in the density of the soil or by the furrowers in the two evaluated crops with average values of 1.9 and 2.3 grains per pod in the 2013/2014 and 2014/2015 crops, respectively. The lack of response for this variable is probably because the number of grains per pod is typically a genetic characteristic and hence did not influence the treatments (Centurion et al., 2006).

For the 2013/2024 crop, the use of the bar provided an increase in the number of pods per plant with an increase in the soil density (Figure 5). However, when utilizing the double disc, the number of pods per plant does not depend on the soil density: an average value of 42.02 pods per plant was observed.

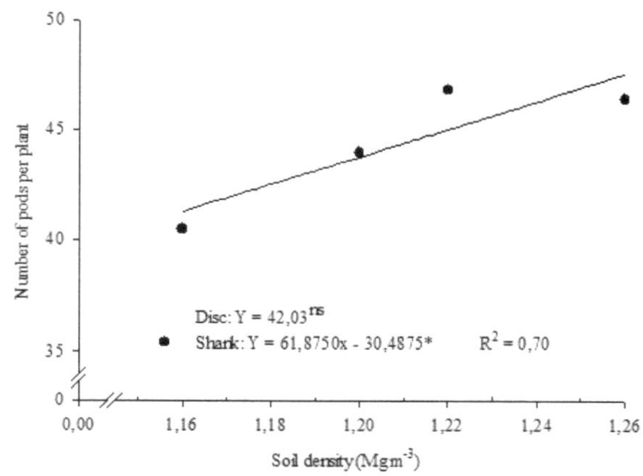

Figure 5. Means number of pods per plant in the 2013/2014 crop in reference to the soil density and furrow openers.

In the 2014/2015 crop, only the furrowing tools had a significant influence (Table 4). The number of pods per plant obtained by the use of the double disc was 15% greater than that presented by the shanks. This result is related to the smaller final stands of the plant obtained by this furrowing tool (Table 3).

Several studies indicate that the increase in the density of the plants reduces the number of pods per plant in the soybean culture. This behaviour occurs because in the largest sowing densities, there is greater competition for light, water and nutrients and less availability of photo-assimilated elements, which reduces the number of pods (Mauad, Silva, Almeida Neto, & Abreu, 2010).

Table 4. Mean number of pods per plant and soy productivity (t ha^{-1}) in reference to the furrow openers.

Furrows	Number of pods per plant	Productivity (t ha^{-1})
	Crop 2014/2015	Crop 2013/2014
Disc	63.8 a	2.10 b
Shanks	55.3 b	2.33 a

Means followed by different lower case letters in a column significantly differ ($p \leq 0.05$) by the Tukey test.

The thousand grain mass was not influenced by the variables of soil density and furrowers in the two agricultural crops. The average observed values were 125.67 and 120.05 in the first and second crops, respectively. Centurion et al. (2006) and Arf et al. (2008) observed that both the number of grains per pod and the thousand grain mass are more related to the genetic characteristics of the variety.

In the two evaluated crops, no influence of the soil density on the grain yield was observed. There was a significant difference only with regard to the furrowing mechanism, with the shank producing approximately 11% more than the double disc in the experiment conducted in the 2013/2014 crop. Taking into consideration that the components of the yield (number of grains per pod and thousand grain mass) did not differ significantly between the furrows, this result was only influenced by the variable number of pods per plant, which on average was 6% greater when using the shank. The average yield observed in the second year of the experiment was 3.57 t ha^{-1}.

The good water supply during the crops (Figure 1) may also have contributed to the lack of response of the tested density levels on productivity. Several authors indicate that soil compacting reduces the root growth of soybean plants. However, it is possible that the reduction of the volume of soil explored by the roots may not be sufficient to restrict the supply of water and nutrients to the aerial parts of the plants, thus maintaining the production of the culture (Centurion et al., 2006; Giarola et al., 2009; Oliveira et al., 2012).

Soy beans are sensitive to temperature and the photo-period, and premature flowering may occur as a result of higher temperatures and fewer sunlight hours. Thus, late sowing may reduce the productivity both of the varieties with a premature cycle and those with a later cycle (Botta et al., 2010; Amorim, Hamawaki, Sousa, Lana, & Hamawaki, 2011). This situation may have generated the greater productivity observed in the 2014/2015 crop in comparison with the 2013/2014 crop.

Working with Rhodic Hapludox under the compacting levels, Beutler et al. (2006) and Oliveira et al. (2012) obtained a reduction in productivity with an increase in the compacting level for soil densities ranging between 1.31 to 1.71 Mg m^{-3} and 1.20 to 1.36 Mg m^{-3}, respectively. However, Secco, Reinert, Reichert, and Silva (2009), working with Rhodic Hapludox, did not verify a significant difference between the compacting levels evaluated with soil densities ranging between 1.38 and 155 Mg m^{-3}.

Conclusion

The application of traffic on the soil increased the resistance to penetration up to a depth of approximately 0.20 m with higher values in the treatment with a greater number of passages.

The use of as hank proved advantageous only in the 2013/2014 crop because it provided greater productivity in comparison with the double disc. In the 2014/2015 crop, the use of the shank resulted in an increase in the depth of the furrow, the mobilized soil area, height of the plants and their final stand, but this was not reflected in productivity gains.

Acknowledgements

The authors would like to thank the CNPq (Research Productivity – PQ) and the Araucaria Foundation for providing financial support for developing this study.

References

Amorim, F. A., Hamawaki, O. T., Sousa, L. B., Lana, R. M. Q., & Hamawaki, C. D. L. (2011). Época de semeadura no potencial produtivo de soja em Uberlândia - MG. *Semina: Ciências Agrárias*, *32*(1), 1793-1802. doi:10.5433/1679-0359.2011v32Suplp1793

Araújo, A. G., Casão Júnior, R., Ralisch, R., & Siqueira, R. (1999). Mobilização de solo e emergência de plantas na semeadura direta de soja (*Glycine max L.*) e milho (*Zea mays L.*) em solos argilosos. *Revista Engenharia Agrícola*, *19*(2), 226-237.

Arf, O., Afonso, R. J., Romanini Junior, A., Silva, M. A. da., & Buzetti, S. (2008). Mecanismos de abertura do sulco e adubação nitrogenada no cultivo do feijoeiro em sistema plantio direto. *Bragantia*, *67*(2), 499-506. doi:10.1590/S0006-87052008000200026

Beutler, A. N., Centurion, J. F., Centurion, M. A. P. C., & Silva, A. P. (2006). Efeito da compactação na produtividade de cultivares de soja em Latossolo vermelho. *Revista Brasileira de Ciência do Solo*, *30*(5), 787-7,94. doi:10.1590/S0100-06832006000500004

Bonini, A. K., Gabriel Filho, G., Secco, D., Souza, R. F., & Tavares, C. (2008). Atributos físicos e requerimento de potência de uma semeadora-adubadora em um

Latossolo sob estados de compactação. *Revista Engenharia Agrícola, 28*(1), 136-144. doi:10.1590/S0100-69162008000100014

Botta, G. F., Tolon-Becerra, A., Lastra-Bravo, X., & Tourn, M. (2010). Tillage and traffic effects (planters and tractors) on soil compaction and soybean (*Glycine max* L.) yields in Argentinean pampas. *Soil & Tillage Research, 110*(1), 167-174. doi:10.1016/j.still.2010.07.001

Botta, G. F., Tolón-Becerra, A., Rivero. D., Laureda, D., Ramírez-Roman, M., Lastra-Bravo, X., & Martiren, V. (2016). Compactión produced by combine harvest traffic: Effect on soil and soybean (*Glycine max* L.) yields under direct sowing in Argentinean Pampas. *European Journal of Agronomy, 74*, 155-163. doi:10.1016/j.eja.2015.12.011

Centurion, J. F., Centurion, M. A. P. C., Beutler, A. N., Rossini, L. A., Freddi, O. S., & Souza Neto, E. L. (2006). Compactação do solo no desenvolvimento e na produção de cultivares de soja. *Científica, 34*(2), 203-209. doi:10.15361/1984-5529.2006v34n2p203+-+209

Gao, W., Whalley, W. R., Tian, Z., Liu, J., & Ren, T. (2016). A simple model to predict soil penetrometer resistance as a function of density, drying and depth in the field. *Soil Tillage & Research, 155*, 190-198. doi:10.1016/j.still.2015.08.004

Giarola, N. F. B., Brachtvogel, E. L., Fontaniva, S., Pereira, R. A., & Fioreze, S. L. (2009). Cultivares de soja sob plantio direto em Latossolo Vermelho compactado. *Acta Scientiarum. Agronomy, 31*(4), 641-646. doi:10.4025/actasciagron.v31i4.851

Kaneko, F. H., Arf, O., Gitti, D. C., Arf, M. V., Ferreira, J. P., & Buzetti, S. (2010). Mecanismos de abertura de sulcos, inoculação e adubação nitrogenada em feijoeiro em sistema plantio direto. *Bragantia, 69*(1), 125-133. doi:10.1590/S0006-87052010000100017

Levien, R., Furlani, C. E. A., Gamero, C. A., Conte, O., & Cavichioli, F. A. (2011). Semeadura direta de milho com dois tipos de sulcadores de adubo, em nível e no sentido do declive do terreno. *Ciência Rural, 41*(6), 1003-1010. doi:10.1590/S0103-84782011000600014

Lima, C. L. R., Reinert, D. J., Reichert, J. M., & Suzuki, L. E. A. S. (2010). Produtividade de culturas e resistência à penetração de Argissolo Vermelho sob diferentes manejos. *Pesquisa Agropecuária Brasileira, 45*(1), 89-98. doi: 10.1590/S0100-204X2010000100012

Mauad, M., Silva, T. L. B., Almeida Neto, A. I., & Abreu, V. G. (2010). Influência da densidade de semeadura sobre características agronômicas na cultura da soja. *Revista Agrarian, 3*(9), 175-181.

Mion, R. L., & Benez, S. H. (2008). Esforços em ferramentas rompedoras de solo de semeadoras de plantio direto. *Ciência e Agrotecnologia, 32*(5), 1594-1600. doi:10.1590/S1413-70542008000500036

Mion, R. L., Benez, S. H., Viliotti, C. A., Moreira, J. B., & Salvador, N. (2009). Análise tridimensional de esforços em elementos rompedores de semeadoras de plantio direto. *Ciência Rural, 39*(5), 1414-1419. doi:10.1590/S0103-84782009005000067

Modolo, A. J., Franchin, M. F., Trogello, E., Adami, P. F., Scarsi, M., & Carnieletto, R. (2013). Semeadura de milho com dois mecanismos sulcadores sob diferentes intensidades de pastejo. *Engenharia Agrícola, 33*(6), 1200-1209. doi:10.1590/S0100-69162013000600013

Moraes, T. M., Debiasi, H., Carlesso, R., Franchini, J. C., Silva, V. R., & Luz, F. B. (2016). Soil physical quality on tillage and cropping systems after two decades in the subtropical region of Brazil. *Soil Tillage & Research, 155*, 351-362. doi:10.1016/j.still.2015.07.015

Naderi-Boldaji, M., & Keller, T. (2016). Degree of soil compactness is highly correlated with soil physical quality index S. *Soil Tillage & Research, 159*, 41-46. doi:10.101/j.still.2016.01.010

Oliveira, P. R., Centurion, J. F., Centurion, M. A. P. C., Franco, H. B. J., Pereira, F. S., Júnior L. S. B., & Rossetti, K. V. (2012). Qualidade física de um Latossolo Vermelho cultivado com soja submetido a níveis de compactação e de irrigação. *Revista Brasileira de Ciência do Solo, 36*(2), 587-597. doi:10.1590/S0100-06832012000200028

Palma, M. A. Z., Volpato, C. E. S., Barbosa, J. A., Spagnolo, R. T., Barros, M. M., & Vilas Boas, L. A. (2010). Efeito da profundidade de trabalho das hastes sulcadoras de uma semeadora-adubadora na patinagem, na força de tração e no consumo de combustível de um trator agrícola. *Ciência e Agrotecnologia, 34*(5), 1320-1326. doi:10.1590/S1413-70542010000500034

Reinert, D. J., Albuquerque, J. A., Reichert, J. M., Aita, C., & Andrada, M. M. C. (2008). Limites críticos de densidade do solo para o crescimento de raízes de plantas de cobertura em Argissolo Vermelho. *Revista Brasileira de Ciência do Solo, 32*(5), 1805-1816. doi:10.1590/S0100-06832008000500002

Santos, A. P., Volpato, C. E. S., & Tourino, M. C. C. (2008). Desempenho de três semeadoras-adubadoras de plantio direto para a cultura do milho. *Ciência e Agrotecnologia, 32*(2), 540-546. doi:10.1590/S1413-70542008000200030.

Secco, D., Reinert, D. J., Reichert, J. M., & Ros, C. O. (2004). Produtividade de soja e propriedades físicas de um Latossolo submetido a sistemas de manejo e compactação. *Revista Brasileira de Ciência do Solo, 28*(5), 797-804. doi:10.1590/S0100-06832004000500001

Secco, D., Reinert, D. J., Reichert, J. M., & Silva, V. R. (2009). Atributos físicos e rendimento de grãos de trigo, soja e milho em dois Latossolos compactados e escarificados. *Ciência Rural, 39*(1), 58-64. doi:10.1590/S0103-84782009000100010

Siczek, A., Horn, R., Lipiec, J., Usowicz, B., & Łukowski, M. (2015). Effects of soil deformation and surface mulching on soil physical properties and soybean response related to weather conditions. *Soil & Tillage Research, 153*, 175-184. doi:10.1016/j.still.2015.06.006

Souza, F. S., Farinelli, R., & Rosolem, C. A. (2007). Desenvolvimento radicular do algodoeiro em resposta à localização do fertilizante. *Revista Brasileira de Ciência do Solo*, *31*(2), 387-392. doi:10.1590/S0100-06832007000200021

Soil Survey Staff. (2014). *Keys to soil taxonomy* (12th ed.). Washington, D.C.: USDA-Natural Resources Conservation Service.

Trogello, E., Modolo, A. J., Scarsi, M., & Dallacort, R. (2013). Manejos de cobertura, mecanismos sulcadores e velocidades de operação sobre a semeadura direta da cultura do milho. *Bragantia*, *72*(1), 101-109.

Valicheski, R. R., Grossklaus, F., Stürmer, S. L. K., Tramontin, A. L., & Baade, E. S. A. S. (2012). Desenvolvimento de plantas e cobertura e produtividade da soja conforme atributos físicos em solo compactado. *Revista Brasileira de Engenharia Agrícola e Ambiental*, *16*(9), 969-977. doi:10.1590/S1415-43662012000900007

2

Mathematical modeling of the drying process of corn ears

Paulo Cesar Corrêa, Fernando Mendes Botelho*, Gabriel Henrique Horta Oliveira, André Luis Duarte Goneli, Osvaldo Resende and Sílvia de Carvalho Campos

*Departamento de Engenharia Agrícola, Centro de Ciências Agrárias, Universidade Federal de Viçosa, Cx. Postal 270, 36571-000, Viçosa, Minas Gerais, Brazil. *Author for correspondence. E-mail: fernando_eaa@yahoo.com.br*

ABSTRACT. The objective of this work was to study and model the drying process of corn ears at different air temperatures. Thermodynamic properties associated with the drying process of this product were also determined. Corn ears with initial moisture content of 0.45 dry basis ($kg_w\ kg_{dm}^{-1}$) were dried until they reached a final moisture content of 0.12 ($kg_w\ kg_{dm}^{-1}$) at temperatures of 45, 55 and 65°C. Traditional models used to describe the drying process of several agricultural products were employed to fit the observed data of the drying process of corn ears. The effective diffusion coefficient (D_{ef}) was determined by means of an analytical solution of Fick's second law. It was concluded that the Logarithmic model was the one that best fit the observed data representing the drying process. D_{ef} values increased with temperature increases, ranging from 5.490×10^{-10} to 1.163×10^{-9} $m^2\ s^{-1}$. Based on the dependence of the drying constant of the Logarithmic model with temperature, thermodynamic properties were determined, concluding that the drying kinetics variation is dependent on the energy contributions of the surrounding environment.

Keywords: *Zea mays*, mathematical models, moisture content.

Introduction

Corn is one of the most important cultivated and consumed cereals in Brazil and in the world due to its potential productivity, chemical composition and nutritional value. Corn ears, still fresh, are often used in Brazil as raw materials for some traditional recipes, such as porridge, pamonha and cakes, besides being consumed in cooked or baked form.

In seed production, aiming higher quality for the final product, producers harvest corn ears in a period near its physiological maturation. This harvest method allows the product to remain for a limited time in the field, avoiding material exposure to climatic events, microorganisms and insect attacks that could depreciate its physical and physiological quality. However, in this process, corn is harvested under conditions of high humidity, requiring it to be dried as soon as it is harvested.

A decrease in the material moisture content reduces its biological activity as well the chemical and physical changes that occur during storage. The moisture content reduction of grain involves simultaneously heat and mass transfer processes, which can substantially change the grain's quality, depending on the method used and on the drying conditions (HALL, 1980). Conservation by means of a drying process is based on the fact that, microorganisms, enzymes and entire metabolic mechanisms, require water for their activities. With a reduction in the water availability, until safe levels of storage are provided, the water activity and speed of chemical reactions are reduced; thus, the microorganism's development is also reduced.

The simulation and theoretical information about each product's behavior during moisture removal are very important for drying systems study, equipment development, optimization and the determination of commercial viability (CORRÊA et al., 2007).

Several methods have been proposed to analyze and simulate the drying process in hygroscopic products. In processes that are able to be modeled (drying, for instance), two types of models can be used: theoretical and empirical. The first models are based on laws and theories, which are difficult to manage due to their complexity and the involvement of several functions and parameters; these are not convenient for computational practices in most situations (MASKAN, 2002). The empirical models normally do not possess a theoretical formulation and are usually obtained through simple mathematical analogies based on experimental data and dimensional and statistical analysis. These models frequently present a good fit for the observed data. However, their use is limited due to their dependence on the experimental conditions in which the data were obtained and the characteristics of the material that was used. They also require several experimental tests and should not be extrapolated (MAYOR; SERENO, 2004). Despite these facts, these models were proposed by researches to simulate several processes in different engineering areas, obtaining satisfactory results (BALA; WOODS, 1984; MIDILLI et al., 2002; PELEG, 1988).

Depending on the material being dried, moisture can move into the interior by different mechanisms. In porous products, like most agricultural products, possible moisture transportation mechanisms include: liquid diffusion, capillary diffusion, diffusion in the surface, hydrodynamic flow, vapor diffusion and thermal diffusion (BROOKER et al., 1992).

The mechanism of moisture diffusion in agricultural products is very complex due to the diversity of chemical compositions and the physical structure of products. In the literature regarding drying involving the moisture diffusion phenomenon, large variations in diffusivity coefficient values are observed, not only due to the product's complexity, but also due to the different prediction methods that are used, the material type, moisture content, the drying process and the methodology that is used (RESENDE et al., 2005).

For the calculation of the effective diffusion coefficient, many researchers have been using the liquid diffusion theory, known as Fick's second law, which establishes a relationship between the diffusion coefficient and the concentration gradient in a given environment. An effective diffusion coefficient can be used when grain is considered to be homogeneous material (IGUAZ et al., 2003).

Drying curves obtained under controlled conditions provide important information regarding the water transportation mechanisms, and they are utilized in the determination of the effective diffusion coefficient (GELY; SANTALLA, 2007).

Having in mind corn processing, the storage of corn ears and the lack of theoretical information about the corn ears' drying processes, the goal of this work was to estimate and model the drying curves and determine the effective diffusion coefficients for corn ears for different drying conditions.

Material and methods

The present work was developed at the Laboratory of Physical Properties and Quality Evaluation of Agricultural Products belonging to the National Grain Storage Training Center – CENTREINAR, located at Federal University of Viçosa, Viçosa, Minas Gerais State, Brazil.

Determination and modeling of drying curves

Corn ears were utilized, (variety UFVM – 100), with an initial moisture content of 0.45 $kg_w\ kg_{dm}^{-1}$ (kg of water by kg of dry matter), dried until a final moisture content of 0.11 $kg_w\ kg_{dm}^{-1}$ was reached.

The product moisture content was measured by a gravimetric process until constant mass was reached and the ears were divided into three pieces. Afterwards, a piece was randomly chosen and dried in an oven at 105 ± 1°C.

Drying the corn ears to obtain the remaining moisture content levels was accomplished in an oven with forced circulation at temperatures of 45, 55 and 65°C, with three repetitions.

The experimental data of the drying process of corn ears were fit to mathematical models expressed by equations one through ten (Table 1). These models are frequently used to describe the drying phenomenon in agricultural products.

To determinate the moisture ratio (*MR*) (equation 11), the equilibrium moisture content of corn ears was estimated through the Modified Chung-Pfost model, according to Corrêa et al. (2005).

Table 1. Mathematical models to obtain the drying curves of agricultural products.

Model name	Model equation	
Page	$MR=\exp(-kt^{\eta})$	(1)
Diffusion approach	$MR=a\exp(-kt)+(1-a)\exp(-kbt)$	(2)
Two-term exponential	$MR=a\exp(-kt)+(1-a)\exp(-kbt)$	(3)
Verna	$MR=a\exp(-kt)+(1-a)\exp(-gt)$	(4)
Newton	$MR=\exp(-kt)$	(5)
Midili	$MR=a\exp(-kt^{\eta})+bt$	(6)
Logarithmic	$MR=a\exp(-kt)+b$	(7)
Henderson and Pabis	$MR=a\exp(-kt)$	(8)
Two-term model	$MR=a\exp(-kt)+b\exp(-gt)$	(9)
Modified Henderson and Pabis	$MR=a\exp(-kt)+b\exp(-gt)+c\exp(-ht)$	(10)

where:

t = drying time, s;

k = drying constant, s^{-1};

a, b, c, g, h, η = model's parameters, dimensionless.

$$MR=\frac{U_t-U_e}{U_0-U_e} \tag{11}$$

where:

U_t = moisture content at a given time, $kg_w\ kg_{dm}^{-1}$;

U_o = initial moisture content, $kg_w\ kg_{dm}^{-1}$;

U_e = equilibrium moisture content, $kg_w\ kg_{dm}^{-1}$.

Statistical analysis

The models were fit to the drying data, and a non-linear regression procedure was performed using the Gauss Newton method to estimate the parameters of the mathematical models, using the program STATISTICA 6.0®. To evaluate the degree of fit of each model to the observed data, the magnitudes of the determination coefficients (R^2), the mean relative error (*MRE*), the standard error of estimate (*SEE*) and the residual distribution (*RD*) plots were used. The mean relative error and standard error of estimate were calculated by the following expressions:

$$MRE = \frac{100}{n}\sum_{i=1}^{n}\left(\frac{\left|M_{exp} - M_{pre}\right|}{M_{exp}}\right) \tag{12}$$

$$SEE=\sqrt{\frac{\sum_{i=1}^{n}\left(M_{exp} - M_{pre}\right)^2}{D_f}} \tag{13}$$

where:

M_{exp} = experimental value of drying process;

M_{pre} = value predicted by the model;

n = number of data points;

D_f = degrees of freedom of the regression model.

Determination of the Effective Diffusion Coefficient (D_{ef})

The effective diffusion coefficient was obtained through the model adjustment of the liquid diffusion (described by Equation 14), to the drying experimental data of corn ears. This equation is the analytical solution of Fick's second law, considering the geometric shape of a cylinder, ignoring the grain volumetric variation and considering the known moisture condition at the grain surface (BROOKER et al., 1992).

$$MR=\frac{U_t-U_e}{U_0-U_e}=\sum_{i=1}^{\infty}\frac{4}{\lambda_n^2}\exp\left[\frac{-\lambda_n^2 D_{ef}\, t}{r^2}\right] \tag{14}$$

where:

MR = moisture ratio, dimensionless;

λ_n = Bessel Equation roots;

D_{ef} = effective diffusion coefficient ($m^2\ s^{-1}$);

R = equivalent radius (m).

The analytical solution of this equation is presented in the form of an infinite series, and therefore, finite terms numbers (n) in truncation are able to determinate the results with satisfactory precision.

The equivalent radius of a product is defined as the radius of a sphere having the equivalent volume of the product. To achieve the average volume for corn ears, a straight cylinder shape was considered. In order to calculate the volume, length (a), width (b) and thickness (c) measurements for the corn ears were made using a digital pachymeter with 0.01 mm precision, considering the cylinder diameter as the average of the width and the thickness, as shown by Equation 15.

$$V=\frac{\pi}{16}a\left(c+b\right)^2 \tag{15}$$

Temperature influence on the drying constant and thermodynamic considerations

To evaluate the temperature influence on the drying constant, the Arrhenius equation was used (Equation 16).

$$k=k_0\exp\left(\frac{E_a}{RT_{abs}}\right) \tag{16}$$

where:

k = drying constant, s^{-1};
k_0 = pre-exponential factor, s^{-1};
E_a = activation energy, kJ mol^{-1};
R = universal gas constant, 8.314 J mol^{-1} K^{-1};
T_{abs} = absolute temperature, in K.

The activation energy value allowed the determination of different thermodynamic parameters, such as the enthalpy, entropy and Gibbs free energy of activation, according to equations 17, 18 and 19 (JIDEANI; MPOTOKWANA, 2009).

$$\Delta H^* = E_a - RT_{abs} \tag{17}$$

$$\Delta S^* = R\left[\ln k - \ln\left(\frac{k_b}{h_p}\right) - \ln T_{abs}\right] \tag{18}$$

$$\Delta G^* = \Delta H^* - T_{abs}\Delta S^* \tag{19}$$

where:

ΔH^* = enthalpy of activation, J mol^{-1};
ΔS^* = entropy of activation, J mol^{-1};
ΔG^* = Gibbs free energy of activation, J mol^{-1};
k_b = Boltzmann constant, 1.38×10^{-23} J K^{-1};
h_p = Planck's constant, 6.626×10^{-34} J s.

Results and discussion

The statistical parameters obtained in order to verify the fit of each model to the observed data during the corn ears' drying at temperatures of 45, 55 and 65°C are presented in Table 2.

In Table 2, we notice that, for the three drying air temperatures, all of the models presented determination coefficients (R^2) greater than 0.99, satisfactory values of standard error of estimate (SEE) and mean relative error (MRE) lower than ten percent, which, according to Mohapatra and Rao (2005), represents a good fit for the experimental data. However, only the Midili and Logarithmic models (models 6 and 7, respectively), presented random distributions of the residual plots for the three drying temperatures. The Logarithmic model was recommended to describe the corn ears' drying phenomenon due to lower parameter numbers. Midili and Kucuk (2003) also recommended the Logarithmic model as the one most suitable for describe unshelled and shelled pistachios using solar energy. Goyal et al. (2007) recommended the model as the one that best represents plums' drying behavior.

Table 3 presents the Logarithmic model's fit coefficients which are designed for the three air drying temperatures for corn ears.

As Table 3 shows, the model coefficients varied with temperature; the coefficient b decreased and the coefficients a and k increased with the increase in temperature. According to Babalis and Belessiotis (2004), the drying constant (k) can be utilized as an approximation for characterizing the temperature effect; it is related to the effective diffusivity of the drying process's falling period and the liquid diffusion that governs this process. The equations that describe these coefficients, correlating to temperature, are shown in Table 4.

The activation energy for the water diffusion of corn ears during the drying process, obtained by the variation of the drying constant with temperature, was 17,644.8 J mol^{-1}. Zogzas et al. (1996) reported that activation energy generally ranges from 12.7 to 110 kJ mol^{-1} for food; therefore, a consistent value was achieved in this study. Furthermore, this value was lower than observed by Doymaz and Pala (2003) with studying corn; they acquired an activation energy-for the variation of effective diffusion coefficients of 29.56 kJ mol^{-1}. It is also slightly inferior to the value that was established by Corrêa et al. (2007), which verified an activation energy of 10.08 kJ mol^{-1} for edible beans, in a temperature range of 25 to 55°C.

The Logarithmic model can be presented with the described coefficients as a function of temperature in the corn ears drying modeling with the temperature range of 45 to 65°C, according to Equation 20.

$$MR = \left\{ \begin{array}{l} (0.01110211T + 0.51609802)\exp\left[\left(-4.6292 - \frac{17644.8}{RT_{abs}}\right)t\right] \\ +(-0.01084858T + 0.47428181) \end{array} \right\} \tag{20}$$

Table 2. Mean relative error (MRE), standard error of estimate (SEE), determination coefficients (R^2) and residual distribution behavior {random (R) or biased (B)}, of the models and drying air temperatures.

Temperature (°C)	Statistical Parameters	Mathematical models									
		1	2	3	4	5	6	7	8	9	10
45	MRE	2.041	1.517	2.021	2.074	2.344	1.669	1.704	2.165	2.113	2.096
	SEE	0.007	0.008	0.008	0.007	0.009	0.007	0.007	0.008	0.008	0.009
	R^2	0.999	0.999	0.999	0.999	0.999	0.999	0.999	0.999	0.999	0.999
	RD	R	R	R	R	R	R	R	R	R	R
55	MRE	1.619	0.512	1.350	0.512	2.058	0.384	0.787	2.033	0.498	2.009
	SEE	0.008	0.003	0.007	0.003	0.009	0.003	0.005	0.009	0.003	0.010
	R^2	0.999	0.999	0.999	0.999	0.999	0.999	0.999	0.999	0.999	0.999
	RD	B	R	B	R	B	R	R	B	R	B
65	MRE	0.806	3.614	0.497	3.614	5.741	0.637	1.468	4.096	3.614	3.615
	SEE	0.005	0.018	0.003	0.018	0.026	0.004	0.007	0.020	0.018	0.020
	R^2	0.999	0.999	0.999	0.996	0.991	0.999	0.999	0.995	0.996	0.996
	RD	R	B	R	B	B	R	R	B	B	B

where:

T = drying temperature, °C;

t = drying time, s.

Table 3. Logarithmic model's fit coefficients for temperatures of 45, 55 and 65°C.

Coefficients	Temperatures (°C)		
	45	55	65
a	1.037059	1.083981	1.259102
k	1.24×10^{-5}	1.51×10^{-5}	1.84×10^{-5}
b	-0.030713	-0.088774	-0.247684

Table 4. Variation of Logarithmic model coefficients with temperature, with the respective determination coefficients (R^2).

Equation	R^2 (%)
$a = 0.01110211T + 0.51609802$	0.999
$b = -0.01084858T + 0.47428181$	0.933
$\ln(k) = -4.6292 - \dfrac{17644.8}{RT_{abs}}$	0.999

Figure 1 represents corn ears' drying curves with observed and estimated values obtained through the Logarithmic model at temperatures of 45, 55 and 65°C.

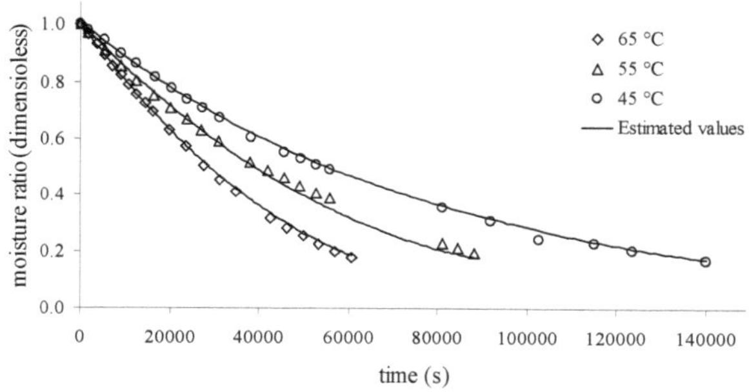

Figure 1. Observed values *versus* estimated values through the Logarithmic model for different temperatures.

Kayacier and Singh (2004) reported that activation energy decreases as the moisture content of the product increases. For drying processes, the lower the activation energy is, the greater the water diffusivity of the product will be.

Enthalpy and entropy of activation decreased, while the Gibbs free energy increased with increased temperatures for the range shown in Table 5. An analysis of the equations used to obtain these thermodynamic parameters (equations 17, 18 and 19) allows us to conclude that this behavior is expected. Bayram et al. (2004), studying the variation in physical properties of soybeans during cooking, also observed this trend; however, they obtained positive values for the entropy of activation. Similar values for entropy were reported by Jideani and Mpotokwana (2009), who evaluated the water absorption of *Vigna subterranea* L. grains.

According to our results, one notices that the water diffusion process for the product during drying responds to sensible heat and requires energy ($\Delta H^* > 0$) in order to promote changes, which is usually observed during a drying process. The evaluation of enthalpy values allows an analysis of the energy contribution that is added to the process which is responsible for altering the water diffusion velocity (BAYRAM et al., 2004).

According to Moyano and Zúñiga (2004), ΔS^* measures the order alterations of a system. Therefore, the values decrease can be explained by the fact that when the product is being dehydrated, moisture content decreases and the movement of water molecules become more restricted (there are less available sites). The positive values for Gibbs free energy of activation ($\Delta G^* > 0$) mean that the phenomenon is not spontaneous and actually requires energy from the environment so that the process can occur. Corn ear diffusion coefficient values at the various temperatures are shown in Table 6.

Table 5. Values of thermodynamic parameters at different temperatures for the drying process of corn ears.

Temperature (°C)	ΔH (J mol^{-1})	ΔS^*(J mol^{-1} K^{-1})	ΔG^*(J mol^{-1})
45	30,132.90	-317.59	13,1174.58
55	30,049.76	-317.85	13,4351.79
65	29,966.62	-318.10	13,7531.53

Table 6. Corn ears diffusion coefficients for each drying temperature.

Temperature (°C)	Effective Diffusion Coefficients (m^2 s^{-1})
45	5.490×10^{-10}
55	7.597×10^{-10}
65	1.163×10^{-9}

According to Table 6, the corn ears diffusion coefficients increased with increasing temperature, and they were consistent with the results of the drying processes of other agricultural products cited in the literature, which, according to Madamba et al. (1996), presented values with magnitudes ranging from 10^{-11} to 10^{-9} m^2 s^{-1}. Doymaz and Pala (2003), observed corn diffusion coefficients ranging from 9.488×10^{-11} to 1.768×10^{-10} m^2 s^{-1} at temperatures of 55 and 75°C, and Mohapatra and Rao (2005) obtained diffusion coefficients for wheat grains ranging from 1.2181×10^{-10} to 2.8611×10^{-10} m^2 s^{-1} for temperatures ranging from 40 to 60°C. Diffusion coefficients can vary depending on air drying temperatures and soybean varieties, as reported by Gely and Giner (2007) for soybeans, or they can vary inside the product, according to Veraverbeke et al. (2003) who studied different parts of the apple.

As the temperature increases, the water viscosity decreases, and since viscosity is a measurement of the fluid's resistance to flow, variations in this property imply that water diffusion alterations in the capillaries of porous media promote fluid movement inside the product. Another factor that can be attributed to this effective diffusion coefficient variation is that, with temperature increase, the molecular vibration level of water molecules also increases, contributing to higher fluid diffusion in the material.

Conclusion

Based on the obtained values, the Midili and Logarithmic models were adequate in describing drying phenomena of corn ears. However, since the Logarithmic model is simpler, it was selected to describe the drying process of corn ears.

The diffusion coefficients for corn ears are 5.490 x 10^{-10}, 7.597 x 10^{-10} and 1.163 x 10^{-9} m^2 s^{-1} for temperatures of 45, 55 and 65°C, respectively.

The activation energy obtained by the variation of drying constants of the corn ears was 17,644.8 J mol^{-1} at a temperature range of 45 to 65°C.

Acknowledgements

The authors thank CNPq and Capes for funding, indispensable for accomplishing this study.

References

BABALIS, S. J.; BELESSIOTIS, V. G. Influence of the drying conditions on the drying constants and moisture diffusivity during the thin-layer drying of figs. **Journal of Food Engineering**, v. 65, n. 3, p. 449 - 458, 2004.

BALA, B. K.; WOODS, J. L. Simulation of deep bed malt drying. **Journal of Agricultural Engineering Research**, v. 30, n. 3, p. 235-244, 1984.

BAYRAM, M.; KAYA, A.; ÖNER, M. D. Changes in properties of soaking water during production of soy-bulgur. **Journal of Food Engineering**, v. 61, n. 2, p. 221-230, 2004.

BROOKER, D. B.; BAKKER-ARKEMA, F. W.; HALL, C. W. **Drying and storage of grains and oilseeds**. Westport: The AVI Publishing Company, 1992.

CORRÊA, P. C.; RESENDE, O.; RIBEIRO, D. M. Isotermas de Sorção das espigas de milho: Obtenção e Modelagem. **Revista Brasileira de Milho e Sorgo**, v. 4, n. 1, p. 126-134, 2005.

CORRÊA, P. C.; RESENDE, O.; MARTINAZZO, A. P.; GONELI, A. L. D. G.; BOTELHO, F. M. Modelagem matemática para a descrição do processo de secagem do feijão (*Phaseolus vulgaris* L.) em camadas delgadas. **Engenharia Agrícola**, v. 27, n. 2, p. 501-510, 2007.

DOYMAZ, I.; PALA, M. The thin-layer drying characteristics of corn. **Journal of Food Engineering**, v. 60, n. 2, p. 173-179, 2003.

GELY, M. C; GINER, S. A. Diffusion coefficient relationships during drying of soya bean cultivars. **Biosystems Engineering**, v. 96, n. 2, p. 213-222, 2007.

GELY, M. C.; SANTALLA, E. M. Moisture diffusivity in quinoa (*Chenopodium quinoa* Willd.) seeds: Effect of air temperature and initial moisture content of seeds. **Journal of Food Engineering**, v. 78, n. 3, p. 1029-1033, 2007.

GOYAL, R. K.; KINGSLY, A. R. P.; MANIKANTAN, M. R.; ILYAS, S. M. Mathematical modelling of thin layer drying kinetics of plum in a tunnel dryer. **Journal of Food Engineering**, v. 79, n. 1, p. 176-180, 2007.

HALL, C.W. **Drying and storage of agricultural crops**. Westport: AVI, 1980.

IGUAZ, A.; SAN MARTIN, M. B.; MATÉ, J. I.; FERNANDEZ, T.; VÍRSEDA, P. Modelling effective moisture diffusivity of rough rice (*Lido cultivar*) at low drying temperatures. **Journal of Food Engineering**, v. 59, n. 2-3, p. 253-258, 2003.

JIDEANI, V. A.; MPOTOKWANA, S. M.; Modeling of water absorption of Botswana bambara varieties using Peleg's equation. **Journal of Food Engineering**, v. 92, n. 2, p. 182-188, 2009.

KAYACIER, A.; SINGH, R. K. Application of effective diffusivity approach for the moisture content prediction of tortilla chips during baking. **Swiss Society of food Science and Technology**, v. 37, n. 2, p. 275-281, 2004.

MADAMBA, P. S.; DRISCOLL, R. H.; BUCKLE, K. A. Thin-layer drying characteristics of garlic slices. **Journal of Food Engineering**, v. 29. n. 1, p. 75-97, 1996.

MASKAN, M. Effect of processing on hydration kinetics of three wheat products of the same variety. **Journal of Food Engineering**, v. 52, n. 4, p. 337-341, 2002.

MAYOR, L.; SERENO, A. M. Modelling shrinkage during convective drying of food materials: a review. **Journal of Food Engineering**, v. 61, n. 3, p. 373-386, 2004.

MIDILLI, A.; KUCUK, H. Mathematical modeling of thin layer drying of pistachio by using solar energy. **Energy Conversion and Management**, v. 44, n. 7, p. 1111-1122, 2003.

MIDILLI, A.; KUCUK, H.; YAPAR, Z. A new model for single-layer drying. **Drying Technology**, v. 20, n. 7, p. 1503-1513, 2002.

MOHAPATRA, D.; RAO, P. S. A thin layer drying model of parboiled wheat. **Journal of Food Engineering**, v. 66, n. 4, p. 513-518, 2005.

MOYANO, P. C.; ZÚÑIGA, R. Enthalpy-entropy compensation for browning of potato strips during deep-fat frying. **Journal of Food Engineering**, v. 63, n. 1, p. 57-62, 2004.

PELEG, M. An empirical model for description of moisture sorption curves. **Journal of Food Science**, v. 53, n. 4, p. 1216-1219, 1988.

RESENDE, O.; CORRÊA, P. C.; GONELI, A. L. D.; MARTINAZZO, A. P.; RIBEIRO, R. M. Contração volumétrica na difusão líquida durante o processo de secagem do arroz em casca. **Revista Brasileira de Armazenamento**, v. 30, n. 2, p. 163-171, 2005.

VERAVERBEKE, E. A.; VERBOVEN, P.; SCHEERLINCK, N.; HOANG, M., L.; NICOLA, B. M. Determination of the diffusion coefficient of tissue, cuticle, cutin and wax of apple. **Journal of Food Engineering**, v. 58, n. 3, p. 285-294, 2003.

ZOGZAS, N. P.; MAROULIS, Z. B.; MARINOS-KOURIS, D. Moisture diffusivity data compilation in foodstuffs. **Drying Technology**, v. 14, n. 10, p. 2225-2253, 1996.

3

Agrometeorological models for groundnut crop yield forecasting in the Jaboticabal, São Paulo State region, Brazil

Victor Brunini Moreto* **and Glauco de Souza Rolim**

*Faculdade de Ciências Agrárias e Veterinárias, Universidade Estadual "Julio de Mesquita Filho", Via de Acesso Prof.Paulo Donato Castellane s/n.°, 14884-900, Jaboticabal, São Paulo, Brazil. *Author for correspondence. E-mai: victor.moreto@hotmail.com*

ABSTRACT. Forecast is the act of estimating a future event based on current data. Ten-day period (TDP) meteorological data were used for modeling: mean air temperature, precipitation and water balance components (water deficit (DEF) and surplus (EXC) and soil water storage (SWS)). Meteorological and yield data from 1990-2004 were used for calibration, and 2005-2010 were used for testing. First step was the selection of variables via correlation analysis to determine which TDP and climatic variables have more influence on the crop yield. The selected variables were used to construct models by multiple linear regression, using a stepwise backwards process. Among all analyzed models, the following was notable: Yield = - 4.964 x [SWS of 2° TDP of December of the previous year (OPY)] – 1.123 x [SWS of 2° TDP of November OPY] + 0.949 x [EXC of 1° TDP of February of the productive year (PY)] + 2.5 x [SWS of 2° TDP of February OPY] + 19.125 x [EXC of 1° TDP of May OPY] – 3.113 x [EXC of 3° TDP of January OPY] + 1.469 x [EXC of 3 TDP of January of PY] + 3920.526, with MAPE = 5.22%, R^2 = 0.58 and RMSEs = 111.03 kg ha^{-1}.

Keywords: crop model, water balance, prediction, production.

Introduction

The groundnut, one of the major oilseeds produced in the world, belongs to the family Leguminosae. The plant has a high energy value, is rich in nutrients and is use to produce a wide range of products derived from the grain, like oil such as peanut butter and 'in natura' peanut (FREITAS et al., 2005).

According to the National Supply Company (CONAB, 2012), the Brazilian production in 2010/2011 compared to 2009/2010, was reduced by 32%, with a 12% reduction in the planted area. These reductions were mainly caused by climatic factors, such as excessive rainfall at the end of the peanut crop cycle.

São Paulo, and the Jaboticabal region in particular, accounts for almost 70% of the entire Brazilian product. This region is distinguished by their average yield, of 3.7 ton. ha^{-1}, which is much higher than the national average of 2,8 t ha^{-1} and the global average of 1.6 ton. ha^{-1} (IEA, 2011).

To estimate the production of an agricultural crop, in Brazil, a survey system is used based on opinions of technicians and economists from of each sector. This survey system is considered a subjective method because it does not allow for a quantitative analysis of the errors involved. An alternative subjective method is the use of crop models, known as agrometeorological models which express the influence of meteorological elements on crop yield.

According to Rossetti (2001), 95% of claims paid by agricultural security entities are related to drought or excess rain. Modeling is a mean to quantify these climates risks, estimate yields and devise strategies to minimize their impacts in a rapid and cost-effective manner.

In São Paulo, the recommended groundnut sowing date, consider climatic condition of each region. Generally, the months of October and November are the most suitable to sowing the groundnut known as the "water crop". The "dry crop" is typically sowed in February (GODOY et al., 1986).

The groundnut cycle ranges from 90 to 115 days for early varieties and from 120 to 140 days for late varieties. Depending on the weather the water requirements range from 500 to 700 mm for both cycles (DOORENBOS; KASSAM, 1979). According to Silva and Rao (2006), in general, water deficits during the growing season cause flowering delays and extend the crop cycle, thereby delaying harvest and reducing yield. The phase of flowering and pod formation is highly sensitive to water stress.

The harvest must be performed in the appropriate season. Harvests at an unsuitable time may lead to considerable losses, both in the quantity and quality of seeds. Earlier harvests result in considerable quantities of immature and poorly formed seeds, whereas late harvests can lead to the greater deterioration of seeds (CARVALHO et al., 1976). Soil moisture can severely damage and reduce the quality of the seeds, most likely because of the occurrence of rain (TOLEDO; MARCOS FILHO, 1977); subsequent to harvest, seed germination may occur with the pod (SAVY FILHO; LAGO, 1985), resulting in the deterioration of the seeds during storage.

Studies of the water balance should be developed to facilitate an understanding of the relationship between culture and climate, which allows for an adjustment of the crop climatic conditions, thereby avoiding disastrous consequences of defective agricultural planning (TUBELIS, 1988). The main climate elements, according to Ometto (1981) and Vianello and Alves (1991), are the precipitation, air temperature, solar radiation, atmospheric moisture, wind and atmospheric pressure.

Doorenbos and Kassam (1979) reported that, to obtain a high yield, a rainfed crop requires approximately 500 to 700 mm of water for the entire period of growth. The indices of drought sensitivity (Ky) for the phases of establishment, vegetative growth, flowering and maturity are 0,2; 0,2; 0,8 and 0,6 respectively, indicating that the flowering period is more sensitive to water deficit (DOORENBOS; KASSAM, 1979).

Groundnuts grow well in environments with average daytime temperatures between 22°C and 28°C (CRAUFURD et al., 2002; DOORENBOS; KASSAM, 1979). The period from sowing to flowering requires 313 GDD (growing degree days) to 360 GDD, depending on the variety (KETRING; WHELESS, 1989).

According to Robertson (1983) yield-climate interactions can be quantified using models and studying the variations and effects of climate on plant performance.

Agricultural simulation models can be understood as empirical or mechanistic mathematical equations that aim to simplify reality and represent the biomass accumulation and plant development as well as estimate their yield according to the influencing factors. Empirical models are those based on regression analyses, whereas mechanistic models quantify and seek to understand the physical and biological interactions of plants with their development and environment (ZHU, 2010).

The knowledge of the space-temporal variability of long-term series of meteorological data can assist in the identification of better areas and periods for sowing for the crops, as well as provide important information on possible climatic trends. The union of the knowledge of space-temporal variability is a fundamental step for reducing climatic risks associated with the agricultural sector (BLAIN, 2009).

Few agrometeorological models are available to estimate the yields of groundnut crops (ESPOSTI, 2002; MARIN et al., 2006). Assunção and Escobedo (2009) developed a model to estimate the crop yield as a function of the water availability. However, as mentioned above, most models estimate but do not forecast the yield. For example, Challinor et al. (2003) indicated that yield estimations and forecasting can be improved if we relate crop models to weather forecasting.

This study aimed to identify the climatic elements in ten-day dataset that influence the annual groundnut yield in the Jaboticabal region. The other objective was to develop agrometeorological model (s) for regional groundnut yield forecasting.

Material and methods

We used agrometeorological data that were obtained from 1990 to 2010 of the

Agrometeorological Station of the Department of Exacts Science - FCAV/Unesp, Jaboticabal, located at latitude 21°14'05", longitude 48°17'09" and an altitude of 600 m.

The mean air temperature and precipitation were used for estimating the ten-day period (TDP) reference evapotranspiration (ETo) according to the Thornthwaite (1948) method. The components of the water balance (WB) proposed by Thornthwaite and Mather (1955), such as the deficit (DEF), water surplus (EXC) and soil storage water (SWS) were calculated using an available water capacity (AWC) of 100 mm.

The groundnut production data were obtained from the Institute of Agricultural Economy (IEA) for a period of 20 years, from 1990 to 2010. The yield data were adjusted as proposed by Prela-Pantano et al. (2011), to remove the technological trend. This adjustment is necessary to minimize effects due to changes in the technological level employed by producers, thereby obtaining the influence of the climate variability on the yield.

A linear correlation analysis (r, Pearson) was performed using a TDP of SWS, DEF and EXC of the harvest year and the same elements of the previous year. The variables with the best correlations were selected for modeling. The models were constructed by multiple linear regression ($Y = a.X_1 + b.X_2 + c.X_3 + \ldots + LC$) where Y is the yield of the harvest year, the independent variables are the climatic elements and LC is the linear coefficient.

In modeling, especially with the use of many independent variables, the main problem is the selection and combination of the variables. A stepwise backwards regression method was used, based on the criterion of accuracy (R^2 adjusted) improvement. A total of 511 regressions were obtained and analyzed by combinatory analysis. Only models shaving at least 0,05 significance (P value) in the parameters and in the regression were selected.

After this step, 28 models were chosen for the region. These models were evaluated using an accuracy analysis based on the mean absolute percentage error (MAPE), precision analysis based on the adjusted coefficient of determination (R^2) and tendency analysis based on the Root Mean Systematic Error (RMSEs) (Equations 1, 2 and 3, respectively). We used a 14 year period (1990 to 2004) for the model calibration and a 6 year period (2005 to 2010) for testing (or validation).

$$MAPE = \frac{\sum_{i=1}^{N}\left(\left|\frac{Yest_i - Yobs_i}{Yobs_i}\right| \times 100\right)}{N} \quad (1)$$

$$R^2\ adjusted = \left[1 - \frac{(1 - R^2) \times (n - 1)}{n - k - 1}\right] \quad (2)$$

$$RMSEs = \sqrt{\frac{\sum_{i=1}^{N}(Yobs_i - Y_{estC})^2}{N}} \quad (3)$$

where:

$Yest_i$: estimation of year 'i' yield; $Yobs_i$: observed yield (without technological trend) in year 'i'; Y_{estC}: yield estimated by simple linear regression between the observed ($Yobs_i$) and estimated ($Yest_i$) yield; N: number of years; n: number of data and k: number of independent variables in the regression.

Results and discussion

The period of analysis showed a wide variability in the crop yield (Figure 1A). This irregularity occurred due to climate variables and changes in the level of technology that was available. A 6° order polynomial adjustment was applied to minimize the effects of the agricultural technology level that was used (Figure 1A) and to clearly demonstrate the effects of climate on the crop yield. The adjusted yields (Figure 1B) are used throughout the remainder of the paper.

The occurrence of heavy rainfall in 1992/1993 season (Figure 2), and the lack of heavy rain in the 1991/1992 season during the final phases of crop development (flowering and fructification) had a considerable negative effect on the groundnut yields in those years. According to Gillier and Silvestre (1996), groundnut flowering and fructification are dependent on certain climatic elements, and a change in the water availability directly and intensely affects the crop yield.

An analyzing of the year's corresponding to the lowest and highest yields at Jaboticabal showed that the distribution of rainfall directly affects the groundnut yield. The 'water crop', which is the more common variety in the region and is normally sowed in mid-November had the lowest yield in 1998 (Figure 3). This low yield occurred due to DEF in January that affected the flowering phase and DEF from March to mid-November that most likely had a negative influence on the fructification and the initial phase of maturity.

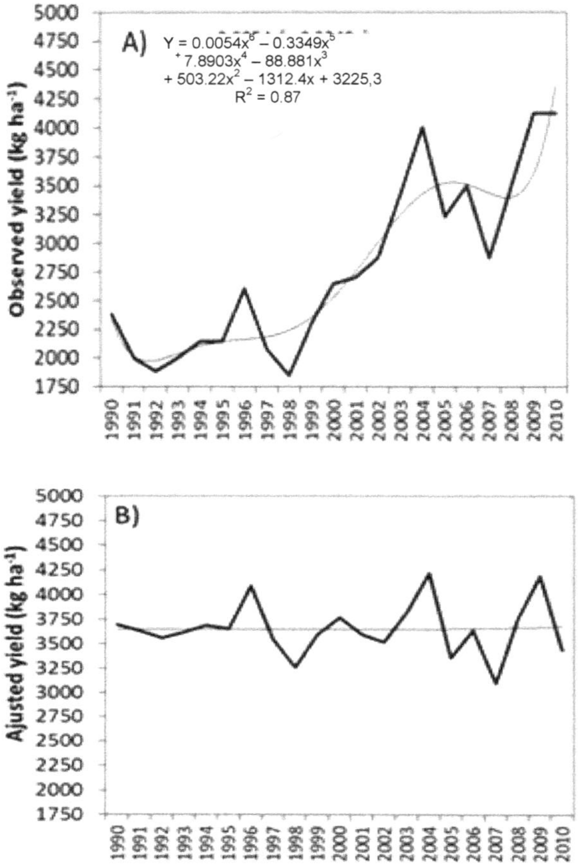

Figure 1. Groundut observed yield (A) and adjusted yield to remove the technological trend (B), in Jaboticabal, São Paulo State, from 1990 to 2010.

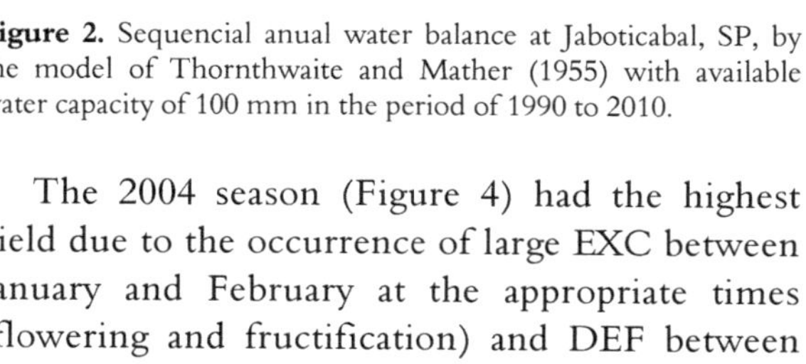

Figure 2. Sequencial anual water balance at Jaboticabal, SP, by the model of Thornthwaite and Mather (1955) with available water capacity of 100 mm in the period of 1990 to 2010.

The 2004 season (Figure 4) had the highest yield due to the occurrence of large EXC between January and February at the appropriate times (flowering and fructification) and DEF between March and December favoring the maturation and harvest of the crop.

A correlation analysis was performed to select elements of the water balance (WB) within the ten-day period (TDP) (independent variables) that had the greatest influence on the crop yield (Table 1). These selected variables were used to construct the agrometeorological models.

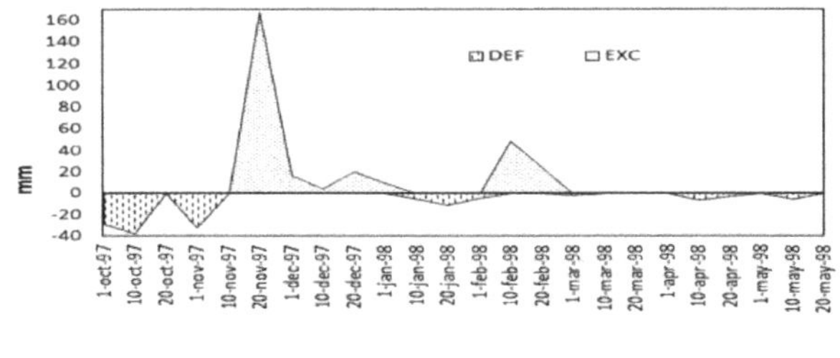

Figure 3. Ten-day period sequential water balance by the method of Thornthwaite and Mather (1955) with available water capacity of 100 mm of the season 1997/1998

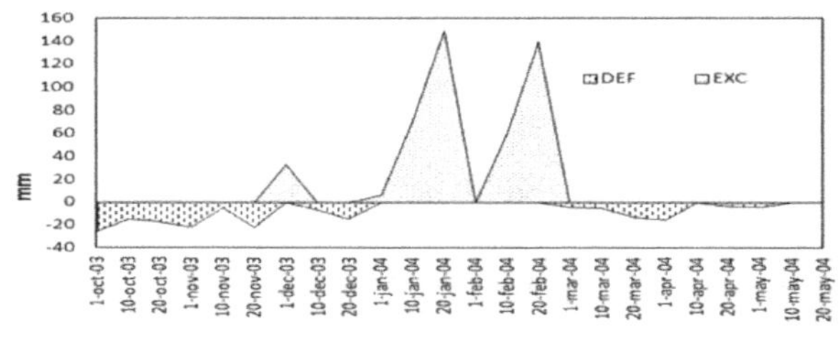

Figure 4. Ten-day period sequential water balance by the method of Thornthwaite and Mather with available water capacity of 100 mm of the season 2003/2004.

Table 1. Selected coefficient of correlation (r) among the most important meteorological variables and yield. Legend: X_1: soil water storage (SWS) of the 2° TDP of December of the previous year (OPY); X_2: SWS of the 2° TDP of November OPY; X_3: water surplus (EXC) of the 1° TDP of February; X_4: SWS of the 2° TDP of February OPY; X_5: water deficit (DEF) of the 2° TDP of December; X_6: EXC of the 1° TDP of May OPY; X_7: EXC of the 3° TDP of January OPY; X_8: EXC of the 3° TDP of January; X_9: yield OPY.

	PROD	X_9	X_1	X_2	X_3	X_4	X_5	X_6	X_7	X_8
PROD	1									
X_9	0.840	1								
X_1	-0.477	-0.213	1							
X_2	-0.448	-0.333	0.451	1						
X_3	-0.438	-0.417	0.348	0.445	1					
X_4	0.449	0.278	-0.355	-0.239	-0.362	1				
X_5	0.452	0.414	0.185	0.012	0.000	0.209	1			
X_6	0.561	0.361	-0.732	-0.480	-0.212	0.210	0.000	1		
X_7	0.674	0.687	-0.359	-0.402	-0.034	0.255	0.000	0.613	1	
X_8	0.701	0.618	-0.290	-0.233	0.000	0.337	0.000	0.406	0.469	1

In the calibration, the models '1' to '7' (Tables 2 and 3) showed the high accuracies having MAPE values found of less than 3,25%. Considering the average yield in the region of 2780 kg ha^{-1}, the observed MAPE corresponds to 90 kg ha^{-1}, or 1.5 sacks of 60 kg ha^{-1}. The independent variable X_6, which represents the EXC of the 1° TDP of the May OPY, was important in all models, except model '5'. This importance can be explained by the fact that the timing of the end of maturation and early harvest period coincides with this variable (Figure 5).

The models were also evaluated with respect to their ability to forecast yields prior to the harvest or before the crop sowing. For this reason, the first four models (Table 2) are most interesting for discussion. The models not only present the best accuracy (MAPE) but they also ignore the variable X_5; therefore these models can forecast the yield prior to the sowing date. Whenever possible the variable X_5, should be discarded because it uses the 2° TDP of December of the productive year, thereby preventing a forecast of the yield.

The second step of the study was to test the selected models using independent data from the period of 2005 to 2010. The first eight models showed high accuracy (Table 3), high precision and low tendency, as evidenced by, MAPE values less than 8,25%, a minimum R^2 of 0.56 and a minimum RMSEs of 110.86 kg ha^{-1}.

Among the top four models (Tables 2 and 3), model '1' showed the best accuracy in the tests, with a MAPE of 0.99% and precision (R^2) of 0.6 (Figure 6b). This model is interesting because it uses just four variables: X_6, X_7, X_8 and X_9 (Table 2). These variables are related to the EXC of 3° TDP of January of the same year and the EXC of the 3° TDP of January and the 1° TDP of May OPY. According to Silva and Rao (2005), this result confirms that in January, the crop is generally at the beginning of the flowering phase, which requires the greatest water supply (Figure 5).

Models '3' and '4' showed good accuracy with MAPE values of 5.22% and 5.73%, respectively. In both models the variable X_9 (Table 2) was not used, so it is possible to forecast the yield using only meteorological data.

Finally, the following models were found to have both high accuracy and high applicability: model '1' (Yield = $23.341.X_6 - 2.315.X_7 + 0.898.X_8 - 0.092.X_9 + 4010.508$), model '2' (Yield = $-5.199.X_1 - 1.118.X_2 + 0.931.X_3 + 2.284.X_4 + 18.641.X_6 - 2.997.X_7 + 1.511.X_8 - 0.033.X_9 + 4076.521$) and model '3' (Yield = $-4.964.X_1 - 1.123.X_2 + 0.949.X_3 + 2.5.X_4 + 19.125.X_6 - 3.113.X_7 + 1.469.X_8 + 3920.526$). These models presented better statistical indices in both, the calibration and tests (Table 3) due to the lower dispersion of the estimate data relative to the observed data (Figures 6, 7 and 8).

Table 2. Coefficients of the agrometeorological models related to linear regression models ($Y= a.X_1+b.X_2+c.X_3+...+LC$) for yiled forecasting of groundnut crop at Jaboticabal region. Legend: X_1: soil water storage (SWS) of the 2° TDP of December of the previous year (OPY); X_2: SWS of the 2° TDP of November OPY; X_3: water surplus (EXC) of the 1° TDP of February; X_4: SWS of the 2° TDP of February OPY; X_5: water deficit (DEF) of the 2° TDP of December; X_6: EXC of the 1° TDP of May OPY; X_7: EXC of the 3° TDP of January OPY; X_8: EXC of the 3° TDP of January; X_9: yield OPY.

Models	CL	X_1	X_2	X_3	X_4	X_5	X_6	X_7	X_8	X_9
1	4010.508						24.341	-2.315	0.898	-0.092
2	4076.521	-5.199	-1.118	0.931	2.284		18.641	-2.997	1.511	-0.033
3	3920.526	-4.964	-1.123	0.949	2.500		19.125	-3.113	1.469	
4	3899.059	-3.995	-1.249	0.702	2.511		23.413	-2.745		
5	5434.651	-10.798	-1.193	0.904	0.039	1.667		-1.510	2.824	-0.231
6	4784.275	-12.168	-0.838	0.534	0.305	41.591	-0.826			
7	4480.624	-10.542	-1.074	1.007	1.185	12.395	-2.046	2.513		
8	4908.069	-8.672	0.802	0.132	-0.205	52.648	18.333		-0.752	-0.134
9	3465.854		-0.761	0.640	2.990	42.965	30.123	-3.419		
10	4312.806	-6.772	-0.413	0.629		53.634	23.893	-2.671		
11	4047.005	-5.837	-0.190	0.727	1.975	51.380	25.110	-2.920		
12	4315.315	-6.653	-0.323	0.557		59.531	25,474	-2.583	-0.462	
Models	CL	X_1	X_2	X_3	X_4	X_5	X_6	X_7	X_8	X_9
13	4042.558	-5.922		0.705	2.011	52.096	25.186	-2.889		
14	3488.693		-0.653	0.554	2.902	50.120	31.845	-3.298	-0.542	
15	4054.817	-5.770	-0.128	0.671	1.931	55.847	26.267	-2.849	-0.346	
16	4052.306	-5.821		0.654	1.952	56.547	26.377	-2.825	-0.363	
17	4067.099	-5.182	0.734		1.305	65.808	28.241	-2.288	-1.160	
18	4340.557	-8.247	1.190	0.056	0.323	66.203	21.096		-1.482	
19	3784.168	-6.193	-0.199	0.525		73.683	28.838	-2.911	-1.101	0.140
20	3040.384	-4.741	-0.038	0.898	3.327	59.072	27.757	-3.784		0.217
21	2907.719					85.647	38.034	-3.208	-2.210	0.241
22	2263.049	-3.258		0.773	4.112	88.946	36.286	-4.166	-1.948	0.396
23	2167.824	-3.329	0.477	0.715	4.295	93.115	37.170	-4.142	-2.090	0.414
24	2036.459				3.577	95.696	40.333	-3.824	-2.982	0.409
25	2278.374	-2.831	1.362		3.507	101.778	38.704	-3.480	-2.866	0.393
26	1518.384			0.718	5.179	96.275	41.723	-4.688	-2.495	0.501
27	1429.390		0.381	0.671	5.344	99.730	42.523	-4.678	-2.619	0.518
28	2167.824	-3.329	0.477	0.715	4.295	93.115	37.170	-4.142	-2.090	0.414

Table 3. Accuracy by the mean absolute percentage error (MAPE), precision by adjusted coefficient of correlation (R^2) and tendency by systematic root mean squared error (RMSEs) for the agrometeorological models presented in Table 2. The phase of calibration used data from 1990 to 2004 and test with 2005 to 2010.

	Calibration			Test		
Models	MAPE (%)	Ajusted R^2	RMSEs (kg ha^{-1})	MAPE (%)	Ajusted R^2	RMSEs kg ha^{-1})
1	3.07	0.56	103.74	0.99	0.60	159.44
2	2.80	0.70	72.13	5.18	0.61	110.86
3	2.79	0.70	72.24	5.22	0.58	111.03
4	2.94	0.66	80.43	5.73	0.59	123.62
5	2.84	0.58	98.60	6.10	0.61	151.55
6	3.25	0.47	126.28	6.42	0.68	194.09
7	3.09	0.56	104.40	7.12	0.56	160.46
8	2.65	0.63	87.72	8.10	0.65	134.82
9	2.56	0.71	68.30	8.21	0.32	104.98
10	2.39	0.02	59.30	8.23	0.46	91.14
11	2.45	0.03	55.22	8.66	0.51	84.88
12	2.40	0.75	58.83	8.75	0.50	90.41
13	2.44	0.77	55.29	8.79	0.44	84.97
14	2.55	0.72	95.84	8.80	0.30	147.59
15	2.44	0.77	54.96	9.03	0.44	84.47
16	2.44	0.77	54.99	9.14	0.52	84.52
17	2.39	0.74	62.37	10.11	0.37	95.86
18	2.64	0.62	89.52	10.18	0.47	137.58
19	2,41	0.76	57.00	10.78	0.33	87.61
20	2.30	0.79	50.63	11.00	0.22	77.81
21	2.55	0.69	74.09	12.64	0.13	113.88
22	2.31	0.81	45.20	14.74	0.10	69.48
23	2.29	0.81	44.84	15.40	0.09	68.91
24	2.48	0.75	58.66	15.45	0.07	90.16
25	2.37	0.78	53.24	16.21	0.08	81.83
26	2.37	0.80	48.52	16.33	0.04	74.58
27	2.36	0.80	48.29	16.88	0.04	74.22
28	2.02	0.81	44.84	25.67	0.09	68.91

The variable X_6 (EXC of the 1° TDP of May OPY) had the greatest influence in all of the models because the adjusted parameter had a higher value than the other variables. The variable X_7 (EXC of the 3° TDP of January OPY) was also important because it was used in all selected models. This variable showed an inverse relationship with the water crop groundnut yield, indicating that excess water is moderately harmful in flowering and fructification. The X_1 (SWS of the 2° TDP of December OPY) also showed inverse relationship with yield, suggesting that water storage during vegetative growth should not be excessively high. These considerations are shown in Figure 5 for both water and dry crops.

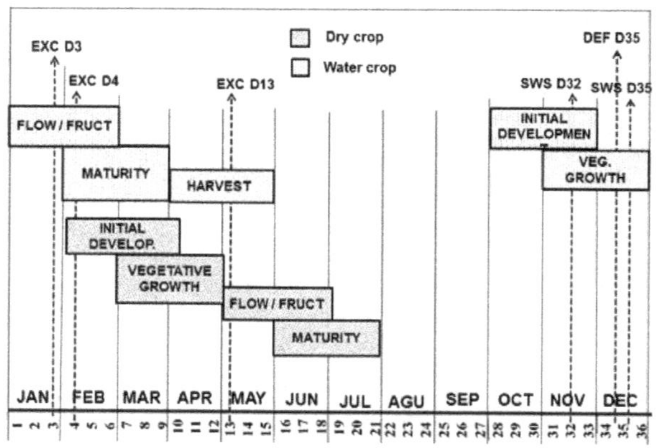

Figure 5. Meteorological variables selected in the models 1, 2 and 3 and the periods of their influence on the water and dry crops of groundnut. Legend: water surplus (EXC), water deficit (DEF) and soil water storage (SWS).

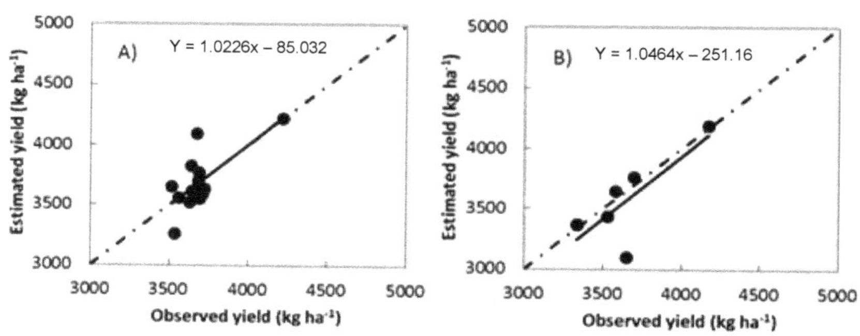

Figure 6. Calibration of the parameters of model '1' using ten-day period data of 1991 to 2004 (A) and test of the parameters, 2005 to 2010 (B).

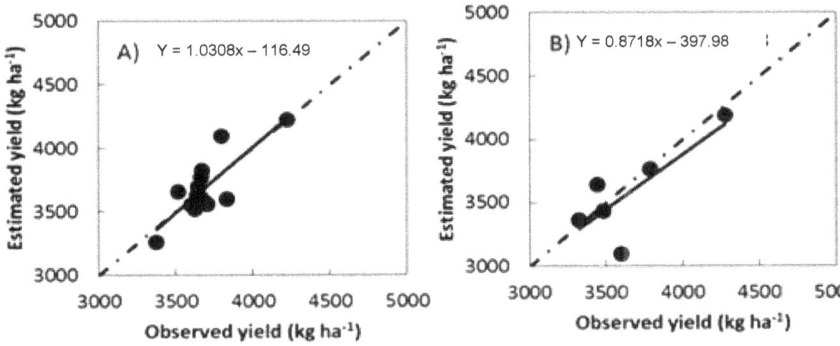

Figure 7. Calibration of the parameters of model '2' using ten-day period data of 1991 to 2004 (A) and test of the parameters, 2005 to 2010 (B).

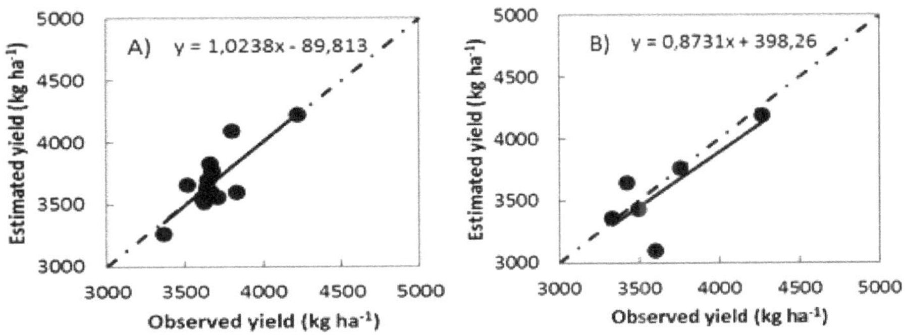

Figure 8. Calibration of the parameters of model '3' using ten-day period data of 1991 to 2004 (A) and test of the parameters, 2005 to 2010 (B).

Conclusion

The results showed that the following meteorological elements had the greatest influence of the groundnut yield at Jaboticabal: a) soil water storage of a 2° ten-day period (TDP) of February, November and December of the previous year (OPY); and b) water surplus of the 3° TDP of January and 1° TDP of May OPY before the seasonal year and the 3° TDP of January and 1° TDP of February of the seasonal year.

Models '1', '2' and '3' were the most accurate. Model '3' uses only meteorological data and, thus does not require using the yield OPY in its calculations.

References

ASSUNÇÃO, H. F.; ESCOBEDO, J. F. Use of an agrometeorological to computing the peanut water requirements. **Revista Irriga**, v. 14, n. 3, p. 325-335, 2009.

BLAIN, G. C. Statistical considerations of eight monthly precipitations series of the state of São Paulo, Brazil. **Revista Brasileira de Meteorologia**, v. 24, n. 1, p. 12-23, 2009.

CARVALHO, N. M.; BUENO, C. R.; SANCHEZ, L. C. Maturação de sementes de amendoim (*Arachis hypogaea* L.). **Científica**, v. 4, n. 1, p. 39-42, 1976.

CHALLINOR, A. J.; SLINGO, J. M.; WHEELER, T. R.; CRAUFURD, P. Q.; GRIMES, D. I. F. Toward a combined seasonal weather and crop yield forecasting system: Determination of the working spatial scale. **Journal of Applied Meteorology**, v. 42, n. 2, p. 175-192, 2003.

CONAB-Companhia Nacional de Abastecimento. Amendoim total (1ª. e 2ª. safra). **Acompanhamento da safra brasileira de grãos**. Brasília: CONAB, 2012. p. 10-13.

CRAUFURD, P. Q.; PRASAD, P. V. V.; SUMMERFIELD, R. J. Dry matter production and rate of change of harvest index at high temperature in peanut. **Crop Science**, v. 42, n. 1, p. 145-151, 2002.

DOORENBOS, J.; KASSAM, A. H. **Yield response to water**. Rome: FAO, 1979. (Irrigation and Drenage, Paper 33).

ESPOSTI, M. D. D. Estimativa da perda de produtividade potencial do amendoim (*Arachis hypogaea* L.) e épocas de plantio na Região Sul do Espírito Santo em função do déficit hídrico. **Revista Ciência Agronômica**, v. 33, n. 2, p. 5-12, 2002.

FREITAS. M.; MARTINS, S. S.; NOMI, A. K.; CAMPOS, A. F. Evolução do mercado brasileiro de amendoim. In: SANTOS, R. C. (Ed.). **O Agronegócio do amendoim no Brasil**. Campina Grande: Embrapa, 2005. p. 16-44.

GILLIER, P.; SILVESTRE, P. Green and chicken manures for the peanut crop (*Arachis hypogeae* L.). **Scientia Agricola**, v. 53, n. 1, p. 88-93, 1996.

GODOY, I. J.; RODRIGUES FILHO, F. S. O.; GERIN, M. A. N. Amendoim. In: **Instituto Agronômico (Campinas)**. Instruções Agrícolas para o Estado de São Paulo. 3. ed. Campinas: IAC, 1986. p. 23. (Boletim, 200).

IEA-Instituto de Economia Agrícola. **Previsões e estimativas das safras agrícolas do estado de São Paulo, ano agrícola 2010/2011**. Informações Econômicas. São Paulo: IEA, 2011. v. 41, n. 4.

KETRING, D. L.; WHELESS, T. G. The Degree-day requirements for phenological development of peanut. **Agronomy Journal**, v. 81, n. 1, p. 910-917, 1989.

MARIN, F. R.; PEZZOPANE, J. R. M.; PEDRO JÚNIOR, M.; GODOY, I. J.; GOUVÊA, J. R. R. F. Estimativa da produtividade e determinação das melhores épocas de semeadura para a cultura do amendoim no Estado de São Paulo. **Revista Brasileira de Agrometeorologia**, v. 14, n. 1, p. 64-75, 2006.

OMETTO, J. C. **Bioclimatologia vegetal**. São Paulo: Agronômica Ceres, 1981.

PRELA-PANTANO, A.; DUARTE, A. P.; SILVA, D. F.; ROLIM, G. S.; CASER, D. V. Yield of maize, and occurrence of enso precipitation in the region of the middle Paranapanema, SP, Brazil. **Revista Brasileira de Milho e Sorgo**, v. 10, n. 2, p. 146-157, 2011.

ROBERTSON, G. W. Guidelines on crop-weather models. Geneva: WMO, 1983. (World Climate Application Programme, 50)

ROSSETTI, L. A. Zoneamento agrícola em aplicações de crédito e seguridade rural no Brasil: aspectos atuariais e de política agrícola. **Revista Brasileira de Agrometeorologia**, v. 9, n 3, p. 386-399, 2001.

SAVY FILHO, A.; LAGO, A. A. Efeito da época de colheita na qualidade de sementes de amendoim. **Revista de Agricultura**, v. 60, n. 2, p. 11-16, 1985.

SILVA, L. C. L.; RAO, V. R. T. Evaluation of methods for stimating the peanut crop coefficients. **Revista Brasileira de Engenharia Agrícola e Ambiental**, v. 10, n. 1, p. 128-131, 2006.

THORNTHWAITE, C. W. An approach towards a rational classification of climate. **Geographical Review**, v. 38, n. 1, p. 55-94, 1948.

THORNTHWAITE, C. W.; MATHER, J. R. The water balance. drexel institute of technology. **Laboratory of Climatology**, v. 8, n. 1, p. 1-104, 1955.

TOLEDO, F. F.; MARCOS FILHO, J. **Manual das sementes**: tecnologia de produção. São Paulo: Agronômica Ceres, 1977.

TUBELIS, A. **Chuva e a produção agrícola**. Sao Paulo: Nobel, 1988.

VIANELLO, R. L.; ALVES, A. R. **Meteorologia básica e aplicações**. Viçosa, UFV, 1991.

ZHU, X. G.; ZHANG, G. L.; THOLEN, D.; WANG Y.; CHIN, C. P.; SONG, Q. F. The next generation models for crops and agro-ecosystems. **Science China Press**, v. 54, n. 3, p. 589-597, 2011.

4

Kinetic modeling of water sorption by roasted and ground coffee

Fernanda Machado Baptestini[1*], Paulo Cesar Corrêa[1], Gabriel Henrique Horta de Oliveira[2], Paulo Roberto Cecon[3] and Nilda de Fátima Ferreira Soares[4]

*[1]Universidade Federal de Viçosa, Departamento de Engenharia Agrícola e Ambiental, Avenida Peter Henry Rolfs, s/n, Campus Universitário, Viçosa, Minas Gerais, Brazil. [2]Instituto Federal de Educação, Ciência e Tecnologia Sudeste de Minas Gerais, Campus de Manhuaçu, Manhuaçu, Minas Gerais, Brazil. [3]Universidade Federal de Viçosa, Departamento de Estatística, Viçosa, Minas Gerais, Brazil. [4]Universidade Federal de Viçosa, Departamento de Tecnologia de Alimentos. Viçosa, Minas Gerais, Brazil. *Author for correspondence. E-mail: fbaptestini@yahoo.com.br*

ABSTRACT. The objective of this study was to model the kinetics of water sorption in roasted and ground coffee. Crude Arabica coffee beans with an initial moisture content of 0.1234 $kg_w kg_{dm}^{-1}$ were used. These beans were roasted to a medium roast level (SCCA # 55) and ground at three particle sizes: coarse (1.19 mm), medium (0.84 mm) and fine (0.59 mm). To obtain the water sorption isotherms and the isosteric heat, different conditions of temperature and relative humidity were analyzed using the dynamic method at 25°C (0.50, 0.60, 0.70, and 0.80 of RH) and 30°C (0.30, 0.40, 0.50, 0.60, 0.70, and 0.80 of RH) and using the static method at 25°C (0.332 and 0.438 of RH). The GAB model best represented the hygroscopic equilibrium of roasted coffee at every particle size. Isosteric heat of sorption for the fine particle size increased with increments of equilibrium moisture content, indicating a strong bond energy between water molecules and the product components. The Gibbs free energy decreased with the increase in equilibrium moisture content and with temperature.

Keywords: particle size, specific surface, isosteric heat.

Introduction

The popularity of coffee is not based on its nutritional value or potential health benefits but is due to its flavor and stimulating effects. Currently, several new coffee products are being developed, such as capsules and sachets. Commercialization and success of these types of products require the correct preservation of ground, roasted coffee beans. Different factors can influence the preservation of coffee and consequently its final quality, including cultivation, harvest and post-harvest processing (for example, grinding, roasting and storage).

The purpose of grinding is to increase the surface area for extraction, as it increases the interface between water and the solid to facilitate the transfer of soluble substances to the brewing water, which affect cup quality (Illy & Viani, 1996). However, grinding promotes cell breakage and enables the release of molecular components, which can make quality preservation difficult (Corrêa et al., 2016a).

Roasting also affects the degree of moisture sorption, with coffee being more (dark roasts) or less (light roasts) hygroscopic according to the degree of roasting (Corrêa, Oliveira, Oliveira, Vargas-Elías, & Baptestini, 2016b). Roasted coffee undergoes important chemical and physical changes that affect its quality and adequacy. Water, oxygen, and light all influence roasted coffee quality during storage (Kreuml, Majchrzak, Ploederl, & Koenig, 2013).

Temperature and relative humidity of the environment surrounding the product are the primary parameters to be considered during storage. They determine the water activity (a_w) of the product and thus the water exchange between the product and the environment (Henao, Queiroz, & Haj-Isa, 2009). Due to its low moisture content, roasted and ground coffee may absorb moisture from the environment, causing clumping and inhibiting handling and movement of the product as well as increasing economic losses (Oliveira et al., 2014). Sorption isotherms are a key metric for assessing water sorption during storage and are affected by coffee particle size and roasting (Oliveira et al., 2014).

Along with water sorption, the thermodynamic sorption properties of a material may be monitored by assessing the a_w. According to Corrêa, Oliveira, and Santos (2012), the thermodynamic properties of agricultural products is a key resource that can provide information for assessing the effect of a_w on product storage as well as food microstructure, and they can aid in the understanding of the properties of the adsorbed water. This information enables the study of the physical phenomena that occur on food surfaces.

Given these reasons, as well as the importance of understanding the hygroscopicity of agricultural products, the objective of this study was to determine the adsorption isotherms of roasted coffee ground at different particle sizes. Additionally, this study sought to determine the thermodynamic properties of water sorption as a function of a_w.

Material and methods

The present work was developed at the Physical Properties and Quality of Agricultural Products Laboratory, belonging to the National Center of Storage Training (CENTREINAR), located at the Federal University of Viçosa, Viçosa, Minas Gerais State, Brazil.

Sample preparation

Crude coffee (*Coffea arabica* L.) were used for these experiments, with average size above sieve number 16 and initial moisture content of 0.1234 $kg_w\ kg_{dm}^{-1}$. A direct burning roaster using GLP gas, with a rotating cylinder at 45 rpm and capacity of 350 g of crude grain was used to roast the beans. The average temperature of roaster was 284°C. Roasting was complete when bean mass temperature reached 235°C, after approximately 16 minutes, which led to 17% weight loss. This characteristic matches the medium roast standard for coffee beans, based on the Agtron number (SCCA # 55). Sample temperature measurement was accomplished with an infrared thermometer.

After roasting, coffee was ground at three particle sizes: coarse (1.19 mm), medium (0.84 mm) and fine (0.59 mm), using milling equipment (Mahlkönig, model K32 S30LAB, Germany), according to Corrêa et al. (2016b).

Sorption isotherms

Moisture content was measured as described by the Regras para análise de sementes (Rules for Seed Testing) (Brasil, 2009). Dynamic and static sorption methods were used to measure the equilibrium moisture content of roasted and ground coffee at different particle sizes.

The following combinations of temperature and relative humidity were used for the dynamic method: 25°C (0.50, 0.60, 0.70, and 0.80 of RH) and 30°C (0.30, 0.40, 0.50, 0.60, 0.70, and 0.80 of RH). These conditions of temperature and humidity were created with an Aminco-Aire atmosphere conditioner unit (Aminco Company).

BOD (Biological Oxygen Demand) chambers kept at 25°C were used for the static method, with desiccators using saturated solutions of potassium carbonate and hydrated magnesium chloride (relative humidities of 0.332 and 0.438, respectively). These saturated solutions were prepared according to techniques cited by Hall (1980).

Samples of 10 g were used and weighed periodically until the mass variation among three consecutive measurements was less than or equal to 0.01 g, signifying hygroscopic equilibrium.

Mathematical models were fitted to the hygroscopic equilibrium data to describe the sorption phenomenon (Table 1).

Table 1. Mathematical models to represent hygroscopic equilibrium curves.

Model name	Model	
Modified Oswin	$U_e = \dfrac{a + bT}{\left[\dfrac{(1-a_w)}{a_w}\right]^{\frac{1}{c}}}$	(1)
Halsey	$U_e = \left[\dfrac{\exp(a - bT)}{-\ln(a_w)}\right]^{\frac{1}{c}}$	(2)
Chung-Pfost	$U_e = a - b\ln\left[(c - T)\ln(a_w)\right]$	(3)
Sigma-Copace	$U_e = \exp\{a - (bT) + [c\exp(a_w)]\}$	(4)
Copace	$U_e = \exp[a - (bT) + (c\,a_w)]$	(5)
Guggnheim-Anderson-de Boer (GAB)	$U_e = \dfrac{U_m C K a_w}{(1 - K a_w)(1 - K a_w + C K a_w)}$	(6)

In which, U_e: equilibrium moisture content; aw: water activity, decimal; a, b e c: parameters of the models and T: temperature, °C; Um: monolayer moisture content, % d.b.; and C and K: parameters of the GAB model which are related to the dimensionless monolayer and multilayer properties, respectively.

The standard deviation of the estimate (SDE), mean relative error (MRE) values, and residual plot were analyzed to assess the goodness of the model fit. The MRE and SDE values of each model were calculated using Equations 7 and 8, respectively:

$$MRE = \frac{100}{n}\sum_{i=1}^{n}\frac{\left|Y_i - \hat{Y}_i\right|}{Y_i} \tag{7}$$

$$SDE = \sqrt{\frac{\sum_{i=1}^{n}\left(Y_i - \hat{Y}_i\right)}{GLR}} \tag{8}$$

In which: MRE = mean relative error, %; SDE = standard deviation of the estimate, % d.b.; Y = observed value; $\hat{Y}$ = estimated value by the model; n = number of observed data; and DF = residue degrees of freedom (number of observed data minus number of model parameters).

Thermodynamic properties

The isosteric heat of sorption (Q^{st}) can be defined as a partial molar property, derived from the isotherm dependence on temperature (Equation 9).

$$\left[\frac{d\ln(P)}{dT}\right]_w = \frac{Q^{st}}{R\,T_a^2} \tag{9}$$

where: P = partial pressure of water vapor; R = water vapor constant, 0.462 kJ kg^{-1} K^{-1}; T_a = absolute temperature, K; Q^{st} = isosteric heat of sorption, kJ kg^{-1}.

The Clausius-Clapeyron equation is often used to establish the isosteric heat of sorption because it provides a rapid computational procedure. From that equation, applied to pure water, the following was obtained:

$$\left\{\left[d\ln\left(\frac{P}{P_0}\right)\right]dT\right\}_w = \left[d\ln\left(\frac{a_w}{dT}\right)\right]_w = \frac{Q_n^{st}}{R\,T_a^2} \tag{10}$$

where, P_o = saturation pressure at temperature T_a; Q_n^{st} = net isosteric heat of sorption, kJ kg^{-1}.

The net isosteric heat of sorption is the difference between the isosteric heat and pure water vaporization energy (L_r). This variable can be calculated according to Equation 10 by plotting the sorption isotherms as ln (a_w), at a specific moisture content, vs. $1/T_a$, and measuring the slope (Wang & Brennan, 1991).

Equation 10 is based on the assumption that Q_n^{st} is invariant with T_a. Although not always true, this assumption has often been accepted (Iglesias & Chirife, 1976). This method requires the measurement of sorption isotherms at more than two temperatures. The isosteric heat may also be calculated from the integrated form of Equation 11 applied to sorption isotherms measured at two temperatures (Wang & Brennan, 1991).

$$Q_n^{st} = R\left\{\left[\frac{T_{a1}\,T_{a2}}{T_{a2} - T_{a1}}\right]\ln\left(\frac{a_{w2}}{a_{w1}}\right)\right\} \tag{11}$$

The Gibbs free energy was calculated using Equation 12. Free energy changes are positive for desorption and negative for adsorption (Atkins, 2000).

$$\Delta G = \pm R\,T_a\,\ln\left(a_w\right) \tag{12}$$

Results and discussion

Table 2 presents the parameters of the models fitted to the equilibrium moisture content of roasted coffee ground to different particle sizes at different temperature and relative humidity conditions. The respective values of the coefficient of determination (R^2), standard deviation of the estimate (SDE) and mean relative error (MRE) are also included.

The coefficient of determination values were higher than 95% for most of the models, but this parameter cannot be used on its own to decide which model best fits the process as it is based on linear models. Low values of SDE indicate good fit of the model as a description of the sorption process (Draper & Smith, 1998), and low values of MRE (lower than 10%) indicate a good fit for practical purposes (Samapundo et al., 2007). Thus, Halsey and Chung-Pfost models are not suitable because values of MRE and SDE at medium particle size are inadequate (Table 2).

The remaining models can be used to predict the sorption isotherms of roasted coffee at different particle sizes. However, the GAB model is superior to empirical models, as its coefficients are correlated to physical changes, allowing a more detailed study of the heat and mass transfer during sorption.

Figure 1 was prepared to verify if the GAB model has biased or random residual plots. The GAB model presents random residual plots, indicating that this model can be used to describe the sorption kinetics of roasted and ground coffee. Thus, the GAB model was chosen to predict sorption isotherms of roasted coffee in different particle sizes.

Table 2. Parameters of the models fitted to equilibrium moisture content of roasted coffee, in different particle sizes, with coefficient of determination (R^2), standard deviation of the estimate (SDE) and mean relative error (MRE), stored at 25 and 30°C.

Models	Particle sizes		
	Coarse	Medium	Fine
Modified Oswin	a = 5.8873	a = 6.6994	a = 6.4785
	b = -0.0303	b = -0.0671	b = -0.0681
	c = 1.5667	c = 1.6029	c = 1.6404
	R^2 = 0.9827	R^2 = 0.9519	R^2 = 0.9643
	SDE (% d.b.) = 0.424	SDE (% d.b.) = 0.674	SDE (% d.b.) = 0.526
	MRE (%) = 6.47	MRE (%) = 9.34	MRE (%) = 8.64
Halsey	a = 1.8672	a = 2.142803	a = 2.160087
	b = 0.0077	b = 0.018049	b = 0.019662
	c = 1.2540	c = 1.282622	c = 1.309824
	R^2 = 0.9759	R^2 = 0.9440	R^2 = 0.9592
	SDE (% d.b.) = 0.501	SDE (% d.b.) = 0.727	SDE (% d.b.) = 0.563
	MRE (%) = 8.35	MRE (%) = 10.85	MRE (%) = 9.66
Chung-Pfost	a = 29.1515	a = 23.4887	a = 22.7647
	b = 5.6263	b = 5.2635	b = 4.7898
	c = -72.1519	c = -20.1356	c = -33.4663
	R^2 = 0.9768	R^2 = 0.9519	R^2 = 0.9532
	SDE (% d.b.) = 0.491	SDE (% d.b.) = 0.674	SDE (% d.b.) = 0.603
	MRE (%) = 7.45	MRE (%) = 10.16	MRE (%) = 8.27
Sigma-Copace	a = -0.8067	a = -0.5718	a = -0.5369
	b = 0.0062	b = 0.0141	b = 0.0151
	c = 1.5640	c = 1.5283	c = 1.4935
	R^2 = 0.9804	R^2 = 0.9493	R^2 = 0.9613
	SDE (% d.b.) = 0.451	SDE (% d.b.) = 0.692	SDE (% d.b.) = 0.548
	MRE (%) = 6.97	MRE (%) = 9.87	MRE (%) = 8.98
Copace	a = 0.3184	a = 0.5280	a = 0.5443
	b = 0.0060	b = 0.0139	b = 0.0150
	c = 2.9193	c = 2.8515	c = 2.7778
	R^2 = 0.9831	R^2 = 0.9536	R^2 = 0.9616
	SDE (% d.b.) = 0.419	SDE (% d.b.) =0.662	SDE (% d.b.) =0.546
	MRE (%) = 5.84	MRE (%) =8.99	MRE (%) = 8.35
GAB 25°C	U_m = 4.851	U_m = 5.263	U_m = 4.123
	K = 0.831	K = 0.791	K = 0.851
	C = 2.510	C = 2.321	C = 2.707
	R^2 = 0.9920	R^2 = 0.9495	R^2 = 0.9798
	SDE (% d.b.) = 0.280	SDE (% d.b.) = 0.676	SDE (% d.b.) = 0.409
	MRE (%) = 3.67	MRE (%) = 6.44	MRE (%) = 7.13
GAB 30°C	U_m = 4.292	U_m = 4.398	U_m = 3.185
	K = 0.882	K = 0.859	K = 0.898
	C = 2.143	C = 1.861	C = 4.030
	R^2 = 0.9220	R^2 = 0.9666	R^2 = 0.9487
	SDE (% d.b.) = 0.447	SDE (% d.b.) = 0.596	SDE (% d.b.) = 0.639
	MRE (%) = 6.57	MRE (%) = 8.33	MRE (%) = 10.07

Figure 2 presents sorption isotherms, estimated by the GAB model, for different particle sizes of coffee, at 25 and 30°C.

According to Oliveira, Corrêa, Santos, Treto, and Diniz (2011), monolayer moisture content indicates the amount of water molecules that are strongly adsorbed at specific sites on the food surface. This value is an important measurement for ensuring foodstuff stability (Lavoyer, Gabas, Oliveira, & Telis-Romero, 2013). Values of this coefficient varied between 3.185 to 5.293% d.b. Minimal deterioration is expected in products with a lower moisture content compared to a monolayer moisture content because the water is strongly bound and does not participate in chemical reactions (Dalgiç, Pekmez, & Belibagl, 2012).

The C constant is associated with the chemical potential differences between the monolayer and superior layers. As observed in Table 2, this constant did not present a defined behavior in the present work, although this trend was reported previously (Oliveira et al., 2011; Goneli, Corrêa, Oliveira, & Afonso Júnior, 2013). According to Goneli et al. (2013), this is due to the heterogeneous nature of coffee products, and variance among layers may lead to this trend.

The constant K corresponds to the difference at the chemical potential between the monolayer and free water. These numbers are, with rare exceptions, close; K is consistently lower than unity (Timmermann, Chirife, & Iglesias, 2001) (Table 2).

Figure 2 shows the effect of temperature on sorption isotherms of roasted coffee. With increased temperature at a constant water activity value, equilibrium moisture content is reduced. According to Palipane and Driscoll (1992), as temperature increases, water molecules become thermodynamically less stable, favoring bond breakage with sorption sites and reducing the moisture content of the product. The excitation, distance, and consequently the attraction of molecules vary with temperature alteration. Thus, the amount of absorbed water changes due to temperature variation at a given relative humidity (Mohsenin, 1986).

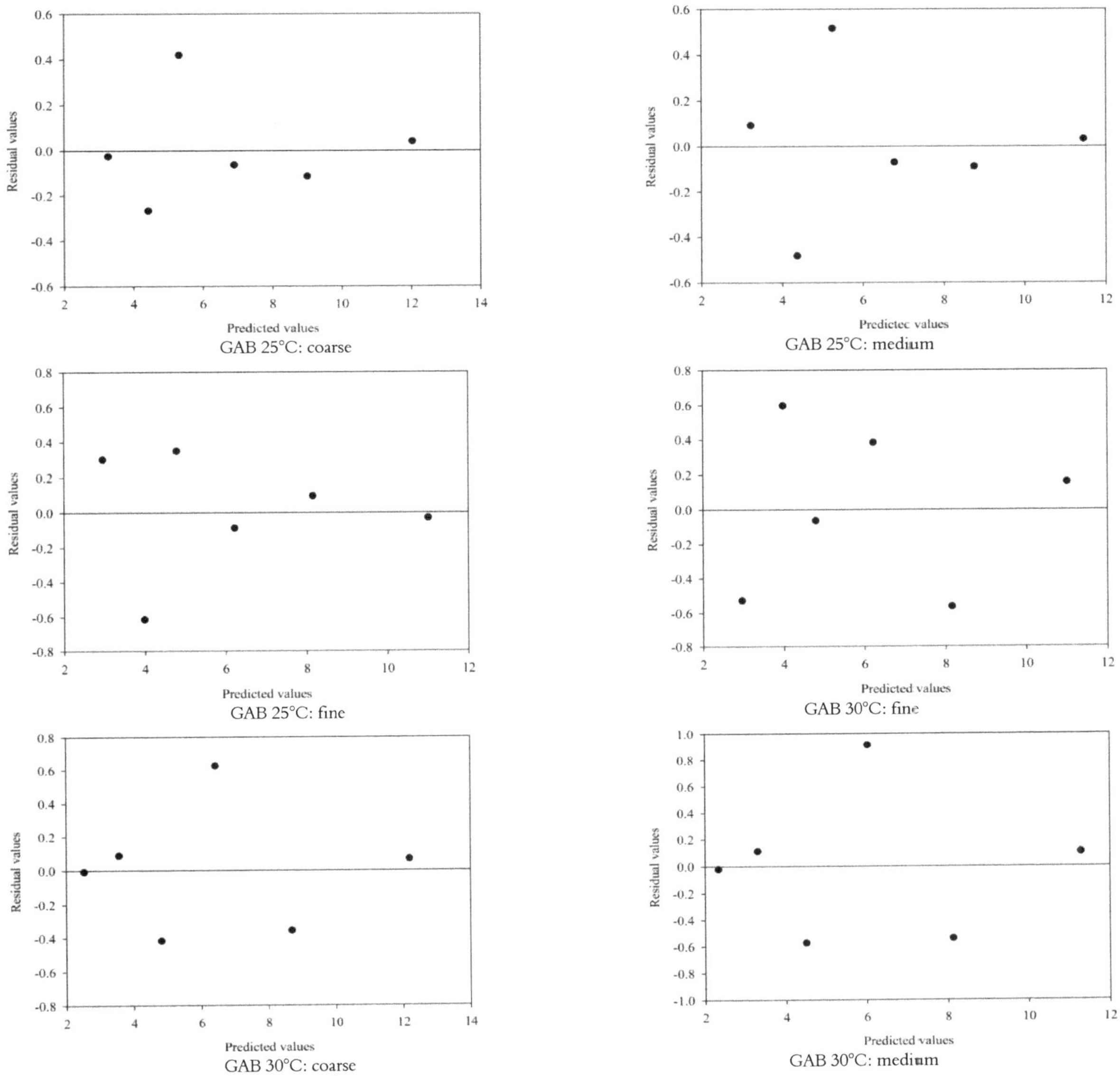

Figure 1. Residual plots of GAB model.

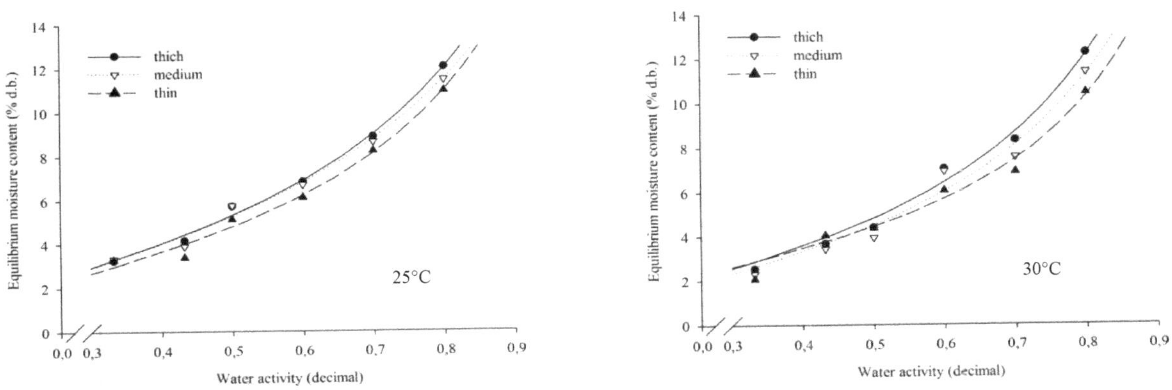

Figure 2. Observed values and sorption isotherms of coffee, estimated by the GAB model, for different particle sizes, at temperatures of 25 and 30°C.

Sorption isotherms of roasted coffee presented a pronounced increase of moisture content in the final values, corresponding to elevated water activity values (Figure 2). Analogous behavior was observed by Corrêa, Afonso Júnior, and Stringheta (2000), and Anese, Manzocco, and Nicoli (2006), suggesting a similar trend for most products of agricultural origin. Furthermore, from Figure 2 it can be concluded that particle size significantly altered the equilibrium moisture content of roasted and ground coffee, especially at higher values of water activity (> 0.80). This trend can be explained by changes to the specific area: lower particle sizes lead to higher specific area and consequently higher interaction with the environment, reaching higher values of equilibrium moisture content.

The moisture content of roasted and ground coffee established by Brazilian law cannot exceed 5.26% (d.b.). Considering this value, regardless of temperature, this limit is surpassed if starting from a water activity of 0.55 (Figure 2), compromising physical, chemical and biological stability of the products.

Observed and estimated values of the integral isosteric heat of sorption, as a function of equilibrium moisture content of roasted coffee samples, in different particle sizes, are shown in Figure 3.

Figure 3 shows an increase of integral isosteric heat of sorption with decrease of equilibrium moisture content, for roasted coffee at medium and coarse particle sizes. This trend was also reported for other agricultural products (Cano-Higuita, Villa-Vélez, Telis-Romero, Váquiro, & Telis, 2015; Ascheri, Moura, Ascheri, & Junior, 2009; Goula, Karapantsios, Achilias, & Adamoupolos, 2008). This is because, at the beginning of sorption, there are highly active polar sites of elevated interaction energy at the product surface, which are covered with water molecules, forming a monomolecular layer. As water molecules chemically link to active sites, sorption starts to occur at less active sites, with lower interaction energy and, consequently, lower isosteric heat of sorption (Pérez-Alonso, Beristain, Lobato-Calleros, Rodríguez-Huezo, & Vernon-Carter, 2006).

An inverse trend was observed for fine particle size in roasted coffee (Figure 3). This same response was observed by Oliveira, Corrêa, Oliveira, Baptestini, and Vargas-Elías (2016). The author explained that water adsorption occurred in vapor form, since the energy released by the process did not reach the value of the latent heat of condensation. The fact that the energy released is lower than that for free water indicates that the bond strength between water molecules and active sites in dry matter of the product is weaker than the bonds between water molecules.

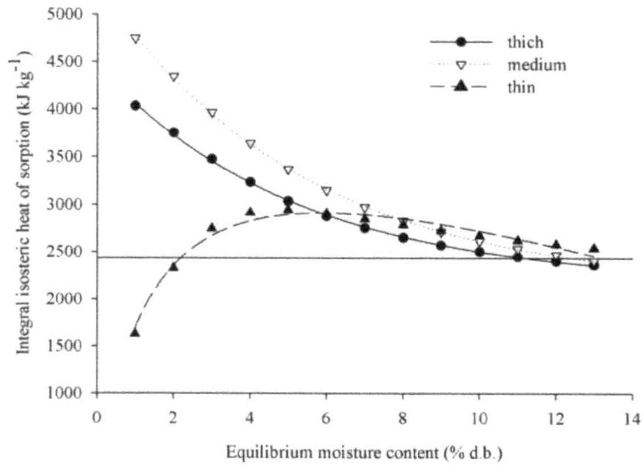

Figure 3. Observed and estimated values of integral isosteric heat of sorption of roasted coffee, for different particle sizes.

The latent heat of vaporization, representing the minimum amount of energy required to evaporate water molecules, was 2436.48 kJ kg^{-1}. Figure 3 shows that integral isosteric heat of sorption was always higher than latent heat of vaporization, until equilibrium moisture content reached 11 and 12% d.b., for coarse and medium particle size, respectively. This trend is because the bonding energy between water molecules and sorption sites is higher than the bonding energy among water molecules at liquid phase (Masuzawa & Sterling, 1968). At these points, it reaches minimum values of 2402.97 and 2108.42 kJ kg^{-1} for coarse and medium particle sizes, respectively, indicating that the liquid isosteric heat is negative. According to Tsami (1991), the more negative the value of heat of sorption, the higher the degree of water binding on the food.

Table 3 presents fitted equations of integral isosteric heat of sorption for roasted and ground coffee in different particle sizes. The proposed equations showed good fitting, with a coefficient of determination above 97%.

Table 3. Fitted equations of integral isosteric heat of sorption for roasted and ground coffee in different particle sizes.

Particle sizes	Models	R^2 (%)
Coarse	$Q^{st} = 2171.2697 + 2286.9769 \times \exp(-0.1931 \times U_e)$	99.94
Medium	$Q^{st} = 2127.2020 + 3197.2904 \times \exp(-0.1884 \times U_e)$	99.97
Fine	$Q^{st} = \frac{4649.1579 \times U_e}{(1.5524 + U_e) - 130.5183 \times U_e}$	97.15

The Gibbs free energy of roasted and ground coffee in coarse, medium and fine particle sizes are presented in Figure 4.

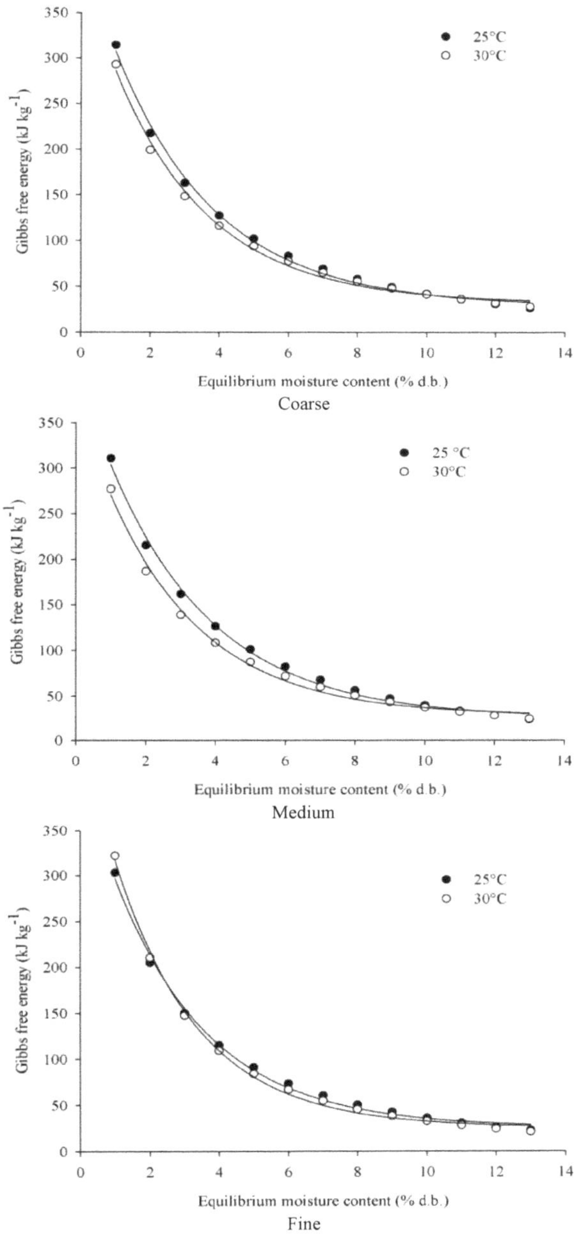

Figure 4. Gibbs free energy of roasted and ground coffee in coarse, medium and fine particle sizes.

Changes of the Gibbs free energy during water sorption between the product and the environment can be defined as required energy to transfer water molecules from the vapor state to a solid surface, or vice versa. This is the amount of work generated by the system during the adsorption or desorption process (Kaleemullah & Kailappan, 2007).

Gibbs free energy values decreased with equilibrium moisture content and temperature increment. Positive values of Gibbs free energy are also expected, corresponding to endothermic reactions, which require energy from the environment to occur (Goneli et al., 2013).

From Figure 4, it is observed that influence of temperature over Gibbs free energy is higher at lower values of equilibrium moisture content: lower temperatures provide higher values of Gibbs free energy. This trend was also observed by Goneli et al. (2013), as they correlated temperature to the vibration of water molecules within the product; at high temperatures this vibration (movement) is higher, leading to lower work required for the product layers to become available to sorption. At high levels of equilibrium moisture content, the temperature influence becomes negligible because the sorption sites are already available.

Table 4 presents fitted equations of Gibbs free energy of roasted and ground coffee in different particle sizes. The proposed equations presented good fitting, with coefficients of determination above 99%.

Table 4. Fitted equations of Gibbs free energy of roasted and ground coffee of different particle sizes.

Particle sizes	Temperature (°C)	Models	R^2 (%)
Coarse	25	$\Delta G = 28.7026 + 393.3855 \times \exp(-0.3442 \times U_e)$	99.83
	30	$\Delta G = 32.7806 + 367.5229 \times \exp(-0.3673 \times U_e)$	99.58
Medium	25	$\Delta G = 25.4401 + 388.7621 \times \exp(-0.3347 \times U_e)$	99.63
	30	$\Delta G = 28.2859 + 349.9071 \times \exp(-0.3675 \times U_e)$	99.53
Fine	25	$\Delta G = 25.7021 + 391.8711 \times \exp(-0.3681 \times U_e)$	99.66
	30	$\Delta G = 25.6747 + 441.0900 \times \exp(-0.4161 \times U_e)$	99.74

Conclusion

The GAB, Modified Oswin, Sigma-Copace and Copace models are suitable to represent the hygroscopic equilibrium of water sorption by roasted coffee, ground to different particle sizes; however, the GAB model was chosen because its coefficients are correlated to physical changes.

Reduction of equilibrium moisture content leads to an increase of released energy by water adsorption at coarse and medium particle sizes, represented by the integral isosteric heat of sorption values.

Isosteric heat of sorption at the fine particle size increased with equilibrium moisture content increments, indicating strong bonding between water molecules and the product.

The Gibbs free energy decreased with equilibrium moisture content and temperature increments.

Acknowledgements

The authors thank CNPq and FAPEMIG for the financial support.

References

Anese, M., Manzocco, L., & Nicoli, M. C. (2006). Modeling the secondary shelf life of ground roasted coffee. *Journal of Agricultural and Food Chemistry*, *54*(15), 5571-5576.

Ascheri, D. P. R., Moura, W. S., Ascheri, J. L. R., & Junior, E. A. F. (2009). Propriedades termodinâmicas de adsorção de água do amido de rizomas do lírio-do-brejo (*Hedychium coronarium*). *Ciência e Tecnologia de Alimentos*, *29*(2), 454-462.

Atkins, W. P. (2000). *Chimie Physique* (6th ed., Chapter 28). Paris, Bruxelles: De Boeck Université.

Brasil. (2009). Ministério da Agricultura, Pecuária e Abastecimento. *Regras para análise de sementes*. Brasília, DF: MAPA/ACS.

Cano-Higuita, D. M., Villa-Vélez, H. A., Telis-Romero, J., Váquiro, H. A., & Telis, V. R. N. (2015). Influence of alternative drying aids on water sorption of spray dried mango mix powders: A thermodynamic approach. *Food and Bioproducts Processing*, *93*, 19-28.

Corrêa, P. C., Afonso Júnior, P. C., & Stringheta, P. C. (2000). Estudo do fenômeno de adsorção de água e seleção de modelos matemáticos para representar a higroscopicidade do café solúvel. *Revista Brasileira de Produtos Agroindustriais*, *2*(1), 19-25.

Corrêa, P. C., Oliveira, G. H. H., & Santos, E. S. (2012). Thermodynamics properties of agricultural products processes. In: I. Arana (Ed.). *Physical properties of foods: novel measurement techniques and applications*. Boca Raton, US: CRC.

Corrêa, P. C., Oliveira, G. H. H., Oliveira, A. P. L. R., Vargas-Elías, G. A., SANTOS, F. L., & Baptestini, F. M. (2016a). Preservation of roasted and ground coffee during storage Part 1: Moisture content and repose angle. *Revista Brasileira de Engenharia Agrícola e Ambiental*, *20*(6), 581-587.

Corrêa, P. C., Oliveira, G. H. H., Oliveira, A. P. L. R., Vargas-Elías, G. A., & Baptestini, F. M. (2016b). Particle size and roasting on water sorption in conilon during storage. *Coffee Science*, *11*(2), 221-233.

Dalçic, A. C., Pekmez, H., & Belibagl, B. K. (2012). Effect of drying methods on the moisture sorption isotherms and thermodynamics properties of mint leaves. *Journal of Food Science Technology*, *49*(4), 439-449.

Draper, N. R., & Smith, H. (1998). *Applied regression analysis. Wiley series in probability and mathematical statistics* (3rd ed.). New York, US: John Wiley & Sons.

Goneli, A. L. D., Corrêa, P. C., Oliveira, G. H. H., & Afonso Júnior, P. C. (2013). Water sorption properties of coffee fruits, pulped and green coffee. *LWT - Food Science and Technology*, *50*(2), 386-391.

Goula, A. M., Karapantsios, T. D., Achilias, D. S., & Adamoupolos, K. (2008). Water sorption isotherms and glass transition temperature of spray dried tomato pulp. *Journal of Food Engineering*, *85*(1), 73-83.

Hall, C. W. (1980). *Drying and storage of agricultural crops*. Westport, US: AVI.

Henao, J. D., Queiroz, M. R., & Haj-Isa, N. M. A. (2009). Umidade de equilíbrio de café cereja descascado baseada em métodos estático e dinâmico. *Revista Brasileira de Engenharia Agrícola e Ambiental*, *13*(4), 470-476.

Illy, A., & Viani, R. (1996). *Express coffee: the chemistry of quality*. London, UK: Academic Press.

Iglesias, H., & Chirife, J. (1976). Prediction of the effect of temperature on water sorption isotherms of food material. *Journal of Food Technology*, *11*(2), 109-116.

Kaleemullah, S., & Kailappan, R. (2007). Monolayer moisture, free energy change and fractionation of bound water of red chillies. *Journal of Stored Products Research*, *43*(2), 104-110.

Kreuml, M. T. L., Majchrzak, D., Ploederl, B., & Koenig, J. (2013). Changes in sensory quality characteristics of coffee during storage. *Food Science & Nutrition*, *1*(4), 267-272.

Lavoyer, F. C. G., Gabas, A. L., Oliveira, W. P., & Telis-Romero, J. (2013). Study of adsorption isotherms of green coconut pulp. *Food Science and Technology*, *33*(1), 68-74.

Masuzawa, M., & Sterling, C. (1968). Gel-water relationshisps in hydrophilic polymers: thermodynamics of sorption of water vapour. *Journal of Polymer Science*, *12*(9), 20-23.

Mohsenin, N. N. (1986). *Physical properties of plant and animal materials*. New York, US: Gordon and Breach Publishers.

Oliveira, G. H. H., Corrêa, P. C., Oliveira, A. P. L. R., Baptestini, F. M., & Vargas-Elías, G. A. (2016). Roasting, grinding, and storage impact on thermodynamic properties and adsorption isotherms of arabica coffee. *Journal of Food Processing and Preservation*. doi:10.1111/jfpp.12779

Oliveira, G. H. H, Corrêa, P. C., Santos, E. S., Treto, P. C., & Diniz, M. D. M. S. (2011). Evaluation of thermodynamics properties using GAB model to describe the desorption process cocoa beans. *International Journal of Food Science & Technology*, *46*(10); 2077-2084.

Oliveira, G. H. H., Corrêa, P. C., Santos, F. L., Vasconcelos, W. L., Calil Júnior, C., Baptestini, F. M., & Vargas-Elías, G. A. (2014). Caracterização física de café após torrefação e moagem. *Semina: Ciências Agrárias*, *35*(4), 1813-1828.

Palipane, K. B., & Driscoll, R. H. (1992). Moisture sorption characteristics of in-shell macadamia nuts. *Journal of Food Engineering*, *18*(1), 63-76.

Pérez-Alonso, C., Beristain, C. I., Lobato-Calleros, C., Rodríguez-Huezo, M. E., & Vernon-Carter, E. J. (2006). Thermodynamic analysis of the sorption isotherms of pure and blended carbohydrate polymers. *Journal of Food Engineering*, *77*(4), 753-760.

Samapundo, S., Devlieghere, F., De Meulenaer, B., Atukwase, A., Lamboni, Y., & Debevere, J. M. (2007). Sorption isotherms and isosteric heats of sorption of whole yellow dent corn. *Journal of Food Engineering*, *79*(1), 168-175.

Timmermann, E. O., Chirife, J., & Iglesias, H. A. (2001). Water sorption isotherms of foods and foodstuffs: BET or GAB parameters? *Journal of Food Engineering*, *48*(1), 19-31.

5

Calibration and testing of an agrometeorological model for the estimation of soybean yields in different Brazilian regions

Leonardo Amaral Monteiro[1*] and Paulo Cesar Sentelhas[2]

*[1]Programa de Pós-graduação em Engenharia de Sistemas Agrícolas, Escola Superior de Agricultura "Luiz de Queiroz", Universidade de São Paulo, Av. Pádua Dias, 11, 13418-900, Piracicaba, São Paulo, Brazil. [2]Departamento de Engenharia de Biossistemas, Escola Superior de Agricultura "Luiz de Queiroz", Universidade de São Paulo, Piracicaba, São Paulo, Brazil. *Author for correspondence. E-mail: leomonteiro@usp.br*

ABSTRACT. This study was designed to calibrate and test an agrometeorological model over 18 growing seasons in three soybean production areas in Brazil: Passo Fundo (Rio Grande do Sul State), Londrina (Paraná State), and Dourados (Mato Grosso do Sul State). The soybean potential yield (Yp) was determined by two methods: estimated using the FAO Agroecological Zone Model or based on the maximum yield published by the Brazilian Institute of Geography and Statistics (IBGE), increased by 10, 20 and 30%. The estimate of actual yield (Ya) was calculated by correcting Yp for the relative water deficit at different growth stages. The results showed that the best performance was obtained when the Yp was represented by the maximum yield increased by a certain percentage. The model showed satisfactory Ya estimates for the three locations, generating R^2 values of 0.64, 0.46 and 0.70, with mean absolute errors (MAE) of 303, 289 and 259 kg ha^{-1} for Passo Fundo, Londrina and Dourados, respectively. In a global analysis, the performance of the model was satisfactory, with an R^2, agreement modified index (d_1), confidence index (C) and MAE of 0.64, 0.52, 0.71 and 300 kg ha^{-1}, respectively.

Keywords: *Glycine max*, potential yield, actual yield, water deficit, simulation models.

Introduction

Agriculture is an important part of the Brazilian economy, in particular the jobs, income and revenues that are generated through the export of agricultural commodities. Today, in a modern and global agriculture, increases in crop yield as well as the assessment of climate risk systems define the capacity of a country to become competitive in the international market (EMBRAPA, 2006).

Soybeans are the most important Brazilian agricultural commodity. To improve the methods for estimating soybean yield and production, there is a need for technologies that make crop-forecasting systems more accurate. According to Oliveira et al. (2003), the official agencies responsible for gathering information about agricultural crop yields are still using subjective methods, such as interviews with farmers and agrochemical/fertilizer sales representatives.

For improved yield estimates, crop simulation models are important tools that make it possible to evaluate the influence of the weather on crop production. According to Confalonieri (2012), these models have been used to identify the best time for sowing, the climate risk associated with the crop and the yield gaps as a function of the interaction between the genotype and the environment under climate change scenarios.

Crop simulation based on agrometeorological models seeks to represent and understand biomass accumulation by plants in their different phenological stages because the impact of climate factors varies with the crop phase (ASSAD et al., 2007; POPP et al., 2003).

To estimate the soybean yield in an accurate way, some authors (ASSAD et al., 2007; BERKA et al., 2003; CONFALONE et al., 2010; MELO et al., 2008) have used agrometeorological models that consider water deficits as the main factor in crop yield reduction. Water deficiency, according to Geerts and Raes (2009), has caused significant impacts on agriculture around the world, mainly for rain-fed crops. In addition, these authors emphasize the importance of soil water availability as a tool for the feasibility of agricultural projects in some regions where water deficiency is substantial.

For soybeans, the sub-periods that are more sensitive to water deficits, and consequently to yield break, are flowering and grain filling (DOORENBOS; KASSAM, 1994). When water deficit occurs during these phases, the potential yield will be reduced to an attainable yield.

According to Pereira et al. (2002), Andarzian et al. (2008) and Lobell et al. (2009), the crop potential yield (Yp) is defined as the productivity obtained by a variety that is completely adapted to the environment of cultivation from sowing/planting through physiological maturity and for which water conditions, plant nutrition and phytosanitary problems are not limiting factors.

Therefore, the interannual variability of Yp is a function of meteorological conditions during the crop cycle. Recent studies showed a significant range in values for soybean Yp, from 3 ton ha^{-1}, as estimated by the DSSAT CROPGRO-Soybean Model for 21 regions in India (BHATIA et al., 2008), to yields close to 8 ton ha^{-1} for some soybean cultivars, as estimated by the STICS crop model in eastern Canada (JÉGO et al., 2010).

Due to the variety of methodologies used to estimate soybean yield, the aim of this paper was to calibrate and test the performance of the FAO agrometeorological model to identify the most suitable methodology to calculate the potential yield. This method was then used to estimate the actual yield during the crop seasons from 1990 to 2007 in Passo Fundo (Rio Grande do Sul State), Londrina (Paraná State) and Dourados (Mato Grosso do Sul State) in southern Brazil.

Material and methods

This study used daily meteorological data obtained during the period from 1990 to 2007 for the variables of maximum and minimum air temperature (°C), sunshine hours and rainfall (mm) in Passo Fundo, Rio Grande do Sul State (28°15'46"S, 52°24'24"W, 684 m), Londrina, Paraná State (23°08'47"S, 51°19'11"W, 610 m) and Dourados, Mato Grosso do Sul State (22°13'18"S, 54°48'23"W, 430 m), which are traditional regions for soybean production in Brazil. These data were obtained from Embrapa Trigo (Passo Fundo, Rio Grande do Sul State), IAPAR (Londrina, Paraná State) and Embrapa Agropecuária Oeste (Dourados, Mato Grosso do Sul State). The weather dataset was converted into a ten-day scale, which divides each month into 3 periods. To estimate the soybean yield, two methods were used to determine Yp. The first method is the Agroecological Zone Method (AZM) (DOORENBOS; KASSAM, 1979), which is shown in equation 1.

$$Yp = \sum_{i=1}^{12} (PPBp \times c_{resp} \times c_{harv} \times c_{LAI} \times c_{hum}) \quad (1)$$

where:

Yp is the final potential yield, calculated as the sum of the 10-day periods;

i is the 10-day period considered, varying from 1 to 12, totaling 120 days;

PPBp is the total potential yield in each 10-day period, expressed as dry matter of a standard crop (kg DM ha^{-1});

c_{resp} is the coefficient for maintenance respiration correction;

c_{harv} is the harvest index;

c_{LAI} is the coefficient for leaf area index correction, and

c_{hum} is the coefficient to add humidity to the grains. All coefficients are dimensionless.

The PPBp value was obtained by the sum of two components, PPBn (standard gross photosynthesis during cloudy periods) and PPBc (standard gross photosynthesis during clear sky conditions), considering that the energy available for photosynthesis is different in each distinct condition and thus affects the plant's energy absorption. The PPBc and PPBn values and all C coefficients were determined according to the methodology described by Pereira et al. (2002).

The second method assumed that Yp would be the highest value observed (YmO) in the historical series published by IBGE (1990-2007), increased by

10, 20 or 30% to minimize factors that are not considered by the agrometeorological model, such as soil conditions and crop management. A similar procedure for potential yield determination was used by Kanemasu (1983). According to this method, Yp is determined by the following equation:

$$Yp = YmO \times (1 + x) \quad (2)$$

where x is the percentage added to the maximum observed yield (10, 20 or 30%).

For the calculation of the actual yield (Ya), the relationship between relative yield break and relative water deficit, which are well correlated variables, was used (PATANÈ et al., 2011).

The actual yield was estimated by the product of the relative water deficit and the water response factor (ky) of each soybean phenological phase, as proposed by Doorenbos and Kassam (1994). The variable ky represents the relationship between the relative yield break and the relative water deficit $(1 - ETa/ETc)$, where ETc and ETa represent the maximum and actual crop evapotranspiration, respectively.

The maximum crop evapotranspiration (ETc) was estimated by the product of the crop coefficient (kc), proposed by Doorenbos and Kassam (1994), for each phenological phase of the soybean crop and the reference evapotranspiration (ETo), which was calculated by the Thornthwaite method using the effective temperature and was parameterized following Camargo et al. (1999). The actual evapotranspiration (ETa) was calculated using the crop water balance according to the method of Thornthwaite and Mather (1955).

The soil water holding capacity was defined as a soil with a medium texture and the capacity to hold 1 mm of water per 1 cm of effective depth. According to the results obtained by Bordin et al. (2008), the roots of the soybean can effectively reach a depth of approximately 50 cm, resulting in a total water holding capacity of 50 mm.

Because excess water can damage the soybean yield in the final phases of the cycle (from R1 to R8), a relative factor of water excess (FEx), which was suggested by Brunini et al. (1982) and parameterized by Camargo et al. (1986), was considered. This factor was calculated by:

$$FEx = 1 - ke \times [1 - \prod (fe)] \quad (3)$$

where ke is the water excess sensitivity coefficient and $\prod$ (fe) is the product of the 10-day water excess index corresponding to the period from the beginning of the flowering to the beginning of physiological maturity.

To calculate the 10-day (fe), the following equation was used:

$$fe = 1 - \left(\frac{EX - ETc}{EX} \right) \quad (4)$$

where EX is the water excess (mm), calculated by the crop water balance.

According to Brunini et al. (1982), the only restriction on using the water excess index is when the ETc is higher than the EX, in which case fe = 1.

The inclusion of the water excess factor (FEx) in the soybean crop simulation model is an important step because, according to Rhine et al. (2010), the occurrence of water excess during the reproductive phase reduces the number of flowers and pods and, thus, the final yield. The complete model used to estimate Ya is presented below:

$$Ya_m = \prod_{i=1}^{m} \left[1 - ky \times \left(1 - \frac{ETa}{ETc} \right) \times Ya_{m-1} \times FEx \right] \quad (5)$$

where:

Ya_m represents the final actual yield of the crop;

m is the phenological phase considered;

ky is the crop response factor to the water deficit, and;

Ya_{m-1} is the actual yield of the previous phenological phase (for the first phenological phase, it is equal to Yp).

The values of the crop coefficients (kc), the water response factor (ky) and the water excess sensitivity coefficient (ke) used in this study were the same for all locations.

Because the current methods of the crop yield survey do not consider the date of sowing, the agrometeorological model was used to simulate different sowing dates, as recommended by the Soybean Crop Agricultural Zone, for the three locations in this study (BRASIL, 2009).

For Passo Fundo and Londrina, the sowing time varied from October 21st to December 11th. In Dourados, the sowing time was set during November only. After the simulations of Yp and Ya for each location and sowing date, the final Ya for each year was calculated by the arithmetic average of all sowing dates.

To evaluate the agrometeorological model with different Yp values, a simple linear regression analysis was carried out for each location. Then, the model was calibrated to adjust kc, ky and ke to minimize the differences between the observed and estimated Ya.

The performance of the agrometeorological model was evaluated based on the statistical indices related to precision, represented by the coefficient of determination (R^2), and accuracy, represented by the agreement index, d_j (WILLMOTT et al., 1985), which was calculated using the following equation:

$$d_j = 1.0 - \frac{\sum_{i=1}^{N} |Oi - Pi|^j}{\sum_{i=1}^{N} (|Pi - Om| + |Oi - Om|)^j} \quad (6)$$

where:

Oi and Om are the actual and mean soybean observed yield, respectively;

Pi is the estimated soybean yield, and

j is an arbitrary positive power.

Usually, the agreement index (d_j) is calculated with j = 2; however, Legates and McCabe Jr. (1999) recommended a modified index of agreement (d_j), calculated with j = 1. The advantage of d_1 is that errors are given their appropriate weighting and are not inflated by their square values. For that reason, d_1 was used.

The confidence index, represented by C (CAMARGO; SENTELHAS, 1997), was calculated between the product of the correlation coefficient (r) and the original agreement index (d_2). The classification of the model performance by the C index is presented in Table 1.

Table 1. Classification for model performance based on the confidence index (C).

C index	Performance
> 0.85	Excellent
0.76 – 0.85	Very Good
0.66 – 0.75	Good
0.61 – 0.65	Reasonable
0.51 – 0.60	Poor
0.41 – 0.50	Very Poor
≤ 0.40	Extremely Poor

Adapted from: Camargo and Sentelhas (1997).

The mean absolute error (MAE, kg ha^{-1}) and the mean percentage error (MPE, %) were also evaluated.

In addition to the analysis using the statistical index, the observed and estimated Ya data were also evaluated using Student's t-test with a 5% probability to determine whether the differences between the data were significant.

Results and discussion

The soybean phenological phases were subdivided according to the classification proposed by Fehr and Caviness (1977). The number of days considered in each phenological phase was: 10 (S –V2); 50 (V2 – R1); 30 (R1 – R5); and 30 (R6 –R8). In Table 2, the kc, ky and ke calibrated values are presented for each phenological phase.

Table 2. Crop coefficient (kc), water response factor (ky) and water excess sensitivity coefficient (ke) calibrated for each phenological phase of the soybean crop.

Phenological Phase	kc	ky	ke
S – V2	0.30	0.2	-
V2 – R1	0.75	0.8	-
R1 – R5/R6	1.15	1.0	0.2
R6 – R8	0.80	-	0.2

These coefficients were calibrated according to the best fit between the estimated and observed Ya to obtain the highest values of R^2 and the smallest values of MAE for each location. Table 3 presents the statistical coefficients obtained by the relationship between the observed and estimated Ya with the different methods to determine Yp.

Table 3. Statistical indices of model performance for estimating actual yield (Ya) in Passo Fundo (RS), Londrina (PR) and Dourados (MS), considering potential yield (Yp) data calculated by the Agroecological Zone Model (AZM) and determined as the maximum observed (YmO) increased by 10% (1.1*YmO), 20% (1.2*YmO) and 30% (1.3*YmO). R^2 = determination coefficient; d_1 and d_2 = Willmott agreement index; confidence index (C); MAE = Mean Absolute Error (kg ha^{-1}); and APE = Average Percentage Error.

Location	Yp	R^2	d_2	d_1	C	MAE	APE
	AZM	0.61	0.78	0.56	0.61	439.1	27.0
Passo Fundo,	1.1*YmO	0.64	0.85	0.58	0.68	314.9	-1.0
RS	1.2*YmO	0.64	0.87	0.63	0.70	303.1	8.0
	1.3*YmO	0.64	0.85	0.63	0.68	343.5	18.0
	AZM	0.46	0.92	0.88	0.63	318.9	-9.8
Londrina,	1.1*YmO	0.45	0.86	0.82	0.57	523.5	-21.0
PR	1.2*YmO	0.46	0.89	0.85	0.61	418.6	-15.0
	1.3*YmO	0.46	0.93	0.89	0.63	289.3	-8.0
	AZM	0.63	0.76	0.55	0.61	556.8	27.0
Dourados,	1.1*YmO	0.70	0.96	0.81	0.80	258.8	1.0
MS	1.2*YmO	0.70	0.93	0.74	0.78	329.8	10.0
	1.3*YmO	0.67	0.83	0.63	0.68	427.2	21.7

As shown in Table 3, the best way to determine Yp for estimating Ya varied among locations. In Passo Fundo, the best results were obtained with the addition of 20% to the maximum observed yield (YmO), while in Londrina, the best statistical indices were obtained with a 30% addition to YmO. In Dourados, the best fit was found with the addition of 10% to the YmO.

For Passo Fundo, the best indices related to the precision (R^2), accuracy (d_1) and confidence (C) were 0.64, 0.63, and 0.70, respectively. The mean absolute error associated with the estimates was 303.1 kg ha^{-1}, which corresponds to approximately 5 bags of 60 kg per hectare, corresponding to a percentage error of 8%.

The best statistical results for Londrina were as follows: $R^2 = 0.46$, $d_1 = 0.89$, and $C = 0.63$. The mean absolute error was 289.3 kg ha^{-1}, which corresponds to almost 5 bags of 60 kg per hectare, corresponding to a mean percentage error of -8%.

In Dourados, the best statistical indices associated with the estimates were as follows: $R^2 = 0.70$, $d_1 = 0.81$, and $C = 0.80$. The mean absolute error was 258.8 kg ha^{-1}, which corresponds to approximately 4 bags of 60 kg per hectare, corresponding to a mean percentage error of 1%.

According to the performance classification of the models proposed by Camargo and Sentelhas (1997) that is presented in Table 1, the calibrated agrometeorological model performance was good for Passo Fundo, reasonable for Londrina and very good for Dourados.

Figure 1a presents the relationship between the observed and estimated Ya, which was calculated considering the potential yield as the highest YmO increased by 20% for Passo Fundo. Figure 1b presents the interannual variability of the soybean actual yield for the same location, reported by IBGE and estimated by the agrometeorological model.

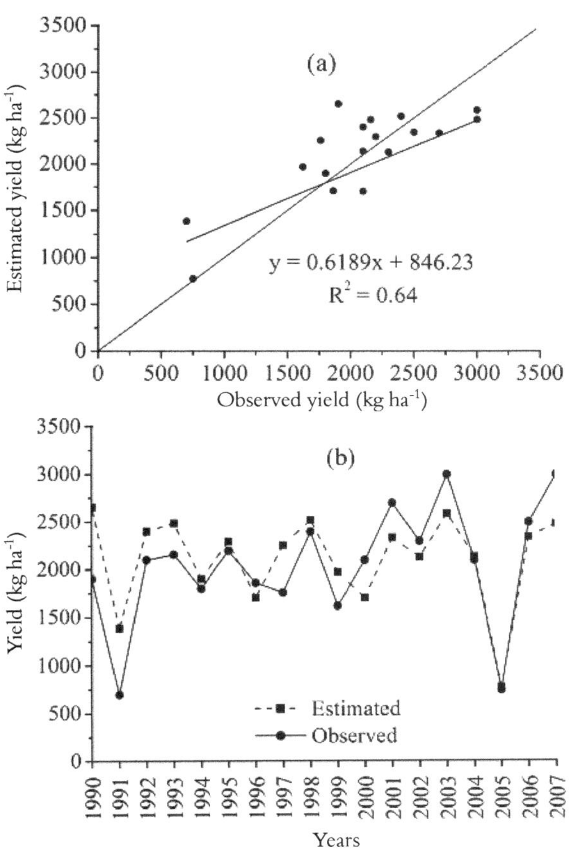

Figure 1. Relationship between the observed and estimated soybean actual yield (Yp = 1.2*YmO) (a) and their interannual variability (b) in Passo Fundo, Rio Grande do Sul State.

In Figures 1a and b, it is clear that the model can reproduce the Ya variability observed in the field, mainly in the critical years 1991 and 2005, when Ya fell to less than 1000 kg ha^{-1} as a result of the intense water deficit observed during the growing season, which was 150 and 200 mm, respectively.

The average soybean yield data obtained by the agrometeorological model in Passo Fundo (2,116 kg ha^{-1}) was similar to the data reported by IBGE for the same period (2,053 kg ha^{-1}); the values were not statistically different. However, the correlation coefficient (r) between the observed and estimated yields was significant (r = 0.80).

For Dourados, the results obtained by the model did not show a clear tendency to overestimate or underestimate high or low yields. Thus, this was the location where the model performed the best. Figure 2a presents the relation obtained by the simple linear regression between the observed and estimated yields in Dourados, and Figure 2b presents the interannual variability.

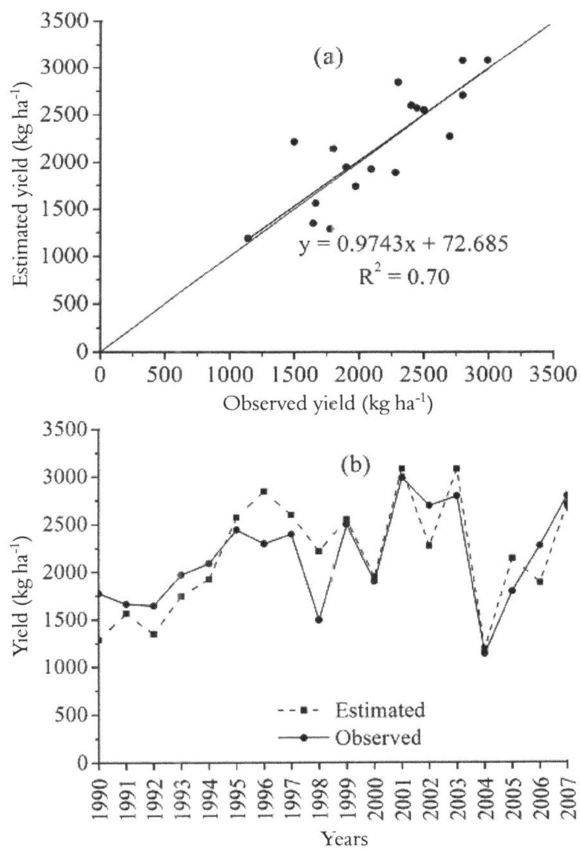

Figure 2. Relationship between the observed and estimated soybean actual yield (YP = 1.1*YmO) and their interannual variability (b) in Dourados, Mato Grosso do Sul State.

During the study period, the average actual yield estimated in Dourados (2170 kg ha^{-1}) was very similar to the observed yield (2150 kg ha^{-1}).These yields were not significantly different but had a significant correlation coefficient (r = 0.84).

The lowest yield in the series for Dourados (Figure 2) was obtained in the 2004-2005 season. This occurred, according to Minuzzi et al. (2010), because of the low percentage of seed germination during this season, which was induced by the high soil temperature and low water availability. It is clear that for years in which the model estimated the yield with high accuracy (for example, 1999, 2000, 2001 and 2004), the climate factors were primarily responsible for yield break. In the years when the errors were higher, soybean yield was affected by other factors primarily related to crop management, such as fertilization and pest and disease control (LOBELL et al., 2009).

Assad et al. (2007) also noted some divergence between the results generated by the FAO model and the average observed yield data for Mato Grosso do Sul. Specifically, for Dourados, the model presented errors ranging from +3 to -17%.

The simple linear regression between the estimated and observed Ya in Londrina was similar to that for Passo Fundo, with a predominance of overestimation (Figure 3a) mainly from 1997 to 2005 (Figure 3b). The average yields were 2,210 and 2,400 kg ha^{-1} for the estimated and observed Ya, respectively. The correlation between the estimated and observed Ya (r = 0.68) was statistically significant.

Most of the differences between the estimated and observed Ya values in Londrina arose because the model was not able to reproduce the technological level adapted in the region, even when Yp was set to 1.3*YmO. Some exceptions, such as 1991 and 2006, were related to specific scenarios for the soybean crop, where economic and cultural contexts limited soybean yield when the model projected an increase.

Using the simulation model SOYGRO for the soybean crop in Londrina, Klosowsky (1997) obtained similar errors, varying from -20 to +20%, among the different sowing dates and growing seasons.

Considering the dataset for the three locations studied (Figure 4), it is possible to observe a satisfactory performance of the model, with a small underestimation of approximately 5%, resulting in a coefficient of determination (R^2) of 0.64, an agreement modified index (d_1) of 0.52 and a confidence index (C) of 0.71. According to Camargo and Sentelhas (1997), the model can be classified as having good performance (Table 3). On average, the absolute error was approximately 5 bags of 60 kg per hectare or 300 kg ha^{-1}.

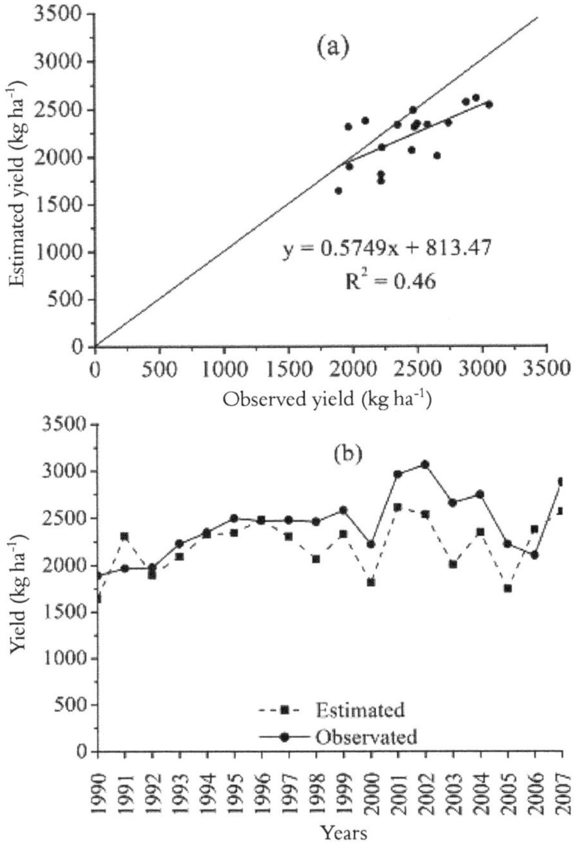

Figure 3. Relationship between the observed and estimated soybean actual yield (YP = 1.3*YmO) and their interannual variability (b) in Londrina, Paraná State.

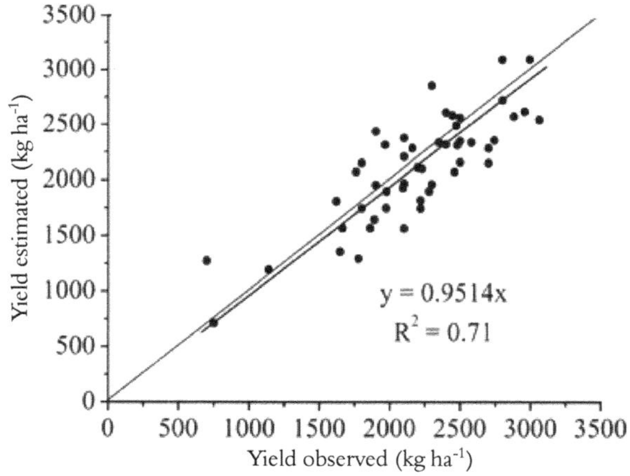

Figure 4. Relationship between observed and estimated soybean actual yield considering the entire dataset for the locations of Passo Fundo (Rio Grande do Sul State), Dourados (Mato Grosso do Sul State) and Londrina (Paraná State).

Based on the results, the use of the agrometeorological model can be recommended for evaluating the climate risks for soybean production in different parts of Brazil because it can reproduce the yield variability observed. However, the determination of Yp should be performed in accordance with the technological level of each location/region to obtain better results (LOBELL et al., 2009). The method for determining the Yp for soybean crops varied among locations. From the methods used, the calculation of Yp by the Agrometeorological Zone Method generated the highest errors in Passo Fundo and Dourados and is not recommended.

The best way to determine Yp for the soybean crop was by adding a percentage to the maximum yield of the historical series. In Passo Fundo, this percentage was 20%, whereas in Londrina, it was 30%, and in Dourados, it was only 10%. The performance of the agrometeorological model for estimating soybean yield also varied among locations. It was classified as good in Passo Fundo, reasonable in Londrina and very good in Dourados. These results are remarkable, considering that there are several others factors that can affect soybean actual yield, such as crop management.

When the entire data set was analyzed as a whole, the performance of the crop model remained good. However, more importantly, the model demonstrated its ability to simulate all ranges of yield variation under very different environments, such as the three locations studied. Based on these results, we recommend the use of agrometeorological models for soybean yield simulation in studies with the objective of evaluating the climate risk associated with water deficit.

Conclusion

The best way to determine Yp for the soybean yield estimate was by adding a percentage to the maximum yield of the historical series. By this way, the used model demonstrated its ability to simulate all ranges of yield variation under very different climatic conditions, showing the potential of this tool for developing studies about climatic risk for soybean production in Brazil as well as for improving the yield monitoring by the state and national governmental agencies.

Acknowledgements

We would like to thank IAPAR (Londrina), Embrapa Trigo (Passo Fundo) and Embrapa Agropecuária Oeste (Dourados) for providing the meteorological data used in this study.

References

ANDARZIAN, B.; BAKHSHANDEH, A. M.; BANNAYAN, M.; EMEM, Y.; FATHI, G.; ALAMI SAEED, K. WheatPot: a simple model for spring wheat yield potential using monthly weather data. **Biosystems Engineering**, v. 99, n. 4, p. 487-495, 2008.

ASSAD, E. D.; MARIN, F. R.; EVANGELISTA, S. R.; PILLAU, F. G.; FARIAS, J. R. B. F.; PINTO, H. S.; ZULLO JUNIOR, J. Sistema de previsão da safra de soja para o Brasil. **Pesquisa Agropecuária Brasileira**, v. 42, n. 5, p. 615-625, 2007.

BERKA, L. M. S.; RUDORFF, B. F. T.; SHIMABUKURO, Y. E. Soybean yield estimation by an agrometeorological model in a GIS. **Scientia Agricola**, v. 60, n. 3, p. 433-440, 2003.

BHATIA, V. S.; SINGH, P.; WANI, S. P.; CHAUHAN, G. S.; RAO, A. V. R.; MISHRA, A. K.; SRINIVAS, K. Analysis of potential yields and yield gaps of rainfed soybean in India using CRCPGRO-Soybean model. **Agricultural and Forest Meteorology**, v. 148, n. 8-9, p. 1252-1265, 2008.

BORDIN, I.; NEVES, C. S. V.; MEDINA, C. C.; SANTOS, J. C. F.; TORRES, E.; SEGUNDO URQUIAGA, S. Matéria seca, carbono e nitrogênio de raízes de soja e milho em plantio direto e convencional. **Pesquisa Agropecuária Brasileira**, v. 43, n. 12, p. 1785-1792, 2008.

BRASIL. Portaria n.° 126 do dia 20 de julho de 2009, Portaria n.° 136 do dia 22 de julho e Portaria n.° 184 do dia 13 de agosto de 2009. Dispõem sobre a aprovação do Zoneamento Agrícola para a cultura da soja nos Estados do Rio Grande do Sul, Mato Grosso do Sul e Paraná, respectivamente. **Diário Oficial da União**, Brasília, 13 de março de 2009. Seção 1, p. 32.

BRUNINI, O.; MIRANDA, M. A. C.; MASCARENHAS, H. A. A. L.; PEREIRA, J. C. V. N. A.; SCHMIDT, N. C. Teste de um modelo agroclimático que relacione o regime pluviométrico com as variações da produtividade agrícola. In: SEGOVIA, R. M. (Ed.). **Determinação do efeito da precipitação pluviométrica na produção agrícola**. Brasília: CFP, 1982. p. 21-46. (Coleção Análise e Pesquisa, 24).

CAMARGO, M. B. P.; BRUNINI, O.; MIRANDA, M. A. C.; Modelo agrometeorológico para estimativa da produtividade para a cultura da soja no estado de São Paulo. **Bragantia**, v. 45, n. 2, p. 279-292, 1986.

CAMARGO, A. P.; SENTELHAS, P. C. Avaliação do desempenho de diferentes métodos de estimativa da evapotranspiração potencial no Estado de São Paulo, Brasil. **Revista Brasileira de Agrometeorologia**, v. 5, n. 1, p. 89-97, 1997.

CAMARGO, A. P.; MARIN, F. R.; SENTELHAS, P. C.; PICINI, A. G. Ajuste da equação de Thornthwaite para estimar a evapotranspiração potencial em climas áridos e superúmidos, com base na amplitude térmica diária. **Revista Brasileira de Agrometeorologia**, v. 7, n. 2, p. 251-257, 1999.

CONFALONE, A. E.; BERNARDES, M. S.; COSTA, L. C.; RIGHI, C. A.; DOURADO NETO, D.; MARTIN, T. N.; MANFRON, P. A.; PEREIRA, C. R. Expolinear model on soybean growth in Argentina and Brazil. **Ciência Rural**, v. 40, n. 5, p. 1009-1016, 2010.

CONFALONIERI, R. Combining a weather generator and a standard sensitivity analysis method to quantify the relevance of weather variables and agrometeorological model outputs. **Theoretical and Applied Climatology**, v. 108, n. 1-2, p. 19-30, 2012.

DOORENBOS, J. KASSAM, A. H. **Yield response to water**. Rome: FAO, 1979. (Irrigation and Drainage Paper, 33).

DOORENBOS, J.; KASSAM, A. H. **Efeito da água no rendimento das culturas**. Traduzido por Gheyi, H. R. et al. Campina Grande: UFPB, 1994. (Irrigação e Drenagem, n. 33).

EMBRAPA-Empresa Brasileira de Pesquisa Agropecuária. **Tecnologias de produção de soja** – Paraná 2007. Brasília: Embrapa Soja, 2006.

FEHR, W. R.; CAVINESS, C. E. **Stages of soybean development**. Ames: Iowa State University of Science and Technology, 1977. (Special Report, 80).

GEERTS, S.; RAES, D. Deficit irrigation as an on-farm strategy to maximize crop water productivity in dry areas. **Agricultural Water Management**, v. 96, n. 9, p. 1275-1284, 2009.

JÉGO, J.; PATTEY, G. J.; BOURGEOIS, G.; MORRISON, M. J.; DRURY, C. F.; TREMBLAY, N.; TREMBLAY, G. Calibration and performance evaluation of soybean and spring wheat cultivars using the STICS crop model in Eastern Canada. **Field Crops Research**, v. 117, n. 2-3, p. 183-196, 2010.

KANEMASU, E. T. Yield and water-use relationships: some problems of relating grain yield to transpiration. In: TAYLOR, H. M.; JORDAN, W. R.; SINCLAIR, T. R. (Ed.). **Limitations to efficient water use in crop production**. Madison: American Society of Agronomy, 1983. cap. 9B, p. 413-417.

KLOSOWSKI, E. S. Estimativa da produtividade de cultivares de soja por meio do modelo SOYGRO para Londrina, estado do Paraná. **Revista Unimar**, v. 19, n. 3, p. 751-765, 1997.

LEGATES, D. R.; McCABE JR., G. J. Evaluating the use of "goodness-of-fit" measures in hydrologic and hydroclimatic model validation. **Water Resources Research**, v. 35, n. 1, p. 233-241, 1999.

LOBELL, D. B.; CASSMAN, K. G.; FIELD, C. B. Crop yield gaps: their importance, magnitudes and causes. **Annual Review of Environment and Resources**, v. 34, n. 3, p. 179-204, 2009.

MELO, R. W.; FONTANA, D. C.; BERLATO, M. A.; DUCATI, J. R. An agrometeorological–spectral model to estimate soybean yield, applied to southern Brazil. **International Journal of Remote Sensing**, v. 29, n. 14, p. 4013-4028, 2008.

MINUZZI, A.; BRACCINI, A. L; RANGEL, M. A. S.; SCAPIM, C. A.; BARBOSA, M. C.; ALBRECHT, L. P. Qualidade de sementes de quatro cultivares de soja, colhidas em dois locais no estado do Mato Grosso do Sul. **Revista Brasileira de Sementes**, v. 32, n. 1, p. 176-185, 2010.

OLIVEIRA, J. C.; FORMAGGIO, A. R.; EPIPHANIO, J. C. N.; LUIZ, A. J. B. Index for evaluation of segmentation (IAVAS): an application to agriculture. **Mapping Sciences and Remoting Sensing**, v. 40, n. 3, p. 155-169, 2003.

PATANÈ, C.; TRINGALI, S.; SORTINO, O. Effects of deficit irrigation on biomass, yield, water productivity and fruit quality of processing tomato under semi-arid Mediterranean climate conditions. **Scientia Horticulturae**, v. 129, n. 4, p. 590-596, 2011.

PEREIRA, A. R.; ANGELLOCI, L. R.; SENTELHAS, P. C. **Agrometeorologia**: fundamentos e aplicações práticas. Guaíba: Ed. Agropecuária, 2002.

POPP, M. P.; DILLON, C. R.; KEISLING, T. C. Economic and weather influences on soybean planting strategies on heavy soils. **Agricultural Systems**, v. 76, n. 3, p. 969-984, 2003.

RHINE, M. D.; STEVENS, G.; SHANNON, G.; WRATHER, A.; SLEPER, D. Yield an nutritional responses to waterlogging of soybean cultivars. **Irrigation Science**, v. 28, n. 2, p. 135-142, 2010.

THORNTHWAITE, C. W.; MATHER, J. R. **The water balance**. Centerton: Drexel Institute of Tecnology, 1955. (Publications in climatology, 8, n .1)

WILLMOTT, C. J.; ACKLESON, S. G.; DAVIES, R. E.; FEDDEMA, J. J.; KLINK, K. M.; LEGATES, D. R.; O'DONNELL, J.; ROWE, C. M. Statistics for the evaluation and comparison of models. **Journal of Geophysical Research**, v. 90, n. 5, p. 8995-9005, 1985.

6

Validation of an experimental prototype of thick-layer fixed bed dryer and determination of the quality of the dried soybean grains

Paulo Carteri Coradi[1*] and Ângelo Francisco Calegare Lemes[2]

*[1]Universidade Federal de Santa Maria, Campus Cachoeira do Sul, Rua Ernesto Barros, 1345, Santo Antônio, 96506-322, Cachoeira do Sul, Rio Grande do Sul, Brazil. [2]Universidade Federal de Mato Grosso do Sul, Campus Chapadão do Sul,Chapadão do Sul, Mato Grosso do Sul, Brazil. *Author for correspondence. E-mail: paulo.coradi@ufsm.br*

ABSTRACT. The aim of this study was to validate an experimental prototype dryer with a thick-layer fixed bed and determine the quality of the dried soybean grains. The drying system was designed as a chamber with four independent drying cells, with a capacity of 164 kilograms, an air ventilation system and heating with electric heaters. This prototype was used for soybeans with an initial water content of 17% (w.b.). The samples were collected from the top, middle and bottom of the four cells to evaluate the quality of the grains. The experimental design was completely randomized factorial (3 x 4 x 3), with three temperatures settings for the drying air (30, 40, and 50°C), four drying cells and three evaluation positions (upper, middle and lower). The grains did not undergo quality changes as a function of the drying cells. In the upper part of the drying cells, there was a reduction in the specific grain mass (851 kg m^{-3}), an increase in electrical conductivity (140.02 μS cm^{-1} g^{-1}), reduction in germination (63%) and oil content (7%) compared with the middle and lower dryer positions. The experimental prototype of the fixed bed dryer allowed a uniform distribution of the heated air in the drying cells, which was effective in the drying operation and in maintaining the quality of the soybean grains.

Keywords: control and optimization of the drying; loss reduction of grain quality in the drying; yield of the drying processes.

Introduction

Agricultural products are harvested at physiological maturity when the water content is high, thus, making them less prone to deterioration and contamination by fungi, insect pests and reduction of matter section (Niamnuy, Nachaisin, Poomsa, & Devahastin, 2012). In the present study, we observed that the water content of the product was higher than that of the product.

Drying is an indispensable pre-processing operation that maintains the initial quality of the agricultural products for storage security. A reduction in the water content of the grains diminishes their biological activity and microbiological, physical and chemical changes during storage. Moreover, the grains subjected to the drying process undergo contraction, which directly affects their physical properties and reduces their physical and chemical quality (Coradi, Fernandes, Helmich, & Goneli, 2016a; Niamnuy, Nachaisin, Poomsa, & Devahastin, 2012).

Artificial drying methods can be classified according to the variation in the heat supply, which is distributed by the air flow. The air imparts heat to the grain mass in a continuous, intermittent or stationary manner (Defendi, Nicolin, Paraíso, & Jorge, 2016; Hashemi, Mowla, & Kazemeini, 2009; Hemis, Choudhary, & Watson, 2012). Several studies have been conducted for developing new dryers and improving the existing ones to obtain a higher yield and maintain the quality of the agricultural products. The main factors involved in this improvement are the temperature and air flow controls, the velocity and pressure of the heated air, and the movement and uniformity of the water content of the product (Khatchatourian, 2012; Sivakumar, Saravanan, Elaya Perumal, & Iniyan, 2016).

Thus, the aim of this study was to validate an experimental prototype for a thick-layer fixed bed dryer and determine the quality of the dried soybean grains. The dryer system consisted of four drying chambers, forced air circulation and electric heating.

Material and methods

The research work was carried out at the Post-Harvest Grain Laboratory, in the experimental area of the Federal University of Mato Grosso do Sul (UFMS), Campus of Chapadão do Sul (CPCS). The designed drying system was composed of a chamber with four independent drying cells for separating the grains according to water content bands, impurities or grain size, or grains from different species, with the aim of increasing the yield and quality of the grains during drying.

The grain dryer was constructed with 14" thick circular shaped metal sheets containing a drying chamber (e) with a height of 2.0-meters and a diameter of 1-meter, with variable static capacity according to the grain type of 164 kg of soybeans (Figure 1).

Ten percent of the total floor area was perforated. In the interior, four mobile drying cells (1.5-meter in height and 0.20-meter in diameter) were built for holding the grain mass during drying. The cells were constructed from 12" metal plates with walls having circular perforations (2 mm in diameter); the top of the cells was left open. The dryer plenum (d) with the function of reducing the drying air speed, was arranged below the drying chamber and had a diameter of 1.5 m and a height of 0.30 m. Two electric resistances (4,000 watts) were installed in the drying chamber – one in the center of the plenum and the other in the front compartment of the plenum (b), which was 1.5 m long and 0.5 m high. A centrifugal fan was connected to the dryer body. The centrifugal fan had the following specifications: straight blades, 20 m^3 min^{-1}, flow, 40 mm static pressure, 0.5 cv power, 220/380 Volts tension, 1,750 rpm and a single-phase, carbon steel structure, cast aluminum rotor with 90 DBA noise, and 35 kg weight coupled to an air expander funnel (1.5 m x 0.5 m, l x h).

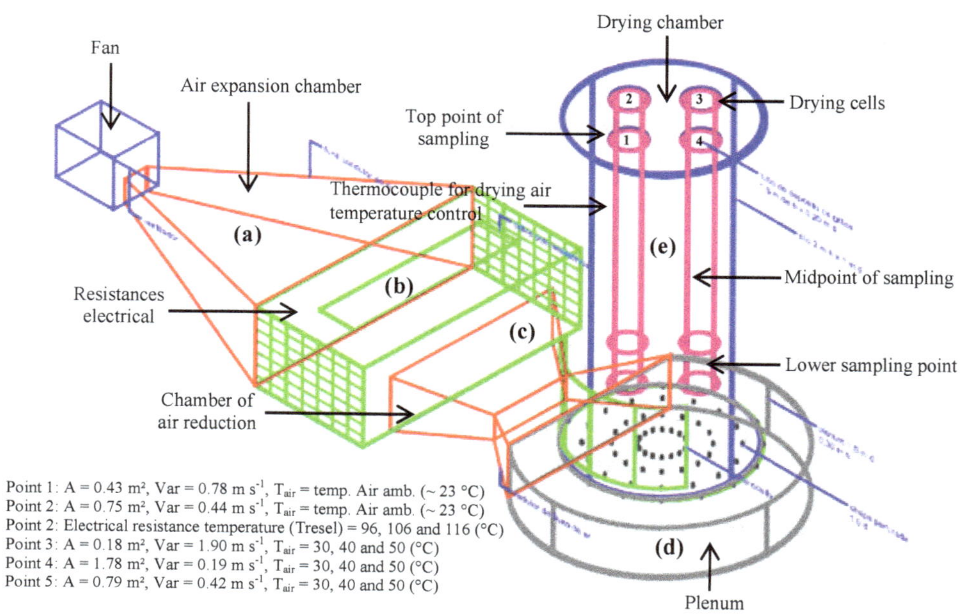

Figure 1. Experimental prototype of grain drying.

The drying system could control the drying air temperature directly using the electrical resistances through an external controller with dimensions of 500 x 400 x 250 mm, channels with open cutout for cable housing, circuit breaker protection for 10A command and 40A force, 24 vcc switching source for solid state relay activation, 40A three phase solid state relay with heat sink, N1040-PR temperature controller for system control, terminal strip for input and output of control cables and force (Figure 1). The electrical resistances used were U-shaped tubular type, manufactured with a 304 stainless steel tube, 11.2 mm diameter, and the iron fins, the total length of 1200 mm, 220 volts and 4500 watts. Depending on the area of the equipment, air flow and dryer heating, we obtained the variations in the drying air velocity and temperature, according to the specifications in Figure 1.

To perform the drying tests, we used soybean grains with initial water contents of 17% (w.b.), collected at a grain storage unit in the municipality of Chapadão do Sul, State of Mato Grosso do Sul, Brazil. A fixed bed with the capacity to hold 164 kg of soybean was loaded on top, and then the air ventilation system and the heating system with electric resistances were activated through an external controller to control the temperature of the drying air at 30, 40, and 50°C. The temperature conditions in the grain mass were monitored during drying using mercury thermometers in the four drying cells. The grains were sampled every hour from each of the four drying cells, and a grain-water sampler was used to determine the water content. At the end of the drying cycle, the samples from the top, middle and lower parts of the four cells were collected to evaluate the grain quality. The experimental design was a completely randomized, factorial (3 x 4 x 3) with three air drying temperatures (30, 40, and 50°C), four drying cells (1, 2, 3, and 4) and three evaluation positions for each drying cell (upper, middle, and lower). To evaluate the quality of the soybean grains, the physical classification, water content, oil content, apparent specific mass, germination, and electrical conductivity were measured. The physical classification of soybean grains was carried out according to the official standard established by the normative instruction no. 9 on March 30, 2007, by the Ministry of Agriculture, Livestock and Food Supply (MAPA) (Brasil, 2007). The water content (% w.b.) was determined by the difference in the initial and final grain mass obtained by oven drying the grains with forced air circulation at 105°C ± 1°C for 24 hours, with three replicates of 15 grams each, according to the Brazilian recommendations (Brasil, 2009). For the analysis of the oil yield, the soybean grains were crushed using a domestic multiprocessor (model, Arno). Subsequently, 5 grams of the sample was weighed, transferred to a filter paper and subjected to a Soxhlet-type extractor, heated at 70°C with 150 mL of solvent hexane PA, for 4 hours. After extraction the vessel containing oil and hexane was brought to a rotary evaporator containing water heated to 70°C and operated at 40 rpm for the separation of solvent and oil. The oil sample was kept in an oven at 65°C for 12 hours, after which the oil was weighed. The oil percentage was determined by multiplying by 100 and dividing by the mass of the initial sample, according to the methodology described by Instituto Adolfo Lutz (2008). The apparent specific mass of the soybean was determined, considering the relationship between the mass and the volume of the grains obtained by weighing with three experimental replicates.

Germination tests were conducted with four subsamples of 50 grains from each treatment, placed in rolls of "Germitest" paper towels and packed in a "Mangesdorf" type germinator, with a constant temperature of 25 ± 2°C. The grain germination results were recorded and interpreted on the 5th and 8th days, according to the criteria established in the Rules for Seed Analysis (Brasil, 2009). The electrical conductivity test in the soybean grains was performed according to a methodology described by Vieira, Penariol, Perecin, and Panobianco (2002). Twenty-five grains and four subsamples from each treatment were used. The samples were soaked in plastic cups with 75 mL of deionized water and maintained in a refrigerated chamber (type B.O.D.) with a controlled temperature of 25 ± 2°C for 24 hours. The solutions containing the grains were slightly agitated to standardize the leachates and immediately read in a portable CD-850 "Instrutherm" portable conductivity meter; the results were divided by the mass of 25 grains and expressed in $\mu S\ cm^{-1}\ g^{-1}$ of grains. The results of the evaluations were analyzed using the analysis of variance and Tukey's test, at 5% probability, to determine the significant differences between the treatments.

Results and discussion

Table 1 shows the results of drying the grains at air temperatures of 30, 40, and 50°C, with a relative humidity of the ambient air varying from 70.5 to 45.5%. The temperature in the grain mass ranged from 26.9 to 33.5°C; the exhaust air temperature ranged from 26 to 32°C, and the relative humidity of the exhaust air ranged from 73 to 59%.

Table 1. Monitoring the drying conditions of soybean grains.

			1-central drying chamber		2-center drying chamber		3-center drying chamber		4-center drying chamber	
$UR_{air\ amb.}$ (%)	$T_{air\ amb.}$ (°C)	$T_{air\ drying}$ (°C)	$T_{air}/UR_{air\ exhaustion}$ (°C/%)	$T_{mass\ grains}$ (°C)	$T_{air}/UR_{air\ exhaustion}$ (°C/%)	$T_{mass\ grains}$ (°C)	$T_{air}/UR_{air\ exhaustion}$ (°C/%)	$T_{mass\ grains}$ (°C)	$T_{air}/UR_{air\ exhaustion}$ (°C/%)	$T_{mass\ grains}$ (°C)
70.5	26.3	30 ±2	26/72	26.9	26/73	26.3	26/73	27.1	26/73	26.9
58.8	28.6	40 ±2	29/67	29.8	29/66	29.4	29/66	30.0	29/66	29.6
45.5	31.2	50 ±2	32/59	33.5	32/59	33.9	32/59	32.1	32/59	31.9

The drying curves, observed in Figure 2A, B, and C, represent the reduction of water contents of the grains from 17 to 11.5% (w.b.) at 30°C, from 17 to 10.3% at 40°C and from 17 to 10.5% (w.b.) at 50°C. Cell 1 showed the greatest reduction (with respect to time) in the water content of the grains at 30 and 40°C, until the end of the drying. The reduction in the water content of the grains at 50°C was similar for all four dryer cells. Moreover, the drying time was two hours less (8 hours) at 50°C compared with the 10 hours at 30 and 40°C.

This behavior is typical of several agricultural products and has been observed by several researchers studying numerous agricultural products (Brooker, Bakker-Arkema, & Hall, 1992; Chua, Hawlader, Chou, & Ho, 2002; Isquierdo et al., 2013; Resende, Oliveira, Chaves, Ferreira, & Bessa, 2014; Siqueira, Resende, & Chaves, 2012). Figure 3A, B, C, and D respectively show that the drying cells 1, 2, 3, and 4 exhibited similar behaviors at the same temperatures and drying at 50°C reduced the drying time to eight hours in all of the four cells. Moreover, drying in cell 1, independent of temperature, resulted in a more pronounced curve, which showed accelerated drying in the first hour, followed by drying at 40°C and 30°C.

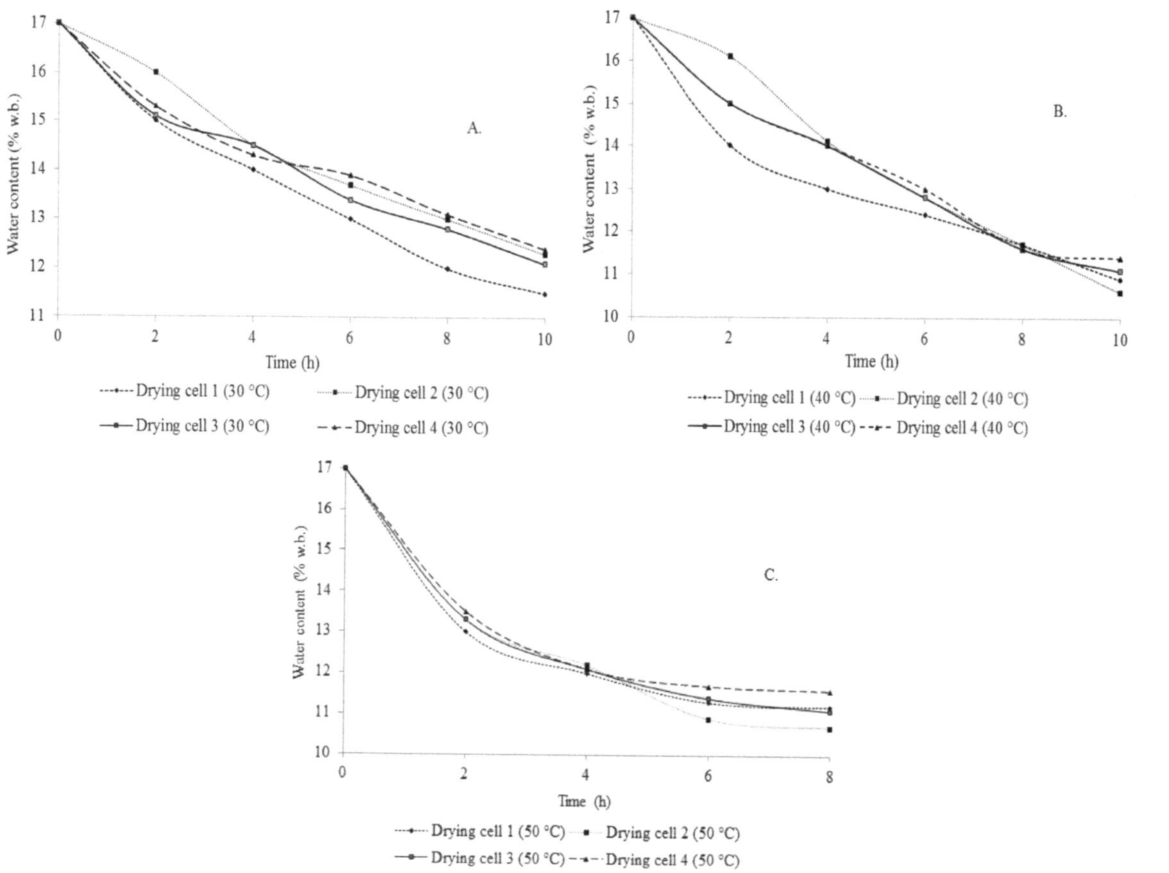

Figure 2. Drying curves of soybean grains as a function of the cells, (A) temperature of 30°C, (B) temperature of 40°C, and (C) temperature of 50°C.

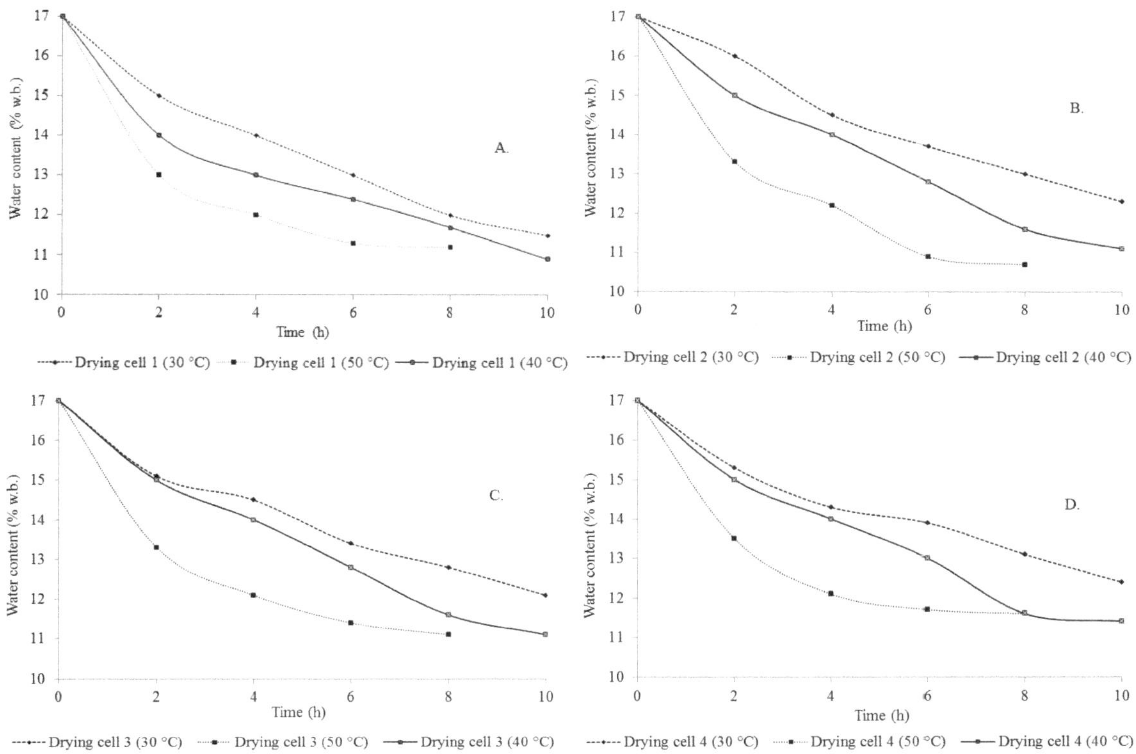

Figure 3. Curves drying soybeans versus temperature, (A) drying cell 1, (B) drying cell 2, (C) drying cell 3, and (D) drying cell 4.

The drying curves show that the system was of the right size for the drying process between the four cells for different air temperatures. The curves represent the same results for all cells at the same drying temperatures and for different temperatures.

Table 2 shows the results of the physical classification of soybean grains. There was an increase in the number of broken grains during drying; however, the breakage was within the maximum limits of tolerance established by Brasil (2007). The broken or fragmented grains could be attributed to the movement of the product in the dryer (load) and the effects of the temperature of the drying air in the grains. Burned grains were not observed in any of the drying treatments; however, the percentage of fermented and burned grains increased at temperatures of 40 and 50°C.

The fermented grains are grains or pieces of grain that have undergone a visible change in the color of cotyledon due to fermentation. These defects are considered light; however, they can evolve with the deterioration process for the burned and moldy grains to a stage classified as 'severe' by the commercial standard (Brasil, 2007). According to Normative Instruction no. 11, which regulates the classification of soybeans, there are no acceptable maximum limits for the fermented grains; however, the total number of damaged grains (sum of grains including burned, moldy, fermented, or germinated should not exceed 4.0% and 6.0% for the classification of Type 1 and Type 2, respectively, for Group I and should not exceed 8.0% for Group 2.

The burned defects are grains or pieces of grains that are visibly fermented and have a brown or dark coloration affecting the pulp. This may include the grains burned by the drying process, which in Group 1 should not exceed 1% for Type 1 and 2% for Type 2, whereas in Group 2 these defects should not exceed 4% as per the Basic Standard (Brasil, 2007). The immature and greenish grains and the presence of foreign matter and impurities are due to the climatic conditions, harvesting, transport and pre-cleaning of the grains.

The total number of defects is the sum of the total percentages of burned, moldy, fermented, germinated, damaged, immature and pimpled grains. The maximum tolerance limits for the total grains damaged, in Group I, are 4.0 and 6.0%, for Types 1 and 2, respectively. In Group II, the limit is 8.0% (Brasil, 2007). The results of the statistical interactions, drying cells, sampling points and air temperatures for the evaluation of water content, specific mass, electrical conductivity test, germination and oil contents are shown in Table 3.

Table 2. Physical classification of soybean grains after drying (%).

Cells	Defects	Air drying temperature (°C)											
		30			40			50			Inicial		
		I	M	S	I	M	S	I	M	S	I	M	S
1	Burned out	-	-	-	-	0.05	-	0.54	0.06	0.96	-	-	-
	Broken	3.32	1.94	1.90	11.7	15.4	20.3	2.26	3.11	2.68	9.11	9.11	9.11
	Fragmented	-	-	-	1.28	0.95	3.07	0.04	0.19	0.18	0.08	0.08	0.08
	Foreign matter	0.16	0.40	0.04	0.17	0.17	0.09	0.51	1.02	0.15	0.18	0.18	0.18
	Immature	0.38	0.92	0.07	0.70	0.49	0.57	-	-	-	0.76	0.76	0.76
	Greenish	-	0.19	-	1.01	0.28	-	-	-	-	0.17	0.17	0.17
2	Burned out	-	-	-	-	-	-	0.12	-	0.16	-	-	-
	Fermented	2.01	-	-	-	-	-	-	-	-	-	-	-
	Broken	0.07	11.9	12.5	21.9	15.4	20.5	2.04	2.42	1.91	9.11	9.11	9.11
	Fragmented	0.08	0.14	0.54	2.46	1.21	3.52	0.13	0.17	0.14	0.08	0.08	0.08
	Foreign matter	-	0.28	0.22	0.23	0.10	0.14	0.11	0.37	0.16	0.18	0.18	0.18
	Immature	-	1.36	1.01	0.67	0.81	0.65	0.29	0.06	0.25	0.76	0.76	0.76
	Greenish	-	0.28	0.46	0.18	0.27	-	-	-	-	0.17	0.17	0.17
3	Burned out	-	-	-	-	-	-	0.27	0.46	0.42	-	-	-
	Broken	9.23	11.7	17.4	17.9	20.4	20.4	1.39	2.52	2.46	9.11	9.11	9.11
	Fragmented	0.35	0.37	0.26	3.02	3.14	0.94	0.31	0.10	0.38	0.08	0.08	0.08
	Foreign matter	0.14	0.03	0.25	0.08	0.09	0.23	1.12	0.36	0.30	0.18	0.18	0.18
	Immature	0.99	0.86	0.72	0.71	0.81	0.62	0.07	-	0.11	0.76	0.76	0.76
	Greenish	0.72	0.29	0.27	0.33	0.31	-	-	-	-	0.17	0.17	0.17
4	Burned out	-	0.04	0.09	0.11	-	-	0.32	0.23	0.18	-	-	-
	Broken	4.89	7.49	11.2	21.5	19.1	20.3	1.61	2.41	2.81	9.11	9.11	9.11
	Fragmented	0.25	0.04	0.15	1.98	1.56	3.03	0.13	0.44	0.25	0.08	0.08	0.08
	Foreign matter	0.14	0.14	0.23	0.22	0.09	0.18	0.89	0.29	0.28	0.18	0.18	0.18
	Immature	0.68	0.81	0.78	0.47	0.91	0.57	0.51	-	0.05	0.76	0.76	0.76
	Greenish	0.22	0.10	0.32	-	0.35	0.24	0.40	-	-	0.17	0.17	0.17

I (Lower), M (Medium), and S (Upper).

Table 3. Analysis of variance of the soybean quality analyzes after drying.

FV	Water content	Specific mass	Electrical conductivity	Germination	Oil content
Cells	0,0001*	0.7287^{NS}	0.5361^{NS}	0.0006*	0.0000*
Points	0,0001*	0.0000*	0.0001*	0.0308*	1.0000^{NS}
Temperatures	0,0001*	0.0027*	0.0000*	0.0000*	0.0000*
Cells*temperatures	0,0001*	0.2136^{NS}	0.4492^{NS}	0.0033*	0.0000*
Cells*points	0,0001*	0.0000*	0.0006*	0.0096*	1.0000^{NS}
Temperatures*points	0,0001*	0.0020*	0.0127*	0.0261*	1.0000^{NS}
Cells*points*temperatures	0,0001*	0.0013*	0.0973^{NS}	0.0036*	1.0000^{NS}

*Significant, ^{NS}Not significant.

With the increase in the drying air temperature, for the same cells and sampling points, the water content of the grains was reduced (Tables 4 and 5). According to Siqueira et al. (2012), the increase in temperature enables a reduction in the drying time of the grains because the difference between the vapor pressure of the drying air and the grain increases with the increase in temperature, providing a greater reduction of water in a shorter period.

The reduction of water in the grain during the drying period is governed by the water reduction rate until it becomes stable because the reduction in free water slows the migration of water from the interior to the surface of the grain (Jangam, Law, & Mujumdar, 2010; Lopes, Steidle Neto, & Santiago, 2014). This behavior indicates that diffusion is the largest probable physical mechanism, which governs the movement of water in the thin layers (Khanchi et al., 2013; Nuthong, Achariyaviriya, Namsanguan, & Achariyaviriya, 2011; Rafiee et al., 2009). Moreover, the drying air temperature is the main indication for accelerating the movement of water in the thin layers.

Thus, regardless of the cell, there was a significant reduction in water content at the upper sampling point compared with the middle and lower points of the cells. Moreover, the difference observed was greater with the increase in temperature of the drying air indicating that the superior points had a higher drying rate, which caused the highest reduction in the grain water content at all drying temperatures and cells.

Table 4. Water content (% w.b.) of soybeans after drying, according to the sampling points in the cells and drying air temperatures[a].

T (°C)	Cell 1			Cell 2			Cell 3			Cell 4		
	L	M	U	L	M	U	L	M	U	L	M	U
30	12.66 Ab	12.80 Ab	11.74 Ba	16.32 Cb	16.16Cb	14.27 Ca	15.98 Bc	14.87 Cb	13.31 Ca	13.21 Aab	14.54 Bb	12.83 Ca
40	14.78 Cb	14.24 Bb	10.90 Aa	15.13 Bb	14.76 Bb	11.19 Ba	14.00 Ac	13.37 Bb	10.66 Ba	14.60 Bc	13.89 Bb	11.18 Ba
50	14.15 Bb	14.14 Bb	12.89 Ca	14.39 Ac	13.10 Ab	8.73 Aa	14.33 Ac	11.90 Ab	9.55 Aa	14.42 Bc	11.15 Ab	9.64 Aa
Initial condition	16.54 Da	16.54 Ca	16.54 Da	16.54 Ca	16.54 Ca	16.54 Da	16.54 Ba	16.54 Da	16.54 Da	16.54 Ca	16.54 Ca	16.54 Da

[a]Means followed by the same lowercase letter in the line do not differ by Tukey test at 1%. Means followed by the same capital letter in the column do not differ by Tukey test at 1%. [b]L (Lower), [c]M (Medium), [d]U (Upper). Coefficient of variation = 1.65%.

Table 5. Water content (% w.b.) of soybeans after drying, depending on the drying cells and the drying air temperatures[a].

	[b]Lower point cells				[c]Medium point cells				[d]Upper point cells			
T (°C)	1	2	3	4	1	2	3	4	1	2	3	4
30	12.66 Aa	16.32 Cb	15.98 Bb	13.21 Aa	12.80 Aa	16.16 Cd	14.87 Cc	14.54 Bb	11.74 Ba	14.27 Cc	13.31 Cb	12.83 Cb
40	14.78 Cb	15.13 Bb	14.00 Aa	14.60 Bab	14.24 Bbc	14.76 Bc	13.37 Ba	13.89 Bab	10.90 Aa	11.19 Ba	10.66 Ba	11.18 Ba
50	14.15 Ba	14.39 Aa	14.33 Aa	14.42 Ba	14.14 Bd	13.10 Ac	11.90 Ab	11.15 Aa	12.89 Cc	8.73 Aa	9.55 Ab	9.64 Ab
Initial condition	16.54 Da	16.54 Ca	16.54 Ba	16.54 Ca	16.54 Ca	16.54 Ca	16.54 Da	16.54 Ca	16.54 Da	16.54 Da	16.54 Da	16.54 Da

[a]Means followed by the same lowercase letter in the line do not differ by Tukey test at 1%. Means followed by the same capital letter in the column do not differ by Tukey test at 1%. [b]L (Lower), [c]M (Medium), [d]U (Upper). Coefficient of variation = 1.43%.

Moreover, although the air distribution occurred uniformly in the dryer, the drying air was released from the equipment necessarily due to the mass of grains in the upper layer as observed in Figure 1. Thus, the increase in the volume of the drying air in the upper part of the equipment caused excessive drying, resulting in a greater reduction of the grain water content.

The lower and middle sampling points of the cells showed homogenous drying independent of air temperature, whereas the upper cell point showed the highest variation in water content, especially at temperatures of 40 and 50°C (Table 3). As the drying air temperature increased, the rate of water removal from the grains also decreased, indicating a greater resistance to heat transfer and mass from the interior of the product to its surface. These results suggest that diffusion is the main mechanism that determines the movement of water (Coradi, Fernandes, & Helmich, 2016b; Jangam et al., 2010; Rafiee et al., 2009; Siqueira, Resende, & Chaves, 2012).The results of the specific mass analysis are shown in Tables 6 and 7. The temperature of the drying air changes the specific mass of the grains. A comparison of the sampling points in the drying cells showed that the specific mass was higher at the midpoints of cells 1 and 2 and the lower points of cells 3 and 4 at the drying air temperature of 30°C. At 40°C, the specific mass was the highest at the lower points of cells 1 and 2 and midpoints of cells 3 and 4. At a drying air temperature of 50°C, the specific mass was highest at the upper point of cell 1, lower points of cells 2 and 3 and the average point of sampling of cell 4. The reduction in the apparent specific mass is associated with the reduction in the water content in response to the increase in the drying air temperature. This causes greater contractions in the internal part of the cotyledons compared with its external dimensions, which contracted to a lesser extent forming voids within the grain (Niamnuy et al., 2012).

By evaluating the drying cells, the specific grain mass varied as a function of drying air temperature and sampling points in the cell. At the drying temperature of 30°C, the specific mass of the grains was higher in cells 3 (lower point) and 1 (mid and upper points), while at 40°C, the specific mass of the grains was higher in cells 4 (lower and medium points) and 3 (upper point). However, at 50°C, the specific mass was the highest in cell 3 (lower point), cell 4 (midpoint), and cell 2 (top point). The treatments that showed the greatest reduction in their masses also showed a significant increase in their electrical conductivity indicating that there was a reduction in the specific mass with a loss in dry matter and consequently quality due to the deterioration of the product.

Table 6. Specific mass (kg m^{-3}) of soybean grains after drying, as a function of sampling points in the cells and drying air temperatures[a].

T (°C)	Cell 1			Cell 2			Cell 3			Cell 4		
	L	M	U	L	M	U	L	M	U	L	M	U
30	889 Aa	892 Aa	868 Ab	869 Aa	873 Aa	861 ABa	892 ABa	877 ABab	861 ABb	872 Aa	857 Cab	851 Bb
40	879 ABa	878 ABab	858 Ab	885 Aa	869 Aa	848 Bb	872 ABab	886 ABa	865 ABb	886 Aa	889 Aba	858 Bb
50	864 Ba	857 Ba	874 Aa	892 Aa	886 Aa	881 Aa	894 Aa	896 Aa	847 Bb	884 Aab	904 Aa	876 Ab
Initial condition	870 ABa	870 ABa	870 Aa	870 Aa	870 Aa	870 ABa	870 Ba	870 Ba	870 Aa	870 Aa	870 BCa	870 Ba

[a]Means followed by the same lowercase letter in the line do not differ by Tukey test at 1%. Means followed by the same capital letter in the column do not differ by Tukey test at 1%. [b]L (Lower), [c]M (Medium), [d]U (Upper). Coefficient of variation = 1.23%.

Table 7. Specific mass (kg m^{-3}) of soybeans after drying, depending on the drying cells and the drying air temperatures[a].

	[b]Lower point cells				[c]Medium point cells				[d]Upper point cells			
T (°C)	1	2	3	4	1	2	3	4	1	2	3	4
30	889 Aa	869 Aa	892 ABa	873 Aa	892 Aa	873 ab	877 ABab	857 Cb	868 Aa	851 ABa	861 ABa	851 Ba
40	879 ABa	885 Aa	872 ABa	886 Aa	878 ABa	869 Aa	886 ABa	889 ABa	858 Aa	848 Ba	865 ABa	858 ABa
50	864 Bb	892 Aa	894 Aa	884 Aab	857 Bb	886 Aa	896 Aa	904 Aa	874 Aa	881 Aa	847 Bb	876 Aa
Initial condition	870 ABa	870 Aa	870 Ba	870 Aa	870 ABa	870 Aa	870 Ba	870 BCa	870 Aa	870 ABa	870 Aa	870 ABa

[a]Means followed by the same lowercase letter in the line do not differ by Tukey test at 1%. Means followed by the same capital letter in the column do not differ by Tukey test at 1%. [b]L (Lower), [c]M (Medium), [d]U (Upper). Coefficient of variation = 1.23%.

Conversely, in some treatments, we verified an increase in the specific mass after drying, which was due to the significant decrease in their water content. This observation concurred with the previous studies in soybean grains by Ribeiro, Corrêa, Rodrigues, and Goneli (2005) and corn by Peplinski, Paulis, Bietz and Pratt (1994), by Resende et al. (2012), Jatropha by Siqueira et al. (2012).

The increase in the temperature of the drying air showed an increase in the amounts of exudates released in the solution due to higher damage to the grain integuments (Tables 8 and 9). Krzyzanowski, Vieira, & França Neto (2009) described seeds with an electrical conductivity greater than 80 µS cm^{-1} g^{-1} at approximately 13% (w.b.) water content as low vigor seeds. According to Vieira et al. (2002), the electrical conductivity is inversely proportional to the water content of the grains; thus, the amount of leachate in the solution increases the quantity of water in the grains. Our results were in agreement with the results mentioned above wherein the highest values of electrical conductivity were observed in the upper sampling points, where there was a greater reduction of grain water content, independent of the cells.

Table 9 shows that the lower sampling point of the drying cells, regardless of the air temperature, showed no difference in the evaluation of the electrical conductivity as compared with the other cells. For the mean sampling point, the electrical conductivity was higher in dry grains in cell 1 at 50°C and cell 2 at 40°C, whereas for the air temperature at 30 °C there was no difference in the electrical conductivity values between the cells. At the upper sampling point, the highest values of electrical conductivity for the air temperature of 50°C were observed in cell 4, with no differences between the cells at temperatures of 30 and 40°C.

According to Vasquez, Bertolin, and Spegiorin (2011), the electrical conductivity test can promote a stratification of different levels of grain vigor. Increased electrical conductivity results in the disorganization of the cell membranes of the grains making them more susceptible to physical damage, such as temperature and humidity, allowing the entry of pathogens and affecting the physiological quality. Resende et al. (2012) and Almeida, Resende, Mendes, Costa, and Rocha (2013) also our study corroborated the results of previous studies, which reported an increase in electrical conductivity with an increase in the drying air temperature in bean grains (Resende et al., 2012; Almeida et al., 2013).

Increasing the temperature of the drying air reduced the percentage of grain germination (Table 10). The highest reduction in the grain germination percentage occurred when they were dried at 50°C; however, this reduction of germination percentage was not significant in the lower point of cell 2 and the higher point of cell 4. However, when the beans were dry at 40°C, there was no significant difference compared to the initial condition. Except for the midpoint of cell 2, a situation similar to the lower drying temperature of 30°C, with a significant reduction of the germination potential of the grains, was verified in cell 1 at the lower and upper points. These data agree with the results of Kowalski and Rajewska (2009) who observed that an increase in drying air temperature reduced the germination in rice grains.

Table 11 shows that at a drying air temperature of 30°C, the percentage germination differed in the cells at the lower and upper points, but at the upper point, only cell 1 differed significantly from the others. Drying at 50°C resulted in a difference in the percentage of grain germination between the cells at the lower and medium points.

However, at 40°C there were no significant differences among cells. Almeida et al. (2013) also observed a decrease in the percentage of germination with an increase in the temperature of the drying air for bean grains. Regardless of the drying position of the grains in the dryer and the temperature of the drying air used, the percentage of germination of the grains was reduced when they reached the lowest water contents after drying, corroborating the results observed by Sanjinez-Argandoña, Branco, Bittencourt, and Munhoz (2011), Niamnuy, Nachaisin, Laohavanich, and Devahastin (2011) and Coradi, Melo, and Rocha (2014). Hegarty (1978) verified that the highest germination value occurred when the grains were dried at a temperature of 45°C both in the shade and prototypes of stationary dryers regulated at temperatures of 35/45°C and 35/45°C. Moreover, the oil content of soybean grains varied significantly compared with the initial oil content due to exposure to various drying air temperatures (Tables 12 and 13).

Drying of the grains caused a reduction in the initial amount of oil in the grains, except for those in cells 3 and 4 at 50 °C. Similar results were observed by Dondee, Meeso, Soponronnarit, and Siriamornpun (2011) who observed that the soybean grains with lower water content had a higher oil yield.

The highest reduction in the oil content was observed in the grains dried at 30°C; except cell 1, which showed the greatest reduction in the amount of oil at 40°C. The four drying cells showed the same oil yield pattern at 30°C, whereas at 40°C, cells 3 and 4 gave a higher yield of oil. The highest yield of oil was obtained from the grains dried at 50°C in cells 1 and 2.

Table 8. Electrical conductivity (μS cm^{-1} g^{-1}) of soybean grains after drying, according to the sampling points in the cells and drying air temperatures[a].

T (°C)	Cell 1			Cell 2			Cell 3			Cell 4		
	L	M	U	L	M	U	L	M	U	L	M	U
30	71.6 Aa	79.0 Aa	77.75 Aa	61.60 Aa	75.52 Aba	77.80 Aba	66.20 Aa	62.45 Aa	71.40 Aa	71.90 Aa	71.22 Aa	77.50 Ba
40	67.0 Aa	73.0 Aa	99.22 Ab	92.85 Ba	102.1 Ba	105.4 Ca	75.72 Aa	85.07 Aab	104.3 Bb	81.85 Aa	81.92 Aa	101.60 a
50	116.1 Ba	132.4 Bb	95.52 Aa	84.82 Aba	94.27 Aba	101.3 Ca	91.35 Aa	118.87 b	113.7 Bab	98.22 Aa	113.5 Ba	140.02 b
Initial condition	74.04 Aa	74.04 Aa	74.04 Aa	74.04 Aba	74.04 Aa	74.04 Aa	74.04 Aa	74.04 Aa	74.04 Aa	74.04 Aa	74.04 Aa	74.04 Aa

[a]Means followed by the same lowercase letter in the line do not differ by Tukey test at 1%. Means followed by the same capital letter in the column do not differ by Tukey test at 1%. [b]L (Lower), [c]M (Medium), [d]U (Upper). Coefficient of variation = 17.01%.

Table 9. Electric conductivity (μcm^{-1} g^{-1}) of soybeans after drying, depending on drying cells and drying air temperatures[a].

	[b]Lower point cells				[c]Medium point cells				[d]Upper point cells			
T (°C)	1	2	3	4	1	2	3	4	1	2	3	4
30	71.63 Aa	61.6 Aa	66.20 Aa	71.90 Aa	79.00 Aa	75.52 ba	62.45 Aa	71.22 Aa	77.75 a	77.80 Ba	71.40 Aa	77.50 ABa
40	67.05 Aa	92.85 Ba	75.72 Aa	81.85 Aa	73.00 Aa	102.1 Bb	85.07 ab	81.92 ab	99.2 Aa	105.4 Ca	104.3 Ba	101.60 Ba
50	116.2 Bb	84.82 Ba	91.35 ab	98.22 ab	132.4 Bb	94.27 Ba	118.8 ab	113.5 ab	95.52 a	101.3 Ca	113.7 ab	140.02 Cb
Initial condition	74.04 Aa	74.04 Aa	74.04 Aa	74.04 Aa	74.04 Aa	74.04 Aa	74.04 Aa	74.04 Aa	74.04 a	74.04 Aa	74.04 Aa	74.04 Aa

[a]Means followed by the same lowercase letter in the line do not differ by Tukey test at 1%. Means followed by the same capital letter in the column do not differ by Tukey test at 1%. [b]L (Lower), [c]M (Medium), [d]U (Upper). Coefficient of variation = 17.01%.

Table 10. Germination (%) of soybean grains after drying, according to the sampling points in the cells and drying air temperatures[a].

T (°C)	Cell 1			Cell 2			Cell 3			Cell 4		
	L	M	U	L	M	U	L	M	U	L	M	U
30	64.00 Bb	78.50 Aa	63.50 Bb	80.50 Aa	83.00 Aa	78.00 Aa	84.00 Aa	73.50 Aa	84.50 Aa	59.50 Ab	84.50 Aa	82.50 Ba
40	77.50 ab	73.00 Ab	85.00 Aa	76.50 Aa	73.00 Ba	82.50 Aa	85.50 Aa	83.00 Aa	81.00 Aa	81.50 Aa	78.00 Aa	84.50 Aa
50	57.00 Ba	53.00 Ba	62.50 Ba	69.00 Aa	65.50 Ba	64.00 Ba	66.00 Ba	53.50 b	65.50 Ba	48.00 Bb	64.00 Ba	71.50 Ba
Initial condition	78.00 Aa	78.00 Aa	78.00 Aa	78.00 Aa	78.00 Aa	78.00 Aa	78.00 Ba	78.00 Aa	78.00 Aa	78.00 Aa	78.00 Aa	78.00 Ba

[a]Means followed by the same lowercase letter in the line do not differ by Tukey test at 1%. Means followed by the same capital letter in the column do not differ by Tukey test at 1%. [b]L (Lower), [c]M (Medium), [d]U (Upper). Coefficient of variation = 9.14%.

Table 11. Germination (%) of soybean grains after drying as a function of drying cells and drying air temperatures[a].

	[b]Lower point cells				[c]Medium point cells				[d]Upper point cells			
T (°C)	1	2	3	4	1	2	3	4	1	2	3	4
30	64.0 Bc	80.5 Aab	84.0 Aa	69.5 Abc	78.5 Aa	83.0 Aa	73.5 Aa	84.50 Aa	63.5 Bb	78.0 Aa	84.5 Aa	82.5 ABa
40	77.5 Aa	76.5 Aa	85.5 Aa	81.50 Aa	73.00 Aa	73.0 ABa	83.0 Aa	78.00 Aa	85.00 Aa	82.50 Aa	81.0 Aa	84.50Aa
50	57.0 Bab	69.0 Aa	66.0 Ba	48.0 Bb	53.00 Bb	65.5 Ba	53.5 ab	64.00 ab	62.5 Ba	64.0 Ba	65.50Ba	71.5 Ba
Initial condition	78.0 Aa	78.0 Aa	78.0 Ba	78.0 Aa	78.00 Aa	78.0 Aa	78.0 Aa	78.00 Aa	78.0 Aa	78.0 Aa	78.0 Aa	78.0 ABa

[a]Means followed by the same lowercase letter in the line do not differ by Tukey test at 1%. Means followed by the same capital letter in the column do not differ by Tukey test at 1%. [b]L (Lower), [c]M (Medium), [d]U (Upper). Coefficient of variation = 9.14%.

Table 12. Oil content (%) of soybean grains after drying, according to the sampling points in the cells and drying air temperature[a].

T (°C)	Cell 1			Cell 2			Cell 3			Cell 4		
	L	M	U	L	M	U	L	M	U	L	M	U
30	9.05 Ca	9.05 Ca	9.05 Ca	9.89 Da	9.89 Da	9.89 Da	9.22 Da	9.22 Da	9.22 Da	9.28 Da	9.28 Da	9.28 Da
40	7.70 Da	7.70 Da	7.70 Da	10.9 Ca	10.9 Ca	10.9 Ca	14.3 Ca	14.3 Ca	14.3 Ca	12.9 Ca	12.9 Ca	12.92 Ca
50	12.17 Ba	12.1 Ba	12.1 Ba	12.3 Ba	12.3 Ba	12.3 Ba	17.9 Aa	17.9 Aa	17.9 Aa	18.0 Aa	18.0 Aa	18.07 Aa
Initial condition	15.44 Aa	15.4 Aa	15.4 Aa	15.4 Aa	15.4 Aa	15.4 Aa	15.4 Ba	15.4 Ba	15.4 Ba	15.4 Ba	15.4 Ba	15.44 Ba

[a]Means followed by the same lowercase letter in the line do not differ by Tukey test at 1%. Means followed by the same capital letter in the column do not differ by Tukey test at 1%. [b]L (Lower), [c]M (Medium), [d]U (Upper). Coefficient of variation = 9.14%.

Table 13. Oil content (%) of soybean grains after drying as a function of the drying cells and drying air temperatures[a].

	[b]Lower point cells				[c]Medium point cells				[d]Upper point cells			
T (°C)	1	2	3	4	1	2	3	4	1	2	3	4
30	9.05 Ca	9.89 Da	9.22 Da	9.28 Da	9.05 Ca	9.89 Da	9.22 Da	9.28 Da	9.05 Ca	9.89 Da	9.22 Da	9.28 Da
40	7.70 Dd	10.9 Cc	14.3 Ca	12.9 Cb	7.70 Dd	10.9 Cc	14.3 Ca	12.92 Cb	7.70 Dd	10.9 Cc	14.3 Ca	12.9 Cb
50	12.1 Bb	12.3 Bb	17.9 Aa	18.0 Aa	12.17 Bb	12.3 Bb	17.9 Aa	18.07 Aa	12.1 Bb	12.3 Bb	17.9 Aa	18.0 Aa
Initial condition	15.4 Aa	15.4 Aa	15.4 Ba	15.4 Ba	15.44 Aa	15.4 Aa	15.4 Ba	15.44 Ba	15.4 Aa	15.4 Aa	15.4 Ba	15.44Ba

[a]Means followed by the same lowercase letter in the line do not differ by Tukey test at 1%. Means followed by the same capital letter in the column do not differ by Tukey test at 1%. [b]L (Lower), [c]M (Medium), [d]U (Upper). Coefficient of variation = 9.14%.

Conclusion

The increase in the drying air temperature to 50°C decreased the quality of soybeans. Different drying cells did not affect the quality of grain. In the upper part of the cells, there was an increase in electrical conductivity (140.02 μS cm^{-1} g^{-1}) and a reduction of the specific grain mass (851 kg m^{-3}), germination (63%) and oil content (7%) compared with the middle and lower dryer positions. The experimental prototype of the fixed bed dryer allowed a uniform distribution of the heated air in the drying cells, being effective in the drying operation and maintaining the quality of the soybean.

Acknowledgements

The authors thank FUNDECT, CNPq and CAPES-Brazil by financial support.

References

Almeida, D. P., Resende, O., Mendes, U. C., Costa, L. M., & Rocha, A. C. (2013). Influência da secagem na qualidade fisiológica do feijão adzuki. *Revista Brasileira de Ciências Agrárias*, *8*(2), 311-315. doi: 10.5039/agraria.v8i2a2164

Brasil. Ministério da Agricultura, Pecuária e Abastecimento. (2007). *Instrução normativa n. 11, de 15 de maio de 2007 para classificação física da soja*. Brasília, DF: SNAD/DNDV/CLAV.

Brasil. Ministério da Agricultura, Pecuária e Abastecimento. (2009). *Regras para análise de sementes*. Brasília, DF: SNAD/DNDV/CLAV.

Brooker, D. B., Bakker-Arkema, F. W., & Hall, C. W. (1992). *Drying and storage of grains and oilseeds*. New York, US: van Nostrand Reinhold.

Chua, K. J., Hawlader, M. N. A., Chou, S. K., & Ho, J. C. (2002). On the study of time-varying temperature drying-effect on drying kinetics and product quality. *Drying Technology*, *20*(8), 1559-1577. doi: 10.1081/DRT-120014052

Coradi, P. C., Melo, E. C., & Rocha, R. P. (2014). Evaluation of electrical conductivity as a quality parameter of lemongrass leaves (*Cymbopogon citratus* Stapf) submitted to drying process. *Drying Technology*, *32*(1), 969-980. doi: 10.1080/ 07373937.2013.879593

Coradi, P. C., Fernandes, C. H. P., Helmich, J. C., & Goneli, A. L. D. (2016a). Effects of drying air temperature and grain initial moisture content on soybean quality (*Glycine max* (L.) Merrill). *Engenharia Agrícola*, *36*(5), 866-876. doi: 10.1590/ 1809-4430-Eng.Agric.v36n5p866-876/2016

Coradi, P. C., Fernandes, C. H. P., & Helmich, J. C. (2016b). Adjustment of mathematical models and quality of soybean grains in the drying with high temperatures. *Revista Brasileira de Engenharia Agrícola e Ambiental*, *20*(4), 385-392. doi: 10.1590/1807-1929/agriambi.v20n4p385-392

Defendi, R. O., Nicolin, D. J., Paraíso, P. R., & Jorge, L. M. D. M. (2016). Assessment of the initial moisture content on soybean drying kinetics and transport properties. *Drying Technology*, *34*(3), 360-371.

Dondee, S., Meeso, N., Soponronnarit, S., & Siriamornpun, S. (2011). Reducing cracking and breakage of soybean grains under combined near-infrared radiation and fluidized-bed drying. *Journal Food Engineering*, *104*(1), 6-13. doi: 10.1016/j.jfoodeng.2010.11.018

Hashemi, G., Mowla, D., & Kazemeini, M. (2009). Moisture diffusivity and shrinkage of broad beans during bulk drying in an inert medium fluidized bed dryer assisted by dielectric heating. *Journal Food Engineering*, *92*(3), 331-338. doi: 10.1016/j.jfoodeng.2008.12.004

Hegarty, T. W. (1978). The physiology of seed hydration and dehydration and relation between water stress and control of germination: a review. *Plant Cell Evironmental*, *1*(1), 101-19. doi: 10.1111/j.1365-3040.1978.tb00752.x

Hemis, M., Choudhary, R., & Watson, D. G. (2012). A coupled mathematical model for simultaneous microwave and convective drying of wheat seeds. *Biosystems Engineering*, *112*(3), 202-209. doi: 10.1016/j.biosystemseng.2012.04.002

Instituto Adolfo Lutz. (2008). *Normas Analíticas do Instituto Adolfo Lutz. Métodos físico-químicos para análises de alimentos* (4a ed., 1a ed. digital). São Paulo, SP: Instituto Adolfo Lutz

Isquierdo, E. P., Borém, F. M., Andrade, E. T., Corrêa, J. L. G., Oliveira, P. D., & Alves, G. E. (2013). Drying Kinetics and quality of natural coffee. *Transactions of the American Society of Agricultural and Biological Engineers*, *56*(3), 1003-1010. doi: 10.13031/trans.56.9794

Jangam, S. V., Law, C. L., & Mujumdar A. S. (2010). *Drying of foods, vegetables and fruits* Singapore: National University of Singapore.

Khanchi, A., Jones, C. L., Sharma, B., Huhnke, R. L., Weckler, P., & Maness, N. O. (2013). An empirical model to predict infield thin layer drying rate of cut switchgrass. *Biomass and Bioenergy*, *58*(1), 128-135. doi: 10.1016/j.biombioe.2013.08.024

Khatchatourian, O. A. (2012). Experimental study and mathematical model for soya bean drying in thin layer. *Biosystems Engineering*, *113*(1), 54-64. doi: 10.1016/j.biosystemseng.2012.06.006

Kowalski, S. J., & Rajewska, K. (2009). Effectiveness of hybrid drying. *Chemical Engineering Process*, *48*(8), 1302-1309. doi: 10.1016/j.cep.2009. 05.009

Krzyzanowski, F. C., Vieira, R. D., & França Neto, J. B. (2009). *Vigor de sementes: conceitos e testes*. Londrina, PR: Associação Brasileira de Tecnologia de Sementes.

Lopes, D. C., Steidle Neto, A. J., & Santiago, J. K. (2014). Comparison of equilibrium and logarithmic models for grain drying. *Biosystems Engineering*, *118*(1), 105-114. doi: 10.1016/j.biosystemseng. 2013.11.011

Niamnuy, C., Nachaisin, M., Laohavanich, J., & Devahastin, S. (2011). Evaluation of bioactive compounds and bioactivities of soybean dried by different methods and conditions. *Food Chemistry*, *129*(3), 899-906. doi: 10.1016/j. foodchem.2011.05.042

Niamnuy, C., Nachaisin, M., Poomsa, N., & Devahastin, S. (2012). Kinetic modeling of drying and conversion/degradation of isoflavones during infrared drying of soybean. *Food Chemistry*, *133*(3), 946-952. doi: 10.1016/j.foodchem.2012.02.010

Nuthong, P., Achariyaviriya, A., Namsanguan, K., & Achariyaviriya, S. (2011). Kinetics and modeling of whole longan with combined infrared and hot air. *Journal Food Engineering*, *102*(3), 233-239. doi: 10.1016/j.jfoodeng.2010.08.024

Peplinski, A. J., Paulis, J. W., Bietz, J. A., & Pratt, R. C. (1994). Drying of high-moisture corn: Changes in properties and physical quality. *Cereal Chemistry*, *71*(2), 129-133.

Rafiee, S. H., Keyhani, A., Sharifi, M., Jafari, A., Mobli, H., & Tabatabaeefar, A. (2009). Thin Layer Drying Properties of Soybean (*Viliamz Cultivar*). *Journal of Agricultural Science and Technology*, *11*(3), 289-300.

Resende, O., Almeida, D. P., Costa, L. M., Mendes, U. C., & Sales, J. F. (2012). Adzuki beans (*Vigna angularis*) seed quality under several drying conditions. *Ciência e Tecnologia de Alimentos*, *32*(1), 151-155.

Resende, O., Oliveira, D. E. C., Chaves, T. H., Ferreira, J., & Bessa, V. (2014). Kinetics and thermodynamic properties of the drying process of sorghum (*Sorghum bicolor* [L.] Moench) grains. *African Journal of Agricultural Research*, *9*(32), 2453-2462. doi: 10.5897/AJAR2014.8808

Sanjinez-Argandoña, E. J., Branco, I. G., Bittencourt, T. U., & Munhoz, C. L. (2011). Influence of temperature and geometry in the drying kinetics of tomato (*Lycopersicum esculentum*). *Food Science and Technology*, *31*(2), 308-312. doi: 10.1590/S0101-20612011000200005

Ribeiro, D. M., Corrêa, P. C., Rodrigues, D. H., & Goneli, A. L. (2005). Análise da variação das propriedades físicas dos grãos de soja durante o processo de secagem. *Food Science and Technology*, *25*(3), 611-617. doi: 10.1590/S0101-20612005000300035.

Siqueira, V. C., Resende, O., & Chaves, T. H. (2012). Drying kinetics of Jatropha seeds. *Revista Ceres*, 59(2), 171-177. doi: 10.1590/S0034-737X2012000200004

Sivakumar, R., Saravanan, R., Elaya Perumal, A., & Iniyan, S. (2016). Fluidized bed drying of some agro products–A review. *Renewable and Sustainable Energy Reviews*, *61*(1), 280-301. doi: 10.1016/j.rser.2016.04.014

Vasquez, G. H., Bertolin, D. C., & Spegiorin, C. N. (2011). Testes de envelhecimento acelerado e de condutividade elétrica para avaliar a qualidade fisiológica de sementes de sorgo (*Sorghum bicolor* (L.) Moench). *Revista Brasileira de Biociências*, *9*(1), 18-24.

Vieira, R. D., Penariol, A. L., Perecin, D., & Panobianco, M. (2002). Condutividade elétrica e teor inicial das sementes de soja. *Pesquisa Agropecuária Brasileira*, *37*(9), 1333-1338. doi: 10.1590/S0100-204X2002000900018

7

Recognition and classification of White Wholes (WW) grade cashew kernel using artificial neural networks

Narendra Veranagouda Ganganagowdar[1] and Hareesha Katiganere Siddaramappa[2]

[1]*Departamento de Engenharia da Computação, Instituto de Tecnologia de Manipal, Universidade de Manipal, Manipal, Karnataka, India-576104.*
[2]*Departamento de Computação Aplicada, Instituto de Tecnologia de Manipal, Universidade de Manipal, Manipal, Karnataka, India 576104.*
**Author for correspondence. E-mail: narendra.vg@manipal.edu*

ABSTRACT. A novel intelligent automated model to recognize and classify a cashew kernels using Artificial Neural Network (ANN). The model primarily intends to work on two phases. The phase one, built with a proposed method to extract features, which includes 16 morphological features and also 24 color features from the input cashew kernel images. In phase two, a Multilayer Perceptron ANN is being used to recognize and classify the given white wholes grades using back propagation learning algorithm. The proposed method achieves a classification accuracy of 88.93%. This study also reveals that the combination of morphological and color features outperforms rather using any one set of features separately to grade cashew kernels.

Keywords: White Wholes (WW) grade cashew kernel images, feature extraction, artificial neural networks, classification.

Introduction

Cashew (*Anarcardium Occidentale* L.) has emerged as an important plantation crop of India and it plays a significant role in Indian economy. Cashew earned foreign exchange equivalent to 43,900 million rupees, from export of 1,32,000 MT of cashew kernels during the year 2011-12 (Saroj & Balasubramanian, 2013). Cashew kernels are obtained from raw cashew nuts, the true seeds of the cashew tree. Cashew are most delicious considered to be among many other edible tree nuts. It is proved that cashew add taste virtually to anything i.e. Ice Creams, Sweets, Chocolates, and various Dishes (Kumar, Rao, & Desai, 2013).

In India cashew farms and processing units are located in the states of Maharastra, Goa, Andhra Pradesh, Orissa, West Bengal, Karnataka, Kerala, Tamilnadu, Madhya Pradesh, and in the eastern regions. In factory, composite lots of raw cashew nuts, after processing, yield the bulk of cashew kernels of varied size and weights, which in turn decide the marketability of the kernels under different grades. Grading of cashew kernels is a prerequisite to meet the requirements of domestic as well as international trades and marketing (Kumar et al., 2013). Cashew kernels are graded depending on their size, shape and color standards specified by the Export Control and Inspection Act, 1963 (Cashew Export Promotion Council of India [CEPCI], 2013). There are as many as 26 cashew kernel grades available in the market ranging from white wholes to pieces of which WW-180, WW-210, WW-240 and WW-320 are the accepted grades in the global market. The important characteristics of the white wholes cashew kernels are as mentioned in the Table 1 (CEPCI, 2013). The current standards for cashew kernels grade are ambiguous, because of count per 454 g size description and general characteristic as mentioned in the Table 1.

Table 1. General characteristic of White Wholes cashew kernels.

White wholes grade	Count per 454 g size description	General characteristic
WW-180	170-180	Cashew kernels are obtained through shelling and peeling cashew nuts. They have the characteristic shape; shall be white, pale ivory or light ash in color reasonably dry and free from insect damage, damaged kernels and black or brown spots. The kernels shall be completely free from testa.
WW-210	200-210	
WW-240	220-240	
WW-320	300-320	

A major problem in sorting or grading of cashew kernels is the use of harmful mechanical equipment or expensive color sorters; still, cashew kernels grading is carried out manually. Manual sorting or grading is based on traditional visual quality inspection performed by trained labours. The problem inherent in this system includes high labor costs, worker fatigue, inconsistency, variability, and scarcity of trained labor. Today, various kinds of cashew kernel grades are provided in the market with different qualities, due to the variability in quality decisions among the graders. Due to the scarcity of trained labor, mechanization at various stages of cashew processing is gaining importance in this country and abroad. This way of grading presents many quality problems and grading is the final opportunity for the quality control (Thakkar et al., 2011). To ascertain the quality, the cashew kernel (white wholes) grade standards have been developed by considering the geometrical and color features.

Computer-based vision technology offers a high level of flexibility and repeatability at relatively low-cost with fairly high throughput with superior accuracy. Computer vision offers an alternative to visual physical inspection, and has been in numerous foods and agricultural commodity sorting or grading systems today. Moreover, this practice is objective, consistent, rapid and economical. Geometrical and color features are the primary information sources for foods and agricultural commodity (i.e. object) inspection, classification, and sorting or grading (Sun, 2008). Computer vision systems have been effectively used to classify or to recognize quality parameters like color and size in several agricultural and food commodities including dry beans (Kumar, Bora, & Lin, 2013), coffee (Soedibyo et al., 2010), soya beans seeds (Namias et al., 2012), peanuts (Chen et al., 2011) and brazil-nuts (Castelo-Quispe et al., 2013) advancement in hardware and image processing makes computer vision a very popular technique for automatic cashew kernel quality inspection of parameters like geometrical and color.

At present, there are no automated systems for cashew kernel (white wholes) grade, to estimate standard quality parameters like color and geometric. Our primary objective of this study is to develop a cost-effective intelligent computer vision system to (i) To extract morphological (i.e. size and shape) and color (i.e. RGB and HCL) features from a cashew kernel image samples of white wholes grades (ii) To recognize and classify the cashew kernel of white wholes grade from an image samples using Multilayer Perceptron Neural Network.

Material and methods

The Methodology of this research involves several tasks such as Data acquisition, Image preprocessing and segmentation, Features extraction and selection, and classification and recognition task using Multilayer Perceptron Neural Network as shown in Figure 1.

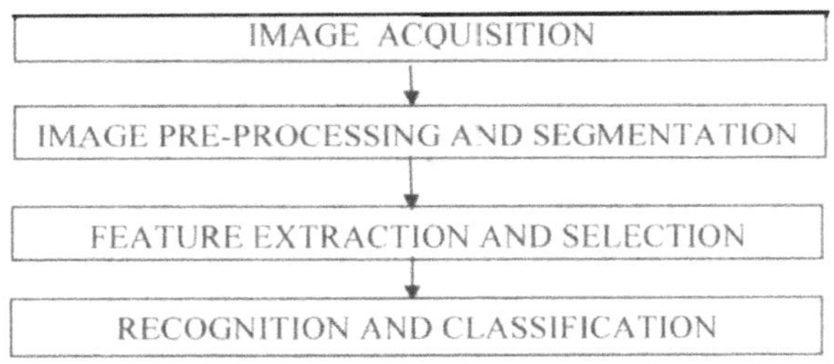

Figure 1. Methodology of the research.

Image acquisition

Samples collection

We have collected the samples of cashew kernel of white wholes grades (i.e. WW-180, WW-210, WW-240 and WW-320) from the various cashew industries located nearby the Manipal, Karnataka. The industries are Achal Cashew Industry Mangalore (ACI), Tirumala Cashew Industry Karkal (TCI), Balaji Cashew Exports Jarkal (BCE), Sathysree Cashew Industry (SCI), Gayathri Cashew Industry Karkal (GCI) and Mahalasa Cashew Exports Hiriyedka (MCE). The mentioned cashew industries were supplying cashew kernels to Domestic as well as international market. The industries follows using Cepci standards (CEPCI, 2013) to grade the cashew kernels manually by skilled labours.

Image acquisition system

The Figure 2 shows an experimental set-up of image acquisition system, used to capture an image and added to database by using two webcams (i) Ball and SmartPC, each of 12M pixels still image resolution), i.e iBall used to capture top view of the cashew kernel and (ii) SmartPC used to capture front view of the cashew kernel, in a color matching cabinet with a proper control of lighting intensity under Artificial Daylight Fluorescent Lamps (D-65) light source. The distance between webcams and each cashew kernel sample with moisture content in the range of 3.5 - 5%, was fixed (10 cm) to regret the effect of the distance on saved images. In order to reduce the influence of surrounding light, a black rubber sheet is used in image acquisition system. We were used Image Acquisition Toolbox (Matlab, 2012), to capture the cashew kernel samples by using an image acquisition system. Images format was 24 bit color JPEG with resolution 640 x 480 pixels. The Image database size of white wholes grades cashew kernel are tabulated in Table 2.

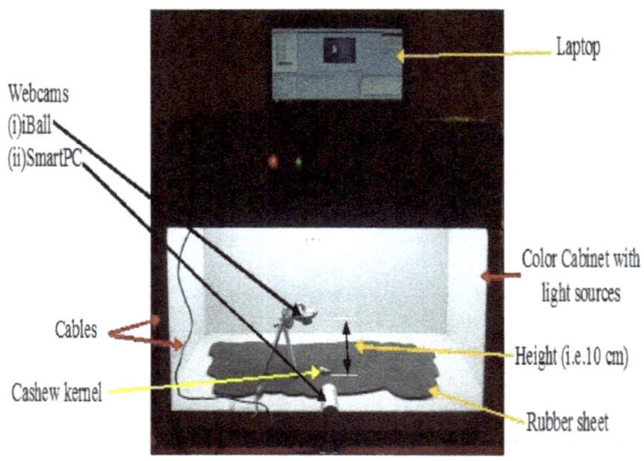

Figure 2. Image acquisition system.

Table 2. Image Database size (in no's) of white wholes grades cashew kernel.

Sl. No.	Cashew Kernel Grades	ACI	TCI	BCE	SCI	GCI	MCE	Total
01	WW-180	150	150	150	150	150	150	900
02	WW-210	150	150	150	150	150	150	900
03	WW-240	150	150	150	150	150	150	900
04	WW-320	150	150	150	150	150	150	900
								Total: 3600

Image pre-processing and segmentation

Image processing Toolbox in the Matlab (2012) is invoked as an image analysis and a processing software to extract the features from the image.

i) Image preprocessing: The median filter has been applied on the acquired image to eliminate noise, which has crept in during the digitization/acquisition process. The median filter replaces the value of a pixel by the median value of the sampling space, which consists of a 5 × 5 centered window on the considered pixel. The use of median filter facilitates the identification of an optimal threshold for the image during segmentation.

ii) Segmentation: This subdivides an image into meaningful non-overlapping region that is used for further analysis. It is assumed that the regions obtained correspond to the physical parts or objects of a scene (3-D) represented by the image (2-D). Image representation involves representing the segmented image as a boundary or a region. Boundary representation is suitable for analysis of morphological (i.e. size and shape) features. In this context, segmenting an image consists of separating the cashew kernel from background of the image. The segmentation carried out by using Otsu's thresholding algorithm. The Otsu's algorithm maximizes the variance between the object and background classes. Pixels that have the values less than the global threshold are classified as cashew kernel and the others are classified as background (resulting in a binary image).

Feature extraction and selection

In this study, morphological and color features were extracted from all the training and testing data sets of cashew kernel images of white wholes grade.

Morphological (i.e. Size and Shape) features of a cashew kernel

Calculation of the length and width is much more complex than that of area and perimeter (Mendoza et al., 2006) especially for food objects, which generally have very irregular shapes. Nevertheless, some measurements for length and width have been developed by researchers and are used in the food industry. The measurements most commonly used are Feret's diameter, by calculating the major axis and the minor axis (Zheng et al., 2006) of an object.

Feret's diameter

Feret's Diameter is also called the caliper length. It is the longest distance between any two points in the boundary of the region of interest (Sun, 2008; Castelo-Quispe et al., 2013) as shown in Figure 3a. All the cashew kernels of white wholes grade are looking like an elliptical shape (Mendoza, et al., 2006; Ganganagowdar, & Siddaramappa, 2011). So we considered the major axis as actual length (L) and minor axis as actual width (W) as well as thickness (T) of cashew kernel as shown in Figure 3b, c and d (Balasubramanian, 2001).

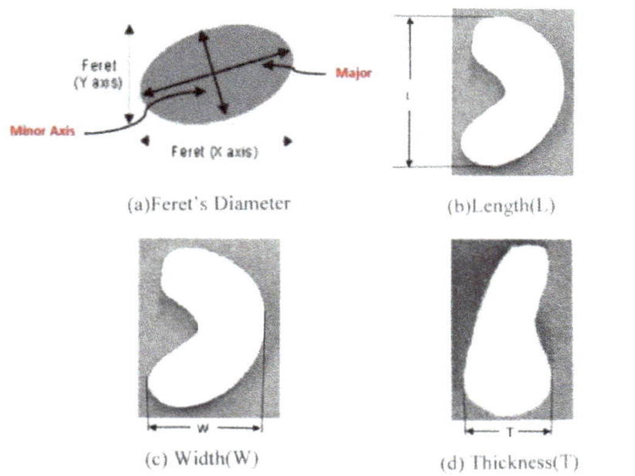

Figure 3. Illustrates the Feret's diameter along with measurement of cashew kernel.

Size

Geometry (i.e. Size) related features deals with the extraction of quantitative information from the previously segmented image regions (Amaral et al., 2009; Chen et al., 2010; Wiwart et al., 2012). After getting the boundary of the selected region, properties such as *area, perimeter, major* and *minor axis length etc.,* can be determined (Liu et al., 2005).

a) The *major axis length* of the cashew kernel is the length (in pixels) of the major axis of the ellipse that has the same normalized second central moments as the region, while the *minor axis length* is the length (in pixels) of the minor axis of the ellipse that has the same normalized second central moments as the region. However, the property of *major and minor axis length* are only supported for a 2-D image input. The *major and minor axis length* can represent the actual length (L) and width (W) as well as thickness (T) of a cashew kernel.

b) The *area* (A) of a region is defined as the actual number of pixels contained within its boundary.

c) The *perimeter* (P) is the contour length of the boundary.

d) Equivalent Diameter (Equation 1): It is the diameter of a circle with the same area as the cashew kernel region.

e)The Aspect ratio is computed by Equation 2

$$Eq = \sqrt{\frac{4 \times Area(A)}{\pi}} \quad (1)$$

Aspect ratio (K) =Major axis length/ Minor axis length (2)

f) Convex area (C): It is the number of pixels in the smallest convex polygon that can contain the cashew kernel region.

g) Solidity(S): The proportion of the pixels in the cashew kernel region are also in the convex hull and computed as Equation 3:

Area (A)/ConvexArea (C) (3)

h) Extent (Ex): The proportion of the pixels in the bounding box which are also in the cashew kernel region.

i) Eccentricity(E): the eccentricity of the ellipse that has the same second-moments as the cashew kernel region. The eccentricity is the ratio of the distance between the foci of the ellipse and its major axis length. The value is between 0 and 1.

j) Roundness (R)= This is given by Equation 4:

$$R = \frac{4 \times \pi \times A}{P^2} \quad (4)$$

k) Compactness (CO): The compactness provides a measure of the object's (i.e. cashew kernel) roundness, according Equation 5.

$$CO = \frac{\sqrt{4 \times A/\pi}}{L} \quad (5)$$

Shape

From the values of major and minor axis length, and area shape factors were derived (Symons & Fulcher, 1988), are as Equation 6, 7, 8 and 9.

Shape factor (SF1) = Major-axis length (L)/Area(A) (6)

Shape factor (SF2) = Area (A)/ Major-axis length $(L)^3$ (7)

Shape factor (SF3) = Area (A)/ [(Major-axis length/ 2) x (Major-axis length/ 2) x π] (8)

Shape factor (SF4) = Area (A)/ [(Major-axis length/ 2) x (Minor-axis length/ 2) x π] (9)

Color

Color provides the basic information for human perception. Color is also elementary information that is stored in pixels to constitute a digital image. Hence color is considered to be one of the most prominent features for object measurements, image understanding and object description (Sun, 2008).

In cashew Industry, color is one of the prominent quality traits of cashew kernels grading by trained labours. So color is playing important role in the cashew kernels grading process (Dhakshinamurthy & Kokila, 2013). In this research, we are making an attempt to use this color information primarily for analysis and grading of cashew kernels using digital image processing technique.

A color space is a method of representing each color in terms of combination of several numeric values. It is a method by which we can specify, create and visualize multiple color. Different color spaces are better for different applications (Sun, 2008). In this approach, we have used different color spaces (i.e. RGB, and HCL), in order to obtain the following color features from a given cashew kernel image.

a) Luminance: Luminance describes the 'achromatic' component of an image. In general, Luminance represents the brightness of an image (Sun, 2008).

b) Chrominance: Chrominance is the color information of an image, separately from the accompanying luminance. Chrominance is usually represented as two color-difference components (Sun, 2008).

RGB color space

RGB color space is an additive color space based on RGB color model. A particular RGB color space is defined by three chromaticities of red, green, and blue additive primaries, and can produce any chromaticity that is the triangle defined by those primary colors. The image is represented in RGB color as a primary color model. The three-dimensional color spaces are normally discretised to 256 levels per axis, which gives over 1.67 million distinct colors (Du & Sun, 2005).

In this study, we have proposed that, (1) First we separate the R, G and B components from a given cashew kernel image. (2) From the R, G and B components, we were extracted the luminance-chrominance by using following Equation 10, 11 and 12.

(1) Luminance (Luma or L for short) is formed as a weighted sum of *linear* RGB components (International Telecommunication Union, 2002).

$$L = 0.2126\,R + 0.7152\,G + 0.0722\,B \tag{10}$$

(2) Chrominance (*chroma* or C for short): Chrominance is usually represented as two color-difference components (International Telecommunication Union, 2002):

$$Cr = R - L \text{ (red – luma)} \tag{11}$$

$$Cb = B - L \text{ (blue – luma)} \tag{12}$$

(3) Also we have measured the hue angle (H^0) from the following Equation 13 (Sun, 2008).

$$H^0 = \begin{cases} \frac{1}{360^0}\left[90^0 - tan^{-1} * \frac{F}{\sqrt{3}} + 0^0\right] & G < B \\ \frac{1}{360^0}\left[90^0 - tan^{-1} * \frac{F}{\sqrt{3}} + 180^0\right] & G > B \end{cases} \tag{13}$$

where:

F = (2*R-G-B)/(G-B)

(4) Finally, we have measured color distance metric by using following Equation 14 (Sun, 2008).

$$\Delta E_{RGB} = \sqrt{(\mu_R - R_0)^2 + (\mu_G - G_0)^2 + (\mu_B - B_0)^2} \tag{14}$$

HCL (Hue, Chrominance and Luminance) color space

In this, we investigates color pixel similarity analysis on a new perceptually uniform color space that we call HCL (Hue, Chroma and Luminance). Such a new color space exploits the advantages of each one of the color spaces: HSL and L*a*b* and discards their drawbacks.

We assume that the chroma and the hue of any color can be defined as a blend of the three chrominance elemental sensations: R-G (from red to green), G-B (from green to blue) and B-R (from blue to red). Based on this assumption and the Munsell color system with the three color attributes closed to human perceptions: hue (H), chroma (C) and luminance (L), we define below a mapping from RGB space to HCL space (Sarifuddin & Missaoui, 2005).

We recall that a color containing a lot of white is brighter than one with less white. A saturated color contains 0% of white and has a maximum value of chroma. An increasing value of white leads to a decreasing value of chroma and a less saturated color. Concretely, a color is saturated if Max (R, G, B) is equal to R, G, or B, and Min (R, G, B) = 0. The saturation of a color is null (i.e., chroma = 0) when Min (R, G, B) = Max (R, G, B). Therefore, we will use the expressions Max (R, G, B) and Min (R, G, B) to compute luminance L (Sarifuddin & Missaoui, 2005).

Color spaces YIQ, YUV, YCrCb, L*u*v* and L*a*b* express Y by Y = 0.299R + 0.587G + 0.114B, whereas color spaces like HSI, HSV, and HSL use Y = I = (R + G + B)/ 3, Y = L = Max (R, G, B) and Y = L = [Max (R, G, B) + Min (R, G, B)]/ 2 respectively.

We define luminance L as a linear combination of Max (R, G, B) and Min (R, G, B) as Equation 15:

$$L = Q.Max\,(R, G, B) + (1 - Q).Min\,(R, G, B)/2 \tag{15}$$

where:

$Q = e^{\alpha\gamma}$ is a parameter that allows a tuning of the variation of luminosity between a saturated hue (color) and a hue containing a great amount of white, with α = (Min (R, G, B)/ Max (R, G, B), $1/Y_0$) and $Y_0 = 100$.

γ is a correction factor whose value (= 3) coincides with the one used in L⋆a⋆b⋆ space. It should be noted that when Min (R, G, B) = 0 and Max (R, G, B) varies between 0 and 255, luminance L takes a value between 0 (black) and 128. When Max (R, G, B) = 255 and in (R, G, B) varies between 0 and 255, luminance takes a value between 128 and 135 (Sarifuddin & Missaoui, 2005).

In a similar way, we define chroma C = Q.Cn, where Cn represents a mixture of three different combinations of R, G, and B components: red-green, green-blue and blue-red. (Sarifuddin & Missaoui, 2005), according Equation 16.

$$C = Q.\,(|R - G| + |G - B| + |B - R|)/3 \quad (16)$$

The hue value can be computed using the following Equation 17:

$$H = \arctan(G - B/R - G) \quad (17)$$

However, hue values (Equation 18) vary between -90^0 and $+90^0$ only. To allow hue values to vary in a larger interval going from -180^0 to 180^0:

$$\begin{aligned} &\text{if } [(R - G) < 0 \text{ and } (G - B) \geq 0], \text{ then } H = 180 + H \\ &\text{if } [(R - G) < 0 \text{ and } (G - B) < 0], \text{ then } H = H - 180 \end{aligned} \quad (18)$$

Or Equation 19 and 20:

$$\begin{aligned} &\text{if } [(R - G) \geq 0 \text{ and } (G - B) \geq 0], \text{ then } H = 2/3H \\ &\text{if } [(R - G) \geq 0 \text{ and } (G - B) < 0], \text{ then } H = 4/3H \\ &\text{if } [(R - G) < 0 \text{ and } (G - B) \geq 0], \text{ then } H = 180 + 4/3H \\ &\text{if } [(R - G) < 0 \text{ and } (G - B) < 0], \text{ then } H = 3/4H - 180 \end{aligned} \quad (19)$$

$$\Delta E_{HCL} = \sqrt{(\mu_H - H_0)^2 + (\mu_C - C_0)^2 + (\mu_L - L_0)^2} \quad (20)$$

Statistical analysis

After features extraction, we have done a statistical analysis of cashew kernel image of white wholes grade using various measures like mean, variance and standard deviation. The color images are recognized by quantifying the distribution of color components throughout the image, change in the each color component with reference to average/mean and difference between the highest and the lowest color values. This quantification is obtained by computing Mean, Variance, Standard Deviation and Range for a given color image. Since these features represent global characteristics for an image, we have adopted Mean, Variance, Standard Deviation and Range color features in this work. The Equation 21, 22, 23 and 24 are used to evaluate Mean, Variance, Standard Deviation and Range of the image samples (Savakar & Anami, 2009; Ganganagowdar & Siddarammappa, 2014).

Mean: The overall brightness of each color component of an image is measured using the mean. This is calculated by Equation 1 of each color component of an image (Savakar & Anami, 2009; Ganganagowdar, & Siddaramappa, 2014).

$$Mean = \mu_x \sum_{x,y=0}^{n-1} x * (f_{x,y}) \text{ and } \mu_y \sum_{x,y=0}^{n-1} y * (f_{x,y}) \quad (21)$$

Variance and Standard Deviation: The variance is a measure of how far perceived brightness is spread out of each color component in an image. The Standard Deviation is the average distance from the mean of the overall perceived brightness and contrast of each color component in a cashew kernel image (Savakar & Anami, 2009; Ganganagowdar, & Siddaramappa, 2014).

$$Variance(\sigma^2) = \sum_{x,y=0}^{n-1} (f_{x,y})(z - \mu)^2 \quad (22)$$

$$Standard\ Deviation(\sigma) = \sqrt{\sigma^2} \quad (23)$$

Range: This gives us the range of maximum and minimum perceived brightness of each color component in an image (Savakar & Anami, 2009; Ganganagowdar, & Siddaramappa, 2014).

$$Range = Max(f_{x,y}) - Min(f_{x,y}) \quad (24)$$

The proposed methods

a) Algorithm 01: Determine the Morphological features of a cashew kernel (Amaral et al., 2009; Chen et al., 2010; Wiwart et al., 2012; Ganganagowdar et al., 2012).

Input: Original 24-bit cashew kernel color image.

Output: 16 Morphological Features

Start

Step 1: Read the image of resolution 640 x 480 pixels.

Step 2: Crop and resize the image.

Step 3: Segment the color of image by using OTSU method.

Step 4: Convert RGB to grayscale image.

Step 5: Remove the noise from image using Median filter (Remove the noise).

Step 6: Binarize the image using Otsu Method.

Step 7: Find the Boundaries.

Step 8: Determine the Geometric properties like major-axis length (L), minor-axis length (W as well as T), Area (A), Perimeter (P), Equivalent diameter (Eq), Convex area (C), Solidity (S), Eccentricity (E), Extent (Ex), Aspect ratio (K), Roundness (R) and Compactness (Co) (Amaral et al., 2009; Chen et al., 2010; Wiwart et al., 2012) of a cashew kernel of white wholes grade from a cashew kernel image using Equation 1, 2, 3, 4 and 5.

Step 9: Determine the Shape properties SF1, SF2, SF3 and SF4 (Amaral et al., 2009; Chen et al., 2010; Wiwart et al., 2012) of a cashew kernel of white wholes grade from a cashew kernel image using Equation 6, 7, 8 and 9.

Stop

We have extracted the 16 Morphological features by using algorithm 1 and all the features are listed in Table 3.

Table 3. Extracted morphological features of individual cashew kernel.

Sl. No.	Features	Sl. No.	Features
01	Major-axis Length (L)	09	Extent (Ex)
02	Minor-axis Length (W T^{-1})	10	Aspect ratio (K)
03	Area (A)	11	Roundness (R)
04	Perimeter (P)	12	Compactness (Co)
05	Equivalent Diameter (Eq)	13	SF1
06	Convex Area (C)	14	SF2
07	Solidity (S)	15	SF3
08	Eccentricity (E)	16	SF4

b) The following algorithm 02 involved in obtaining the 24 color features (Savakar & Anami, 2009; Ganganagowdar & Siddaramappa, 2011; Ganganagowdar, & Siddaramappa, 2014).

Algorithm 02: RGB and HCL Color features extraction from a cashew kernel image. Input: Original 24-bit cashew kernel color image.

Output: 24 color features.

Start

Step 1: Read the image of resolution 640 x 480 pixels.

Step 2: Crop and resize the image.

Step 3: Segment the color of image using OTSU method.

Step 4: Separate the RGB components from the image.

Step 5: Compute Mean, Variance, Standard deviation and Range for each RGB components, using the Equation 21, 22, 23 and 24.

Step 6: Compute Luminance, Chrominance, Hue angle(H) and Color distance metric (ΔE_{RGB}) for cashew kernel image from RGB components using the Equation 10, 11, 12, 13 and 14.

Step 7: Obtain the HCL components from RGB components using the Equation 15, 16, 17, 18 and 19.

Step 8: Compute Mean, Variance, Standard Deviation and Range from HCL components using the Equation 21, 22, 23 and 24.

Step 9: Compute Color distance metric (ΔE_{HCL}) for cashew kernel image from HCL components using the Equation 20.

Stop

We have extracted 24 color features by using the algorithm 02 and all the features are listed in Table 4.

Features selection

Feature selection can be defined as a process that chooses a minimum subset of M features from the original set of *N* features. Feature selection reduces the dimensionality of feature space, removes redundant, irrelevant, or noisy data (Novaković et al., 2011). In many classification problems, it is difficult to learn good classifiers before removing the unwanted features due to the huge size of the data.

Table 4. Extracted color features of individual cashew kernel.

Sl. No.	RGB Features	Sl. No.	HCL Features
1	Mean of Red (μR)	15	Mean of Hue (μH)
2	Standard deviation of Red (σR)	16	Standard deviation of Hue (σH)
3	Range of Red (rangeR)	17	Range of Hue (rangeH)
4	Mean of Green (μG)	18	Mean of Chroma (μC)
5	Standard deviation of Green (σG)	19	Standard deviation of Chroma (σC)
6	Range of Green (rangeG)	20	Range of Chroma (rangeC)
7	Mean of Blue (μB)	21	Mean of Luminance (μL)
8	Standard deviation of Blue (σB)	22	Standard deviation of Luminance (σL)
9	Range of Blue (rangeB)	23	Range of Luminance (rangeL)
10	Luminance (L)	24	Color distance metric (ΔEHCL)
11	Chrominance (Cr)		
12	Chrominance (Cb)		
13	Hue angle(H^0)		
14	Color distance metric (ΔERGB)		

Reducing the number of irrelevant/redundant features can drastically reduce the running time of the learning algorithms and yields a more generic classifier. This helps in getting a better insight into the underlying concept of a real-world classification problem. We have obtained features subset using Sequential Floating Forward Selection (SFFS) algorithm with classification rate (CCR) of minimum: 0.70 and maximum: 0.99, from both morphological and color features. The feature subsets obtained are (L, W, A, Eq, C, S, E, SF1) and (μ_R, rangeR, σ_B, rangeB, Cr, Cb, H^0, ΔE_{RGB}, μ_H, σ_C, rangeC, rangeL, ΔE_{HCL}).

Recognition and classification

In this, the MLP neural network with back propagation is established under Matlab (2012) environment to train the training set. The training and testing set are normalized using the formula $x_i = (x_i - \mu_x)/\sigma_x$, where x_i is a feature (Sun, 2008). There were 750 training data set and 150 test data set, for each white wholes grade and, set includes eight morphological features (L, W, A, Eq, C, S, E, SF1) and thirteen color features (μ_R, rangeR, σ_B, rangeB, Cr, Cb, H^0, ΔE_{RGB}, μ_H, σ_C, rangeC, rangeL, ΔE_{HCL}). The network structure consists of three layers. The number of nodes of input layer (N_i) is the number of input parameters, and therefore, the number of nodes are 21. We tested the four white wholes grades at the same time, and therefore, the number of nodes of the output layer (N_o) is 4. The number of nodes in the hidden layer (N_h) were calculated using the Equation $N_h = N_i + N_o$ (Sun, 2008).

Results and discussion

Morphological features

We have estimated morphological features of the white wholes grades cashew kernel such as length, width, equivalent diameter, area, convex area, eccentricity etc. in pixel counts using proposed algorithm 1 and the clustered column chart is used to compare average values across cashew kernel grades as shown in Figure 4. We observed that, there is a distinct differences between WW-180, WW-210, WW-240 and WW-320 cashew kernels grades in terms of length (186.5, 166.2, 162.3 and 147.7), width (101, 92.8, 90.4 and 83.8), equivalent diameter (124.4, 113.7, 111 and 103), area (12188.7, 10164.4, 9716.5 and 8356) and convex area (14688.7, 12035, 11488 and 9822.4).

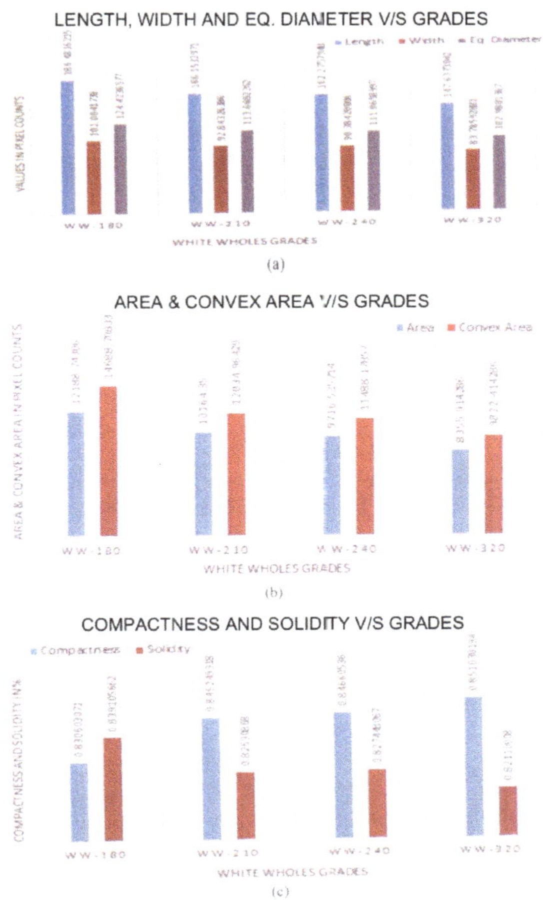

Figure 4. The Clustered column chart for White wholes grades.

Color features

RGB

Features are relating to the color of the entire cashew kernel were extracted from the cashew kernel image. The white whole cashew kernel grade color was characterized by the means (μ_R, μ_G, μ_B), standard deviations (σ_R, σ_G, σ_B) and range of the red, green and blue color components We were measured the luminance and chrominance from cashew kernel image representing the amount of brightness and contrast (Sun, 2008). Also we were extracted the hue angle (H°) from cashew kernel image. Features representing the amount of color and spatial distribution of particular white wholes grade were extracted from the cashew kernel image, using the algorithm 2 and summarized RGB color features are listed in Table 5. The average hue angle of particular white wholes grade ranged from green to blue are shown in Figure 5c. And also average color distance metric as well as average Cr and Cb are shown in Figure 5b and a. Some of the observations noted empirically are listed in Table 5.

Table 5. The RGB color space summary of white wholes grades cashew kernel.

Cashew kernel grades	RGB Color features summary
WW-180	i) $200.8 < \mu_R < 230$, $202.7 < \mu_G < 231.7$ and $198 < \mu_B < 229$ ii) Max (R, G, B) – Min (R, G, B) > 34 iii) R < G, B < G and B < R iv) 37.4 < Hue Angle < 50.5 v) 202 < L < 231.15
WW-210	i) $202.4 < \mu_R < 225$, $204 < \mu_G < 226.4$ and $199.8 < \mu_B < 227.3$ ii) Max (R, G, B) - Min (R, G, B) > 27.5 iii) R < G and G > B iv) 9.13 < Hue Angle < 48.7 v) 203.5 < L < 226.2
WW-240	i) $141.6 < \mu_R < 228.8$, $153.8 < \mu_G < 231.8$ and $113.7 < \mu_B < 228.9$ ii) Max (R, G, B) – Min (R, G, B) > 118 iii) R < G and G > B iv) 23 < Hue Angle < 52.5 v) 148.3 < L < 231
WW-320	i) $199.4 < \mu_R < 232$, $202.5 < \mu_G < 233$ and $198 < \mu_B < 232$ ii) Max (R, G, B) – Min (R, G, B) > 35 iii) R < G and G > B iv) 46.7 < Hue Angle < 56.8 v) 201.5 < L < 233

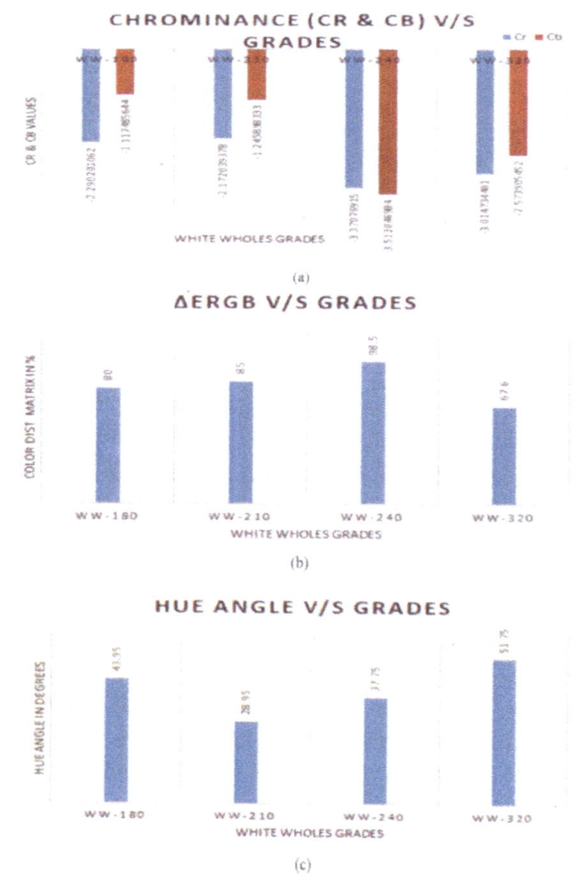

Figure 5. Clustered column bar chart for white wholes cashew kernel grades.

HCL

The features are relating to HCL color coordinates: Hue (H), Chroma (C) and Luminance (L). The color features of the white whole cashew kernel grades was measured by using the algorithm 2 and listed in the Table 6. From the color features set, we have observed that, some of features are contributing more. The features are hue (refer Table 6 and Figure 6a and b), chroma, (refer Table 6 and Figure 6c), luminance (refer Table 6 and Figure 6d) and color distance metric (ΔE_{HCL}) (refer Table 6).

The average color feature of HCL features are as follows.

i) The Hue angle of WW cashew kernel grades (w-180, w-210, w-240, w-320) are (12°, 11°, 21°, 19°).

ii) The Luminance of WW cashew kernel grades (w-180, w-210, w-240, w-320) are (103, 99, 89, 102).

iii) The color difference metric values of WW cashew kernel grades (w-180, w-210, w-240, w-320) are (21.24, 21.14, 31.3, 21).

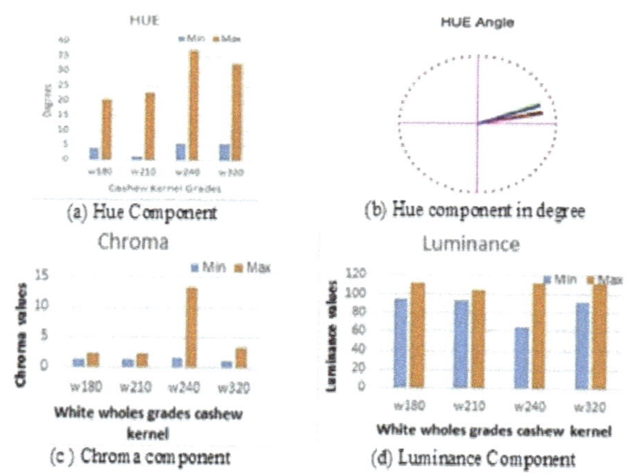

Figure 6. HCL components of white wholes grades cashew kernel.

Table 6. Summarized values of HCL color features.

Cashew kernel grades	Minimum and maximum values of Hue, Chroma, Luminance and color distance metric			
WW-180	$4^0 < H < 20.3^0$	$1.25 < C < 2.16$	$94.5 < L < 111.34$	$16.31 < \Delta E_{HCL} < 26.2$
WW-210	$1^0 < H < 22.7^0$	$1.23 < C < 2.15$	$93.35 < L < 104.64$	$17.53 < \Delta E_{HCL} < 24.8$
WW-240	$5.6^0 < H < 36.9^0$	$1.45 < C < 13.4$	$65 < L < 112.5$	$15.15 < \Delta E_{HCL} < 47.5$
WW-320	$5.8^0 < H < 32.3^0$	$1.10 < C < 3.24$	$91.5 < L < 112.25$	$14.5 < \Delta E_{HCL} < 27.6$

Recognition and classification

The proposed method is implemented using Intel (R) Core (TM) i5-3230M Laptop with 4GB RAM and 2.60 GHz CPU. We applied a MLP neural network of learning rate:0.3, momentum:0.2, no of epochs: 500, Error: 0.001 and sigmoid activation function with 2 hidden layers. The input layer of the ANN has 21 neurons because the data set

contains 21 parameters and the output layer of ANN has 4 neurons. We trained a network with total samples of random manner were 2480 (WW-180:615; WW-210:675; WW-240:630 and WW-320:560) and tested the four grades of white wholes (see Table 8 and 9). The MLP neural network model 21-25-25-4 were evaluated (see Table 7).

Table 7. Recognition accuracy of white wholes grades cashew kernel.

Neural Network Architecture	White wholes grades accuracy (%) WW-180	WW-210	WW-240	WW-320	Accuracy average (%)
21-25-25-4	84.49	83.68	89.45	98.10	88.93

Table 8. Confusion matrix of trained the neural network model (21 – 25 – 25 – 4).

Grades	WW-180	WW-210	WW-240	WW-320	Total	Success rate in %
WW-180	496	82	9	0	587	84.49
WW-210	97	554	11	0	662	83.68
WW-240	23	38	602	10	673	89.45
WW-320	0	0	5	553	558	98.10
		Total			2480	88.93

Table 9. Tested result of white wholes grades cashew kernel with neural network model (21 – 25 – 25 – 4).

Grades	No. of Samples	Correctly classified	Wrongly Classified	Remarks
WW-180	100	90	08	06 →WW-210; 1 →WW-240; 01 →WW-320
WW-210	100	87	13	06 →WW-180; 7 →WW-240
WW-240	100	93	07	02 →WW-180; 01 →WW-210; 04 →WW-320
WW-320	100	96	04	02 →WW-210; 02 →WW-240

Conclusion

Following conclusion were drawn from the study.

i) According to the characteristics of the images, extract morphological and color characteristic parameters as the recognition bases; the color characteristic parameters are more significant to the white wholes grades of cashew kernels.

ii) Recognizing and testing the four (WW-180, WW-210, WW-240 and WW-320) cashew kernels grades, and the results show that the accuracy rate of comprehensive recognition reaches 88.93%, which indicates that the method is feasible and the study provides a theoretical and practical basis for follow-up studies.

Acknowledgements

The author would like to thank the Zonal Agricultural Research Station, Brahmavar, Udupi Taluka, Udupi District. Karnataka State and Directorate of Cashew Research Puttur, Mangalore, Karnataka for provided the information about cashew kernels grading.

References

Amaral, L. A., Rocha O., Goncalves C., Ferreira, A., & Ferreira E. C. (2009). Application of Image analysis to the prediction of EBC barley kernel weight. *Industrial Crops and Products, 30*(3), 366-371.

Balasubramanian, D. (2001). Physical properties of raw Cashew nut. *Journal of Agricultural Engineering and Research, 78*(3), 291-297.

Cashew Export Promotion Council of India (CEPCI). (2013). *Cashew kernels*. Kollam, Kerala: Government of India. Retrieved from http://www.cepci.org

Castelo-Quispe, S., Banda-Tapia, J. D., Lopez-Paredes, Ḿ. N., Barrios-Araniba, D., & Patino-Escarcin, R. (2013). Optimization of Brazil-nuts classification process through automation using color spaces in computer vision. *International Journal of Computer Information Systems and Industrial Management Applications, 5*(1), 623-630.

Chen, H., Wang, J., Yuan, Q., & Wan, P. (2011). Quality classification of peanuts based on image processing. *Journal of Food, Agriculture and Environment, 9*(3-4), 205-209.

Chen, X., Xun, Y., Li, W., & Zhang, J. (2010). Combining discriminant analysis and neural networks for corn variety identification. *Computers and Electronics in Agriculture, 71*, 48-53.

Dhakshinamurthy, B., & Kokila, J. R. L. (2013). Performance valuation high speed color sorter for cashew kernels. *Acta Agrophysica, 20*(4), 543-553.

Du, C. J., & Sun, D. W. (2005). Comparison of three methods for classification of pizza topping using different color space transformations. *Journal of Food Engineering, 68*(3), 277-287.

Ganganagowdar, N. V., & Siddaramappa, H. K. (2011). Cashew kernels classification using color features. *International Journal of Machine Intelligence, 3*(2), 52-57.

Ganganagowdar, N. V., & Siddaramappa, H. K. (2014). Computer Vision System to Estimate cashew kernel (White wholes) grade Geometric and Color parameters. *Electronic Journal of Polish Agricultural Universities, 17*(4), 1-28.

Ganganagowdar, N. V., Shetty, D. K., & Siddaramappa, H. K. (2012). Computer Vision system for cashew kernel area estimation (p. 1-6). In *Third International Conference on Computing Communication & Networking Technologies,* Coimbatore, India: IEEE. doi: 10.1109/ICCCNT.2012.6395942

International Telecommunication Union. (2002). *Recommendation ITU-R BT.709-5. Parameter values for the HDTV standards for production and international programme exchange. ITU Radiocommunication.* Retrieved from https://www.itu.int/dms_pubrec/itu-r/rec/bt/R-REC-BT.709-5-200204-S!!PDF-E.pdf

Kumar, J. A., Rao, P. R., & Desai, A. R. (2013). Cashew kernel classification using machine learning approaches. *Journal of the Indian Society of Agricultural Statistics, 67*(1), 121-129.

Kumar, M., Bora, G., & Lin, D. (2013). Image processing technique to estimate geometric parameters and volume of selected dry beans. *Journal of Food Measurement and Characterization,* 7(2), 81-89.

Liu, Z. Y., Cheng, F., Ying, Y. B., & Rao, X. Q. (2005). Identification of rice seed varieties using neural network. *Journal of Zhejiang University. Science. B, 6*(11), 1095-1100.

Matlab (2012). *The MathWorks,* Inc., Natick, Massachusetts, United States

Mendoza, F., Dejmek, P., & Aguilera, J. M. (2006). Calibrated color measurements of agricultural foods using image analysis. *Post Biology and Technology, 41:* 285-295.

Namias, R., Gallo, C., Craviotto, R. M., Arango, M. R., & Granitto, P. M. (2012). Automatic grading of green intensity in soybean seeds (p. 96-104). In *13th Argentine Symposium on Artificial Intelligence.* Buenos Aires, ARG.

Novaković, J., Strbac, P., & Bulatović, D. (2011). Toward optimal feature selection using ranking methods and classification algorithms. *Yugoslav Journal of Operations Research, 21*(1), 119-135.

Sarifuddin, M., & Missaoui, R. (2005). A new perceptually uniform color space with associated color similarity measure for content-based image and video retrieval (p, 1-8). In *Workshop on multimedia information retrieval.* Retrieved from https://www.researchgate.net/profile/Sarifuddin_Madenda/publication/228906385_A_new_perceptually_uniform_color_space_with_associated_color_similarity_measure_for_content-based_image_and_video_retrieval/links/562a20a308ae04c2aeb1551a.pdf

Saroj, P. L., & Balasubramanian, D. (2013). Cashew Industry in India-a sustainable road map. *Indian Horticultue, 58*(1), 9-15.

Savakar, D. G., & Anami, B. S. (2009). Recognition and classification of food grains, fruits and flowers using machine vision. *International Journal of Food Engineering, 5*(4), 14-34.

Soedibyo, D. W., Seminar, K. B., Ahmad, U., Subrata, I. D. M. (2010). *The Development of Automatic Coffee Sorting System Based on Image Processing and Artificial Neural Network.* Bogor: Asian Federation for Information Technology.

Sun, D. W. (2008). Computer vision technology for food quality evaluation. *Food Science and Technology, International Series,* 63-70.

Symons, S. J., & Fulcher, R. G. (1988). Determination of wheat kernel morphological variation by digital image analysis, I Variation in eastern Canadian milling quality wheats. *Journal of Cereal Science, 8*(3), 211-218.

Thakkar, M., Bhatt, M., & Bhensdadia, C. K. (2011). Performance evaluation of classification techniques for computer vision based cashew grading system. *International Journal of Computer Applications, 18*(6), 9-12.

Wiwart, M., Suchowilska, E., Lajszner, W., & Graban, Ł. (2012). Identification of hybrids of spelt and wheat and their parental forms using shape and color descriptors. *Computers and electronics in Agriculture, 83,* 68-76.

Zheng, C., Sun, D. W., & Zheng, L. (2006). Recent developments and applications of image features for food quality evaluation and inspection–a review. *Trends in Food Science & Technology,* 17(12), 642-655.

8

Mechanical properties of the rachis from macaw palm bunches

Zenil Ricardo Cunha Rodrigues de Oliveira[1]; Fábio Lúcio Santos[2*], Domingos Sarvio Magalhães Valente[1], Francisco de Assis de Carvalho Pinto[1] and Nara Silveira Velloso[2]

*[1]Departamento de Engenharia Agrícola, Universidade Federal de Viçosa, Av. P.H. Rolfs, s/n, Campus Universitário, 36570-900, Viçosa, Minas Gerais, Brazil. [2]Departamento de Engenharia, Universidade Federal de Lavras, Lavras, Minas Gerais, Brazil. *Author for correspondence. E-mail: fabio.santos@deg.ufla.br*

ABSTRACT. Numerical implementation methods such as the finite element method can be used in machine design to study the dynamic behavior of the plant. Input parameters for using the finite element method should be set based on the geometrical, physical and mechanical system under study. Thus, the aim of this study was to determine the mechanical properties of the rachis of macaw palm in the green maturation stage at an age of 180 days and at fruit maturation. The water content was calculated using the oven method of 105 ± 3°C for 24 hours. The modulus of elasticity for the rachis was determined by means of mechanical compression testing. The Poisson's ratios were determined from the results of the longitudinal and transverse deformation of rachis test bodies. The shear tests were performed in two scenarios. In the first one, a blade was used without a tensioner, and in the second scenario, a blade was fixed with a tensioner, characterized by a chamfer at 45°. It was concluded that the modulus of elasticity of macaw palm rachis in the green maturity stage was higher than the values for the mature stage of ripeness.

Keywords: *Acrocomia aculeate*; modulus of elasticity; Poisson's ratio; shear test.

Introduction

The macaw palm (*Acrocomia aculeata*) is a native palm of tropical America, occurring from Central America to the extreme south of the American continent (Motoike et al., 2013). In Brazil, it is considered one of the most dispersed palm trees; tolerant of drought and low temperatures, this palm is found along in practically all regions of the country. The adult palm is arboreal, between 10 and 15 m in height, with a high number of spines in its trunk (Montoya, Motoike, Kuki, Oliveira, & Honório, 2015).

Mature macaw palm fruits are slightly flattened globes of yellowish-brown coloring, with a diameter varying from 3 to 6 cm. The composition of fruit by weight is approximately 20% bark, 40% pulp, 33% endocarp and 7% almond. The oils are extracted from two parts of the fruit, the almond and the pulp. Oil withdrawn from the former represents approximately 15% of total oil withdrawn from the plant and has notable uses in the pharmaceutical, food and cosmetic industries. Meanwhile, oil withdrawn from the latter has greater potential for biodiesel production (Nobre, Trogello, Borghetti, & David, 2014).

In relation to productivity, the macaw palm is a promising source of vegetable oil for industries producing fuel (Navarro-Días et al., 2014) and cosmetics (Silva, Silva, & Parente, 2009). In addition to the oil, the processing of macaw palm fruits generates co-products with high added value. There is also the possibility of the endocarp being used in the production of charcoal (Motoike et al., 2013; Nobre et al., 2014).

Different principles can be employed for the mechanized harvesting of agricultural products. In relation to macaw palm cultivation, it is believed that cutting and mechanical vibration of the palm bunches are promising working principles for harvesting machines. Cutting refers to a shear across the cross section of the rachis, promoting the separation of bunch from the plant and, afterwards, contributing to the detachment of the fruits. Generally, a Malaysian knife is used to make this operation achievable for manual harvesting (Motoike et al., 2013). Techniques based on the principle of mechanical vibration depend on knowledge of parameters such as frequency, amplitude and time of vibration. They are used in the mechanized harvesting of different crops such as coffee (Coelho, Santos, Pinto, & Queiroz, 2015) and orange (Castro-Garcia, Blanco-Roldán, Ferguson, González-Sánchez, & Gil-Ribes, 2017). This type of harvesting provides a better selection of fruits when one has knowledge about the dynamic behavior of the plant to be harvested, such as natural frequencies and vibration modes (Santos, Queiroz, Pinto, & Santos, 2010).

Currently, the operation of macaw palm harvesting is carried out using extractive methods, such as collecting fruits that fall on the ground or cutting mature bunches by means of scythes adapted to the height of the plant. In both forms of harvesting, the fruits come into contact with the soil, leaving them susceptible to both the mechanical damages of the fall and the action of microorganisms. This leads to degradation of the pulp, resulting in an oil product with low quality (Ciconini et al., 2013).

As the macaw palm's potential as a source of energy increases (Gonçalves, Batista, Rodrigues, Nogueira, & Santos, 2013), the demand for scientific data that provide knowledge about the plant and its fruits, oil composition, and physical and mechanical properties increases as well. Using this data, it would be possible to expand extractive exploitation to an industrial scale with rational and sustainable management.

In addition, it will be possible to improve industrial processes and design machines and equipment for use in the harvest and post-harvest operations based on knowledge of the physical and mechanical properties of structures that compose the bunches. In this way, fieldwork for the operator will be made safer, while reducing the contingent of labor and production costs associated with extraction.

In this context, and considering the potential of the crop, the aim of this work was to determine and evaluate the mechanical properties of rachis from macaw palm bunches in order to constitute a basis of knowledge for the design of harvesting and post-processing machines.

Material and method

Research was carried out from November 2015 to February 2016 using macaw palm bunches collected at the Active Germplasm Bank (BAG), located at the Experimental Farm in Araponga, Minas Gerais State, Brazil, at 20° 40' South latitude and 42° 31' West longitude. Samples of macaw palm rachis came from four sites: BGP 12 - Ibiá - Araxá; BGP 13 - Sítio Paraíso - Belo Horizonte; BGP 31 - Três Marias; and BGP 53 - Lavras - São João Del Rey.

Samples were always collected in the morning, and tests performed during the same day as collection. The ambient temperature of the laboratory was maintained at 20°C.

The water content, Poisson's ratio, shear strength, modulus of elasticity and compressive strength were determined for the samples at each stage of maturation. Immature fruits were considered to be in their maturation stage at 180 days and mature when natural detachment of fruits occurred.

Water content

To obtain the water content of the rachis samples, the direct method (Brasil, 2009) was used. First, the empty crucibles that would be used were weighed. Subsequently, the crucibles were weighed with the addition of a volume of 1 cm^3 chopped samples of macaw palm rachis (Figure 1). As established by the direct method, three replicates were then dried in an oven at 105 ± 3°C for 24 hours. Finally, the crucibles were weighed with the dried samples and the results recorded.

Figure 1. Crucibles containing chopped samples of macaw palm rachis with a volume of 1 cm^3.

Poisson's ratio

Poisson's ratio was determined using a machine universal testing INSTRON 3360 Series Dual Column Table Frames with Bluehill 3 software. The tests were performed employing four test bodies (Figure 2) at each stage of maturation. Test specimen composition was standardized at 12 cm of length, 2.50 cm of width and 0.50 cm of height.

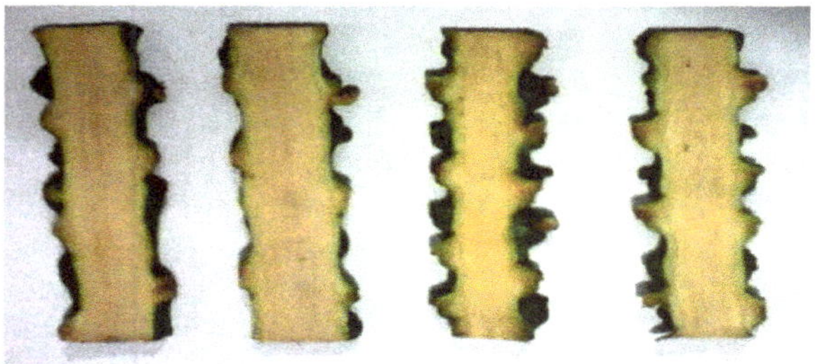

Figure 2. Test specimens of macaw palm rachis used to determine the Poisson's ratio.

The rachis test specimens were fixed directly to the universal testing machine by means of jaws of the equipment and subject to traction. The system displaced the sample vertically by 3.0 mm. The deformation was then measured transversally and longitudinally using a caliper rule of 0.01 mm resolution.

The Poisson coefficient was calculated by Equation 3.

$$\mu = \frac{-\Delta d}{\Delta l} \qquad (3)$$

where: μ = Poisson's ratio; Δd = specific transversal contraction, mm mm^{-1}; and Δl = specific longitudinal elongation, mm mm^{-1}.

Shear test

The shear test was also accomplished using the universal testing machine INSTRON 3360, configured to perform a compression routine with a velocity of 15 mm min.$^{-1}$. For this test, specimens were prepared from the upper third of the rachis (Figure 3) using lengths of 50 mm and actual cross sections of the rachis with average diameters of 25 mm.

Figure 3. Test specimens of macaw palm rachis prepared for shear test.

The shear tests were performed considering two conditions. In the first, a blade without a tensioner was fixed to the upper jaw of the testing machine, while on the lower jaw, a flat circular plate (Figure 4) was used. In order to allocate and support the specimens during the tests, an apparatus was developed and positioned over the flat circular plate (Figure 5).

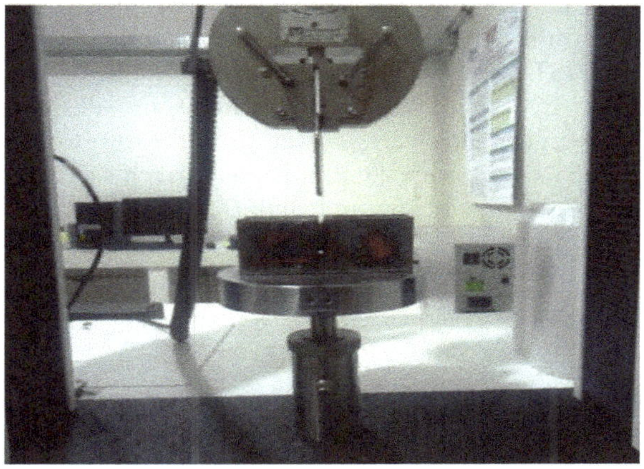

Figure 4. Set up used for the shear test of macaw palm rachis.

For the second condition, the tests were performed using a blade with a tensioner, which was characterized by a 45° chamfer at its end (Figure 5).

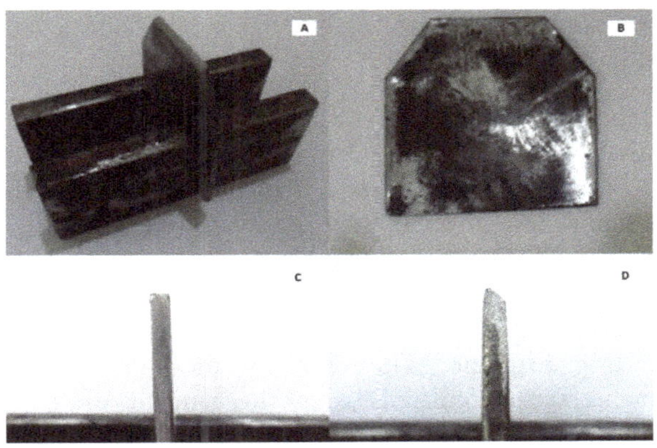

Figure 5. Apparatus developed for the shear test (A), blade used for test (B), without tensioner (C), and blade with tensioner and 45° chamfer at the end (D).

The test specimens were inserted into the fixed part of the apparatus. During the test, the downward movement of the upper jaw of the universal testing machine, where the test blade was fixed, provided a shearing force in the cross-section of the specimens; this force was monitored until the test specimen cut itself off.

From the maximum shear force and the cross-sectional area of the test specimen, shear stress of the macaw palm rachis was calculated (Equation 4).

$$\tau = \frac{F}{S} \quad (4)$$

where: τ = shear stress, MPa; F = shear force, N; and S = cross-sectional area of the test specimen, mm^2.

Compression test

The modulus of elasticity of the rachis was determined through a compression test, employing an INSTRON 3360 universal testing machine and using five cylindrical test specimens for all sites. These specimens were sourced from the lower third of the rachis, with a 20 mm diameter (rachis circular cross section) and 15 mm length (Figure 6).

Figure 6. Test specimens of macaw palm rachis prepared for compression tests.

The tests were performed by means of compression, and the samples were placed between two parallel flat circular plates (Figure 7) and subjected to constant deformation at small magnitudes on the two opposite faces of the specimen. The constant deformation rate was 15.0 mm $min.^{-1}$, monitored by Bluehill 3 software and managed by the computer coupled to the universal testing machine. The elasticity modulus of the rachis was calculated using the tangent method. Additionally, compressive strength was determined.

Statistical analysis

The data for the water content, Poisson's ratio, shear test, modulus of elasticity and compressive strength of the samples (test specimens) were submitted to analysis of variance, according to a completely randomized design and considering two stages of maturation (rachis from immature and mature fruits). The stage of maturation effect on the behavior of mechanical properties was studied using the Tukey test at a significance level of 5%. Statistical analysis was completed by the computer software Assistat Statistic version 7.7 beta (Silva & Azevedo, 2016).

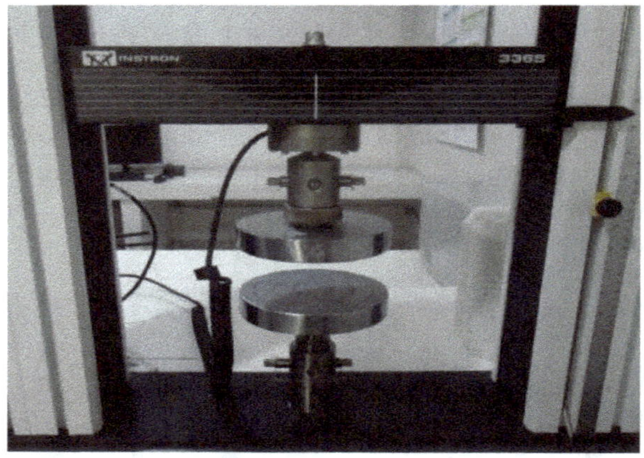

Figure 7. Equipment used for compression test of macaw palm rachis.

Result and discussion

Water content

After analyzing the average water content of the macaw palm rachis for the BGP 12, BGP 13, and BGP 53 sites, no significant differences were observed between the stages of maturation from immature to mature fruits (Table 1). However, for the BGP 31 site only, it is possible to confirm that from the immature to mature maturation stages, the average water content showed a slight decrease.

Table 1. Average water content (% w.b.) of the macaw palm rachis in immature and mature stages of maturation.

Sites	Average water content (%w.b.)	
	Immature	Mature
BGP 12	78.20 aA	80.55 aA
BGP 13	76.34 aA	78.10 abA
BGP 31	77.84 aA	72.31 bB
BGP 53	82.72 aA	81.30 aA

Averages followed by the same letters (lowercase) in the columns do not significantly differ from each other by Tukey test at 5% probability. Averages followed by the same letters (uppercase) in the rows do not significantly differ from each other by Tukey test at 5% probability.

Comparing the average water contents for the same stage of maturation, it can be observed that there are no significant differences between sites in the immature stage. However, for the mature stage, it can be stated that the BGP 31 site has lower water content than the BGP 12 and BGP 53 sites.

With the physiological maturation of the fruits, the water content remained unchanged from the immature to the mature stages, which was expected due to the need for a water supply to maintain the fruits in bunches. Water content can influence the required shear strength to cut biological structures such as the macaw palm rachis. Taghijarah, Ahmadi, Grahderijani, and Tavakoli (2011) reported that when cutting sugarcane stalks with an average water content of 75.3% (% w.b.) and cross-sectional area of

453 mm², a shear strength of 3.64 MPa was necessary.

Poisson's ratio

The Poisson's ratios for stages of maturation and sites are presented in Tables 2 and 3, respectively.

Table 2. Poisson's ratio of macaw palm rachis in immature and mature stages of maturation.

Stages of maturation	Poisson's ratio
Immature	0.29 ± 0.14 a
Mature	0.31 ± 0.14 a

Averages followed by the same letter do not differ statistically from each other by Tukey test at 5% probability.

Table 3. Poisson's ratio of macaw palm rachis for the sites.

Sites	Poisson's ratio
BGP 12	0.32 ± 0.27 a
BGP 13	0.33 ± 0.27 a
BGP 31	0.35 ± 0.27 a
BGP 53	0.22 ± 0.27 a

Averages followed by the same letter do not differ statistically from each other by Tukey test at 5% probability.

There was no significant difference in the Poisson ratios in relation to the stage of maturation (Table 2) or site (Table 3). However, it can be observed that the Poisson ratios at the mature stage were higher than those found in the immature stage, meaning that for a given longitudinal deformation, the mature rachis of macaw palm presented greater deformation in the longitudinal direction when compared to the rachis of macaw palm in the immature stage.

Although there is no significant difference in the results, the BGP 31 site was the one that obtained the highest value of Poisson coefficient, 0.35, and the BGP 53 site obtained the lowest value, 0.22. This indicates that for the same longitudinal deformation, the BGP 31 site will deform more easily in the same direction than the other sites.

Comparing the results obtained with the fibrous biological materials analyzed by Ghavami and Marinho (2005), which found a Poisson coefficient of 0.26 for *Guadua angustifolia*, this value is below the value found for macaw palm rachis. In addition, Coelho et al. (2015), in a study of the "Catuaí Vermelho" coffee variety, obtained a Poisson's coefficient of 0.35 for coffee stems in the green and cherry stages of maturation. These results are by magnitude close to those obtained in the present work for macaw palm rachis.

Shear stress

The results for rachis shear stress are presented in Table 4. Two conditions were evaluated: the first with a cutting blade without a tensioner and the second with a cutting blade with 45° chamfer, characterizing a tensioner.

Table 4. Results for shear stress of macaw palm rachis with cutting blades without tensioner and a cutting blade with 45° chamfer.

Cutting blade	Shear stress (MPa)
Without tensioner	2.79 ± 0.54 a
With tensioner	0.90 ± 0.54 b

The averages followed by the same letter do not differ statistically from each other by the Tukey test at 5% probability.

For different cutting blades, a magnitude of 2.79 MPa was observed for the cutting blade without tensioner, while 0.90 MPa was observed for the cutting blade with 45° chamfer. Using a cutting blade with a tensioner required one-third of the applied force to cut the rachis of the macaw palm.

The results for shear stress considering the different stages of maturation are presented in Table 5.

Table 5. Shear stress of rachis at immature and mature stages of maturation.

Stages of maturation	Shear stress (MPa)	
	Without tensioner	With tensioner
Immature	2.55 ± 0.64 a	0.37 ± 0.17 b
Mature	3.02 ± 0.64 a	1.44 ± 0.17 a

The averages followed by the same letter do not differ statistically from each other by the Tukey test at 5% probability.

It was observed that there was no significant difference between the maturation stages. However, while using a tensioner and cutting blade with 45° chamfer, there was a significant difference between the stages of maturation (Table 5). Over time, the magnitude of shear tends to increase, which implies that the application of greater force is necessary to perform the same cut. For a cutting blade with a tensioner, it was observed that the shear stress magnitude increased approximately 3.9 times, reaching 1.44 MPa, for the mature stage.

The shear stress for the sites is shown in Table 6.

Table 6. Shear stress of the macaw palm rachis for the sites.

Sites	Shear stress (MPa)	
	Without tensioner	With tensioner
BGP 12	2.78 ± 1.62 a	0.86 ± 3.09 a
BGP 13	2.85 ± 1.62 a	0.92 ± 3.09 a
BGP 31	2.36 ± 1.62 a	0.96 ± 3.09 a
BGP 53	3.17 ± 1.62 a	0.87 ± 3.09 a

The averages followed by the same letter do not differ statistically from each other by the Tukey test at 5% probability.

From the results, it was observed that there was no significant difference between the evaluated sites or for the types of cutting blades used.

In a study carried out for bamboo species *Guadua angustifolia*, Ghavami and Marinho (2005) found an average shear stress equal to 2.02 MPa. This magnitude is very close to the results obtained for macaw palm rachis, which is also a fibrous biological material, for a cutting blade without tensioner. It can be inferred that the result found by Ghavami and Marinho (2005) is in the interval between the values obtained from the different cutting blades analyzed in this research.

Compression test

Modulus of elasticity

For the modulus of elasticity, significant differences were observed between the stages of maturation evaluated (Table 7).

Table 7. Modulus of elasticity of the macaw palm rachis in immature and mature stages of maturation.

Maturation stages	Modulus of elasticity (MPa)
Immature	51.97 ± 9.18 a
Mature	37.93 ± 9.18 b

The averages followed by the same letter do not differ statistically from each other by the Tukey test at 5% probability.

The modulus of elasticity for the immature stage was higher than that of the mature stage. There was a greater elastic deformation by the macaw palm rachis in the mature stage for the same tension applied.

Regarding the sites, no significant differences were determined for the modulus of elasticity (Table 8).

Table 8. Modulus of elasticity of different sites on macaw palm rachis.

Sites	Modulus of elasticity (MPa)
BGP 12	39.83 ± 18.23 a
BGP 13	39.40 ± 18.23 a
BGP 31	54.72 ± 18.23 a
BGP 53	45.85 ± 18.23 a

The averages followed by the same letter do not differ statistically from each other by the Tukey test at 5% probability.

Lobão, Lúcia, Moreira, and Gomes (2004) obtained a modulus of elasticity of 18342 MPa from compression tests in a study developed for the fiber of *Eucalyptus grandis* wood, with an average density of 0.88 g mL^{-1}. This magnitude is approximately 420 times greater than that reported for the macaw palm rachis, and it can be inferred that the macaw palm rachis is much more flexible than the *E. grandis* wood, which is mechanized from harvester heads that use knives to perform of the tree cutting (Leite et al., 2014). Therefore, these results suggest that the use of cutting blades assemble in a harvester head could be a solution for the harvesting of macaw palm bunches.

Compressive strength

In Table 9, the results for compressive strength are presented.

Table 9. Compressive strength of macaw palm rachis in immature and mature stages of maturation.

Stages of maturation	Compressive strength (MPa)
Immature	6.81 ± 1.03 a
Mature	3.34 ± 1.03 b

The averages followed by the same letter do not differ statistically from each other by the Tukey test at 5% probability.

The compressive strength of the immature stage was approximately twice that of the mature stage. Magnitude of compressive strength evidence the difficulty of compressing a certain body. In the case under study, the macaw palm rachis in the green maturation stage had a higher resistance to compression than the macaw palm rachis belonging to the mature maturation stage.

Regarding the values of compressive strength, no significant differences were found between sites (Table 10).

Table 10. Compressive strength of macaw palm rachis between sites.

Sites	Compressive strength (MPa)
BGP 12	4.86 ± 2.88 a
BGP 13	4.54 ± 2.88 a
BGP 31	5.98 ± 2.88 a
BGP 53	4.93 ± 2.88 a

The averages followed by the same letter do not differ statistically from each other by the Tukey test at 5% probability.

The compressive strength magnitude found by the present work, 5.10 MPa, is very close to the results found by Magalhães, Braunbeck, and Pagnano (2004) for the compressive strength of sugarcane, which obtained an average value of 4.9 MPa. These similar magnitudes are due to the fibrous structural similarities of sugarcane and macaw palm rachis.

Conclusion

The water content of the macaw palm rachis only showed a significant difference between maturation stages for the BGP 31 site. The Poisson coefficient was 0.29 for the immature stage of maturation and 0.31 for the mature stage. The cutting blade without a tensioner exhibits about three times greater magnitude of shear stress than the cutting blade with a 45° chamfer. The modulus of elasticity at the immature stage of maturation was higher than for the mature stage, at 51.97 MPa and 37.93 MPa respectively. The compressive strength for the immature stage of maturation was approximately twice that of the mature stage.

Acknowledgements

The authors gratefully acknowledge funding *Fundação de Amparo a Pesquisa do Estado de Minas Gerais (FAPEMIG)* and *Conselho Nacional de Desenvolvimento Científico e Tecnológico (CNPq)*.

References

Brasil. (2009). Ministério da Agricultura, Pecuária e Abastecimento. *Regras para análise de sementes*. Secretaria de Defesa Agropecuária. Brasília, DF: Mapa/ACS.

Castro-Garcia, S., Blanco-Roldán, G. L., Ferguson, L., González-Sánchez, E. J., & Gil-Ribes, J. A. (2017) Frequency response of late-season 'Valencia' orange to selective harvesting by vibration for juice industry. *Biosystems Engineering, 155*(1), 77-83, doi: 10.1016/j.biosystemseng.2016.11.012.

Ciconini, G., Favaro, S. P., Roscoe, R., Miranda, C. H. B., Tapeti, C. F., Miyahira, M. A. M., ... Naka, M. H. (2013). Biometry and oil contents of *Acrocomia aculeata* fruits from the Cerrados and Pantanal biomes in Mato Grosso do Sul, Brazil. *Industrial Crops and Products, 45*(1), 208-214. doi: 10.1016/j.indcrop.2012.12.008.

Coelho, A. L. F., Santos, F. L., Pinto, F. A. C., & Queiroz, D. M. (2015). Determinação das propriedades geométricas, físicas e mecânicas do sistema fruto-pedúnculo-ramo do cafeeiro. *Revista Brasileira de Engenharia Agrícola e Ambiental, 19*(1), 286-292. doi: 10.1590/1807-1929/agriambi.v19n3p286-292.

Ghavami, K., & Marinho, A. B. (2005). Propriedades físicas e mecânicas do colmo inteiro do bambu da espécie *Guadua angustifolia. Revista Brasileira de Engenharia Agrícola e Ambiental, 9*(1), 107-114. doi: 10.1590/S1415-43662005000100016.

Gonçalves, D. B., Batista, A. F., Rodrigues, M. Q. R. B., Nogueira, K. M. V., & Santos, V. L. (2013). Ethanol production from macaúba (*Acrocomia aculeata*) presscake hemicellulosic hydrolysate by *Candida boidinii* UFMG14. *Bioresource Technology, 146*(1), 261–266. doi: 10.1016/j.biortech.2013.07.075.

Leite, E. S., Minette, L. S., Fernandes, H. C., Souza, A. P., Amara, E. J., & Lacerda, E. G. (2014). Performance of the harvester on eucalyptus crops in different spacing and slopes. *Revista Árvore, 38*(1). doi: 10.1590/S0100-67622014000100009.

Lobão, M. S., Lúcia, R. M. D., Moreira, M. S. S., & Gomes, A. (2004). Caracterização das propriedades físico-mecânicas da madeira de eucalipto com diferentes densidades. *Revista Árvore, 28*(6), 889-894.

Magalhães, P. S. G., Braunbeck, O. A., & Pagnano, N. B. (2004). Resistência à compressão e remoção de folhas da cana-de-açúcar visando à colheita mecânica. *Engenharia Agrícola, 24*(1), 177-184. doi: 10.1590/S0100-69162004000100020.

Montoya, S. G., Motoike, S. Y., Kuki, K. N., Oliveira, C. M., & Honório, I. G. (2015). Registro da presença e danos causados por coleópteros em macaúba. *Pesquisa Florestal Brasileira, 35*(82), 159-162. doi: 10.4336/2015.pfb.35.82.763.

Motoike, S. Y., Carvalho, M., Pimentel, L. D., Kuti, K. N., Paes, J. M. V., Dias, H. C. T., & Sato, A. Y. (2013). *A cultura da macaúba*: implantação e manejo de cultivos racionais. Viçosa, MG: UFV.

Navarro-Días, H. J., Gonzalez, S. L., Irigaray, B., Vieitez, I., Jachmanián, I., Hense, H., & Oliveira, J. V. (2014). Macauba oil as alternative feedstock for biodiesel: Characterization and ester conversion by the supercritical method. *Journal of Supercritical Fluids, 93*(1), 130-137. doi: 10.1016/j supflu.2013.11.008.

Nobre, D. A. C., Trogello, E., Borghetti, R. A., & David, A. M. S. S. (2014). Macaúba: palmeira de extração sustentável para biocombustível. *Colloquium Agrariae, 10*(2), 92-105. doi: 10.5747/ca.2014.v10.n2.a112.

Santos, F. L., Queiroz, D. M., Pinto, F. A. C., & Santos, N. T. (2010). Analysis of the coffee harvesting process using an electromagnetic shaker. *Acta Scientiarum. Agronomy, 32*(3), 373-378. doi: 10.4025/actasciagron.v32i3.6782.

Silva, F. A. S., & Azevedo, C. A V. (2016). The Assistat Software Version 7.7 and its use in the analysis of experimental data. *African Journal of Agricultural Research, 11*(39), 3733-3740. doi: 10.5897/AJAR2016.11522.

Silva, B. P., Silva, G. M., & Parente, J. P. (2009). Chemical properties and adjuvant activity of a galacto glucomannan from *Acrocomia aculeata. Carbohydrate Polymers,* 75(1), 380-384. doi: 10.1016/j.carbpol.2008.09.031.

Taghijarad, H., Ahmadi, H., Grahderijani, M., & Tavakoli, M. (2011). Shearing characteristics of sugar cane (*Saccharum officinarum* L.) stalks as a function of the rate of the applied force. *Australian Journal of Crop Science, 5*(6), 630-634.

9

Thermodynamic properties of water desorption of forage turnip seeds

Kelly Aparecida de Sousa[1*], Osvaldo Resende[1], André Luis Duarte Goneli[2], Thaís Adriana de Souza Smaniotto[1] and Daniel Emanuel Cabral de Oliveira[1]

*[1]Instituto Federal de Educação, Ciência e Tecnologia Goiano, Câmpus Rio Verde, Rodovia Sul Goiana, Km 01, Zona Rural, 75901-970, Rio Verde, Goiás, Brazil. [2]Departamento de Engenharia Agrícola, Faculdade de Ciências Agrárias, Universidade Federal da Grande Dourados, Dourados, Mato Grosso do Sul, Brazil. *Author for correspondence. E-mail: kellyapsousa@yahoo.com.br*

ABSTRACT. The purpose of this study was to determine the thermodynamic properties of the process of water sorption in forage turnip seeds. The equilibrium moisture content of forage turnip seeds was determined by the gravimetric-dynamic method for different values of temperature and water activity. According to the results, increasing the moisture content increases the energy required for the evaporation of water in forage turnip seeds, and the values of integral isosteric heat of desorption, within the moisture content range of 3.33 to 11.30 (% d.b.), varies from 4,222.70 to 2,870.34 kJ kg^{-1}. With the elevation in the equilibrium moisture content, there is an increase in differential entropy and Gibbs free energy, which has positive values, demonstrating non-spontaneity in the process of desorption in the seeds. The theory of enthalpy-entropy compensation can be satisfactorily applied to the sorption phenomenon, and the process of water desorption of forage turnip seeds is controlled by enthalpy.

Keywords: hygroscopic equilibrium, isosteric heat, gibbs free energy, compensation theory.

Introduction

Forage turnip (*Raphanus sativus* L.) is a plant originally from East Asia and Europe, belonging to the Brassicaceae family. The cultivation of forage turnip has become advantageous because its seeds have significant levels of non-edible oils that are easily extracted. In addition, the forage turnip has productive potential in times of little utilisation or fallowing arable lands and can also be used as vegetal cover in the intervals between crops of other species.

The quality of the grains is an important parameter for marketing and processing and can significantly affect the value of the product. Despite all current available technology, the qualitative and quantitative losses arising during the post-harvest process are not yet fully controlled, and during storage, the grain mass is constantly subjected to external factors. These factors can be physical, such as temperature and air relative humidity, chemical, such as the oxygen supply, and biological, such as bacteria, fungi, insects and rodents (BROOKER et al., 1992).

The moisture content of the agricultural products is extremely important to show the interactions between water molecules and the compounds of the product (GHODAKE et al., 2007), and it is directly related to the drying and storage processes of these agricultural products.

With high levels of water, the risk of fungal and insect infestation becomes greater, seriously impairing the germination power and vigour of the seeds, deteriorating the quality of the stored fruits and seeds and increasing post-harvest losses. However, a reduction in moisture content promotes economic losses due to mass loss of the product (YAZDANI et al., 2006) while reducing the biological activity of the product and chemical and physical changes that occur during storage.

The thermodynamic properties of water desorption are crucial for evaluating the design of the equipment involved in several processes of preservation, drying, storage, packaging and mixing of food (AVIARA et al., 2004; ASCHERI et al., 2009; KAYA; KAHYAOGLU, 2006). They are easily calculated because they relate the equilibrium moisture content of the product with the surrounding medium at a certain temperature and water activity. They are determined through sorption isotherms and allow the experimental results to be read in accordance with the theory (CHEMKHIA et al., 2004; KAYA; KAHYAOGLU, 2006).

According to Wang and Brennan (1991), to remove the water associated with hygroscopic material, the energy required is greater than that used to vaporise the same amount of free water under the same conditions of pressure and temperature. The additional energy, due to the binding forces between the water and the surface of the adsorbent substance, is known as the isosteric heat of desorption in drying processes and is a good parameter for estimating the minimum amount of heat required to remove a given amount of water from the product.

Differential enthalpy is used to design driers capable of providing heat above the latent heat of vaporisation of pure water to dry the material at low levels of moisture content (TSAMI et al., 1990). Differential entropy is related to the number of sorption sites for a given energy level inherent in the biological material (MADAMBA et al., 1996) and is used to obtain information regarding energy rationalisation during the processing, dissolution and crystallisation of the product (AVIARA et al., 2002).

Integral quantities allow the qualitative interpretation of the sorption phenomenon. They describe the degree of disorder and disordered motion for the water molecules as well as the energy involved in the sorption processes. These functions derive from the first law of thermodynamics applied to pure adsorbent material (RIZVI, 1986).

The viability and range of a chemical reaction is best determined by measuring the changes in Gibbs free energy, i.e., water vaporisation during the dehydration process (TELIS et al., 2000).

Recently, studies are being conducted to determine the thermodynamic parameters, such as enthalpy, entropy, Gibbs free energy and isosteric heat, for different oilseeds products, such as sesame seed (*Sesamum indicum* L.) (KAYA; KAHYAOGLU, 2006), peanut (*Arachis hypogaea* L.) (CORRÊA et al., 2007), macadamia (*Macadamia integrifolia* Maid. and Bet.) (DOMINGUEZ et al., 2007) and cocoa beans (*Theobroma cacao* L.) (OLIVEIRA et al., 2011).

Therefore, this study aimed to determine the thermodynamic properties of forage turnip seeds for the process of water desorption.

Material and methods

This study was conducted in the Laboratory of Postharvest of Vegetable Products from the Federal Institute of Education, Science and Technology of State Goias (Instituto Federal de Educação, Ciência e Tecnologia Goiano – IF Goiano) - Rio Verde Câmpus, located in the municipality of Rio Verde, Goias State.

We used manually collected forage turnip seeds with a moisture content of approximately 36% d.b.

To obtain the moisture content of seeds at hygroscopic equilibrium, the dynamic-gravimetric method was used. The desorption of the product in a thin layer was conducted under different controlled conditions of temperature (25, 30, 35, and 40°C) and relative humidity (from 20 to 80%) until the product reached its equilibrium moisture content under the specified air condition.

The environmental conditions for the testing of hygroscopicity tests were provided through an atmospheric conditioning chamber. The samples, each containing 10 g of product, were surrounded by a permeable fabric (cheesecloth type) to allow air flow through the seeds and were placed in the apparatus. The temperature and relative humidity of the air were monitored by a psychrometer installed close to the trays containing the samples.

During the process, the samples were periodically weighed. The hygroscopic equilibrium was reached when the mass changes remained approximately invariable over three consecutive weighings at intervals of 24 hours.

The moisture content was determined gravimetrically, with the oven at 105±1°C for 24 hours, in two repetitions (BRASIL, 2009).

The value of the water activity of forage turnip was obtained using the Copace equation shown below:

$$Xe = \exp\left\{1.335028^{**} - \left(0.016934^{**} \times T\right) + \left(1.979293^{**} \times a_w\right)\right\} \quad (1)$$

**Significant at 1% by t test.

where:

Xe: equilibrium moisture content,% d.b.;

a_w: water activity, decimal;

T: temperature, °C.

The net isosteric heat of desorption (or differential enthalpy) for each equilibrium moisture content was calculated using the Clausius-Clayperon equation (IGLESIAS; CHIRIFE, 1976):

$$\frac{\partial \ln(a_w)}{\partial T}=\frac{\Delta h_{st}}{RT_a^2} \quad (2)$$

where:

T_a: absolute temperature, K;

Δh_{st}: differential enthalpy, kJ kg^{-1};

R: universal gas constant for water vapour 0.4619 kJ kg^{-1} K^{-1}.

Integrating Equation 2 and assuming that the liquid isosteric heat of sorption is independent of the temperature, the net isosteric heat of desorption for each equilibrium moisture content was obtained according to the following equation (WANG; BRENNAN, 1991):

$$\ln(a_w)=-\left(\frac{\Delta h_{st}}{R}\right)\times\frac{1}{T_a}+C \quad 3)$$

where:

C: model coefficient.

The integral isosteric heat of desorption was obtained by adding the value of the latent heat of vaporisation of free water according to Equation 5 to the value of the net isosteric heat of sorption:

$$Q_{st}=\Delta h_{st}+L=a\times\exp(-b\times Xe)+c \quad (4)$$

where:

Q_{st}: integral isosteric heat of sorption, kJ kg^{-1};

a, b and c: coefficients of the model.

The latent heat of vaporisation of free water (L), in kJ kg^{-1}, required to calculate Q_{st} was obtained with the average temperature (T) in the studied range, in °C, using the following equation:

$$L = 2502.2 - 2.39\times T \quad (5)$$

where:

L: latent heat of vaporisation of free water, kJ kg^{-1}.

The differential entropy of sorption was calculated using the Gibbs-Helmholtz equation shown below (RIZVI, 1995):

$$\Delta S=\frac{\Delta h_{st}\Delta G}{T_a} \quad (6)$$

where:

ΔS: differential entropy of sorption (kJ $kg^{-1}K^{-1}$);

ΔG: Gibbs free energy (kJ kg^{-1}).

The Gibbs free energy can be calculated through the following equation:

$$\Delta G=-R\times T\times \ln(aw) \quad (7)$$

The effect of the changes in the sorption of water on free energy is usually followed by changes in the values of enthalpy and entropy. Thus, when we substitute Equation 7 in Equation 6 and rearrange the resulting equation, we have the following:

$$\ln(aw)=\frac{\Delta h_{st}}{R\times T_a}-\frac{\Delta S}{R} \quad (8)$$

The values of the differential enthalpy and entropy of sorption were calculated using Equation 8. The calculated values of enthalpy (Δ_{hst}) and entropy (ΔS) were correlated with the following equation (BERISTAIN et al., 1996):

$$\Delta h_{st}=T_B(\Delta S)+\Delta G_B \quad (9)$$

where:

T_B: isokinetic temperature (K);

ΔG_B: Gibbs free energy at isokinetic temperature (kJ kg^{-1}).

The isokinetic temperature is the temperature at which the reactions in a series occur at the same rate. Given that enthalpy and entropy are correlated, the compensation theory is assumed to be valid for the sorption (BERISTAIN et al., 1996). To confirm the existence of this compensation, the isokinetic temperature was compared with the harmonic mean of the temperatures used to determine the sorption isotherms, according to Equation 10 (KRUG et al., 1976a and b):

$$T_{hm}=\frac{n}{\sum\left(\frac{1}{T}\right)} \quad (10)$$

where:

T_{hm}: harmonic mean of temperature (K);

n: number of temperatures used.

According to Krug et al. (1976a and b), the linear chemical compensation or compesantion theory exists only if the isokinetic temperature (T_B) is different from the harmonic mean of the temperature (T_{hm}). An approximate confidence interval, (1-α) 100%, for the isokinetic temperature was calculated with the following equation:

$$\hat{T}_B = \bar{T}_B \pm t_{m_p-2,\alpha/2}\sqrt{Var(\hat{T}_B)} \quad (11)$$

where:

$$\hat{T}_B = \frac{\sum(\Delta h_{st} - \overline{\Delta h_{st}})(\Delta S - \overline{\Delta S})}{\sum(\Delta S - \overline{\Delta S})^2} \quad (12)$$

and,

$$Var(\hat{T}_B) = \frac{\sum(\Delta h_{st} - \overline{\Delta G}_B - \hat{T}_B \Delta S)^2}{(m_p - 2)\sum(\Delta S - \overline{\Delta S})^2} \quad (13)$$

where:

m_p: number of pairs of data on enthalpy and entropy;

$\overline{\Delta h_{st}}$: mean enthalpy, kJ kg^{-1};

$\overline{\Delta S}$: mean entropy kJ kg^{-1}.

If the harmonic mean of the temperature is in the estimated interval of the isokinetic temperature T_B, the relationship between the values of sorption differential enthalpy and entropy reflect only experimental errors and not the existence of chemical and physical factors that guide the compensation theory (BERISTAIN et al., 1996). For T_B, we adopted a confidence interval of 95% for the range of data studied.

Results and discussion

The mean values of the moisture content of hygroscopic equilibrium of the forage turnip seeds obtained by desorption at temperatures of 25, 30, 35, and 40°C and water activity from 0.27 to 0.82 (decimal) are shown in Table 1:

Table 1. Mean values of equilibrium moisture content (% d.b.) of forage turnip, obtained by desorption process, as a function of temperature (°C) and water activity (decimal, ± 0.03).

Temperature (°C)	Water activity (decimal)									
	0.27	0.30	0.37	0.40	0.45	0.52	0.67	0.75	0.77	0.82
25	-	5.04	-	-	6.60	6.95	-	10.88	-	-
30	-	-	4.57	-		6.36	8.42	-	10.21	-
35	-	-	4.36	-	-	5.83	7.92	-	-	11.30
40	3.33	-	-	3.61	-	5.69	7.54	-	-	-

Table 1 shows that for the same temperature, with variation in the water activity, there is a significant increase in the equilibrium moisture content for forage turnip seeds.

Figure 1 presents the curves of the neperian logarithm of water activity (decimal) calculated using the Copace equation for the specific values of equilibrium moisture content (% d.b.) as a function of the values of the inverse of the absolute temperature (K) for forage turnip seeds.

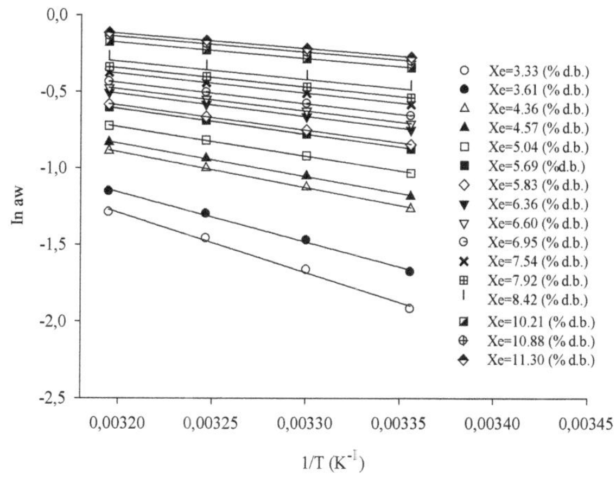

Figure 1. Curves of the neperian logarithm of the water activity (decimal) for specific values of equilibrium moisture content (% d.b.) as a function of the values of 1 T^{-1} (K^{-1}) for forage turnip seeds.

The values of differential enthalpy (Δh_{st}), in kJ kg^{-1}, for forage turnip seeds as a function of the equilibrium moisture content (% d.b.) were calculated according to Equation 3 and are represented for each condition by the slopes of the lines in Figure 1. These values are shown in Table 2.

Table 2. Linear equations of the neperian logarithm of water activity (decimal) as a function of the inverse absolute temperature (1 T^{-1} K^{-1}) for different values of the equilibrium moisture content (% d.b.) of forage turnip seeds.

Xe (% d.b.)	Linear equation	Xe (% d.b.)	Linear equation
3.33	$\ln(a_w) = 11.166 - 3893.1 \times \frac{1}{T}$	6.60	$\ln(a_w) = 4.1212 - 1438.3 \times \frac{1}{T}$
3.61	$\ln(a_w) = 9.2127 - 3240.7 \times \frac{1}{T}$	6.95	$\ln(a_w) = 3.9549 - 1373.4 \times \frac{1}{T}$
4.36	$\ln(a_w) = 6.5434 - 2324.4 \times \frac{1}{T}$	7.54	$\ln(a_w) = 3.7271 - 1283.1 \times \frac{1}{T}$
4.57	$\ln(a_w) = 6.109 - 2170.9 \times \frac{1}{T}$	7.92	$\ln(a_w) = 3.6047 - 1233.8 \times \frac{1}{T}$
5.04	$\ln(a_w) = 5.3855 - 1911.1 \times \frac{1}{T}$	8.42	$\ln(a_w) = 3.4648 - 1176.5 \times \frac{1}{T}$
5.69	$\ln(a_w) = 4.7159 - 1664.5 \times \frac{1}{T}$	10.21	$\ln(a_w) = 3.1147 - 1028.6 \times \frac{1}{T}$
5.83	$\ln(a_w) = 4.6056 - 1623.1 \times \frac{1}{T}$	10.88	$\ln(a_w) = 3.0212 - 987.68 \times \frac{1}{T}$
6.36	$\ln(a_w) = 4.2528 - 1489.1 \times \frac{1}{T}$	11.30	$\ln(a_w) = 2.9706 - 965.19 \times \frac{1}{T}$

For the integral isosteric heat of desorption (Q_{st}), in kJ kg^{-1}, the value of the latent heat of vaporisation of free water (L), which represents the minimum amount of energy required for water evaporation, calculated for the average temperature of 32.5°C, was added to Δh_{st}, resulting in the value of 2,424.52 kJ kg^{-1}.

Figure 2 shows the values of integral isosteric heat of desorption (Q_{st}) as a function of the equilibrium moisture content (% d.b.), estimated according to Equation 10.

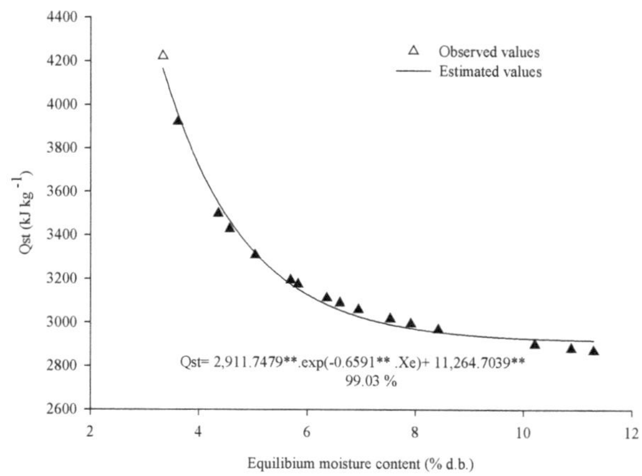

Figure 2. Estimated and experimental values of the integral isosteric heat of desorption as a function of the equilibrium moisture content. **Significant at 1% by t test.

Figure 2 shows that the reduction in the moisture content increases the energy required to remove water from the product, represented by the values of integral isosteric heat of desorption (Q_{st}), as observed for different vegetables such as quinoa (TOLABA et al., 2004), food groups (KUMAR et al., 2005), beans (RESENDE et al., 2006), sweet potato (FASINA, 2006), nutritionally enriched cactus (CAMPOS et al., 2009), the leaves and stems of dehydrated cilantro (SILVA et al., 2010), sweet corn (OLIVEIRA et al., 2010), cassava flour (AYALA-APONTE, 2011) and dehydrated passion fruit pulp (CATELAM et al., 2011). Brooker et al. (1992) claimed that the amount of energy required to remove water from seeds with low moisture content is greater than that required for moist products. According to Aviara and Ajibola (2002), these results confirm the observation that the binding force between the water molecules and dry matter is significantly reduced in products with higher water content.

The values of the integral isosteric heat of desorption for forage turnip seeds in the moisture content range from 3.33 to 11.30 (% d.b.) ranged from 4,222.70 to 2,870.34 kJ kg^{-1}.

Knowledge of the magnitude of the integral isosteric heat of sorption, for a given amount of water, provides an indication of the state of the water sorbed into the product and serves as a measure of the physical, chemical and biological stability under a given storage condition (MCMINN; MAGEE, 2003).

Thus, the elevated value of the integral isosteric heat of sorption for low values of moisture content can be explained by the differences in the binding forces of the water with the adsorbent surface of a product. In the initial stages of sorption, i.e., with a low water content, there are highly active polar sites of sorption with an elevated energy of interaction on the adsorbent surface, and these polar sites are covered with water molecules forming a monomolecular layer (AL-MUHTASEB et al., 2004). As water molecules chemically bind to active sorption sites, the sorption begins to occur at less active sites with a high moisture content, lower interaction energy and, consequently, lower isosteric heat of sorption (WANG; BRENAN, 1991).

Figure 3 shows the values of the Gibbs free energy for each temperature as a function of the moisture content of the forage turnip seeds, calculated using Equation 7.

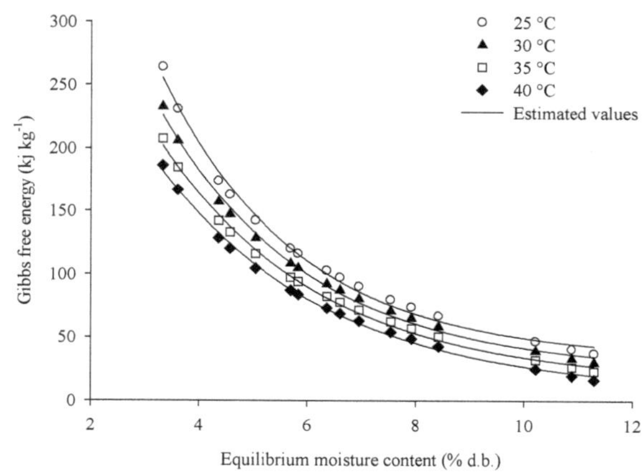

Figure 3. Gibbs free energy as a function of the equilibrium moisture content of forage turnip seeds.

From a thermodynamic standpoint, the Gibbs free energy of a product is the indicative parameter of affinity between the product and water because it provides information on the spontaneity when the Gibbs free energy is negative, entropy is positive and enthalpy is negative or on the non-spontaneity when entropy is negative and enthalpy is positive and thus the Gibbs free energy is necessarily positive (MCMINN et al., 2005; TELIS et al., 2000).

The change in the Gibbs free energy during the water exchange between a product and its medium is the energy required to transfer the water molecules from the gas state to a solid surface or vice versa. This amount may be considered as the measure of the work performed by the system to conduct the desorption or adsorption process.

Equilibrium is achieved when the gradient of Gibbs free energy is zero (NAYAK; PANDEY, 2000).

Ho and McKay (2003) explain this phenomenon by the higher energy of the adsorbate particles at higher temperatures. The 'escape' rate of the adsorbent surface would be higher, causing a reduction in the adsorption efficiency of the material.

Figure 3 shows that with increased equilibrium moisture content, there is an increase in the value of Gibbs free energy. Positive ΔG values are a characteristic of an endergonic reaction, in which the reaction requires an input of energy from the medium for water desorption from the seeds to occur. These results agree with those found by Goneli et al. (2010), who obtained smaller available sorption sites for high values of the equilibrium moisture content of okra seeds. Therefore, less work was necessary to make those sites accessible, with low values of Gibbs free energy, which was positive for all studied temperatures from 10 to 50°C.

Figure 3 shows that the phenomenon's behaviour for each temperature can be described by the following equation:

$$\Delta G = a \times \exp(-b \times Xe) + c \quad (14)$$

where:

a, b and c: coefficients of the equation.

Table 3 presents the equations fit to the values of Gibbs free energy as a function of the equilibrium moisture content (% d.b.) for each temperature tested during desorption in the forage turnip seeds.

Table 3. Exponential regressions of Gibbs free energy (kJ kg^{-1}) as a function of the equilibrium moisture content (% d.b.) of forage turnip seeds.

Temperature (°C)	Equation	R^2 (%)
25	$\Delta G = 33.3458^{**} + 829.9038^{**} \exp\left(-0.3959^{**} \times X_e\right)$	99.41
30	$\Delta G = 23.4438^{**} + 680.5346^{**} \exp\left(-0.3640^{**} \times X_e\right)$	99.54
35	$\Delta G = 14.4671^{**} + 583.0741^{**} \exp\left(-0.3405^{**} \times X_e\right)$	99.63
40	$\Delta G = 6.1274^{NS} + 514.0807^{**} \exp\left(-0.3223^{**} \times X_e\right)$	99.69

**Significant at 1% by t test, ^{NS}Not Significant.

Figure 4 shows the values of the differential entropy of desorption in kJ kg^{-1} K^{-1} as a function of the equilibrium moisture content (% d.b.).

Figure 4 shows a strong dependence of the differential entropy values on the moisture content, similar to the trend of the integral isosteric heat of sorption. There was a reduction in the values of differential entropy, with a tendency to stabilise at higher moisture contents. According to Madamba et al. (1996), these results indicate that up to a certain moisture content, the sorption processes are reversible.

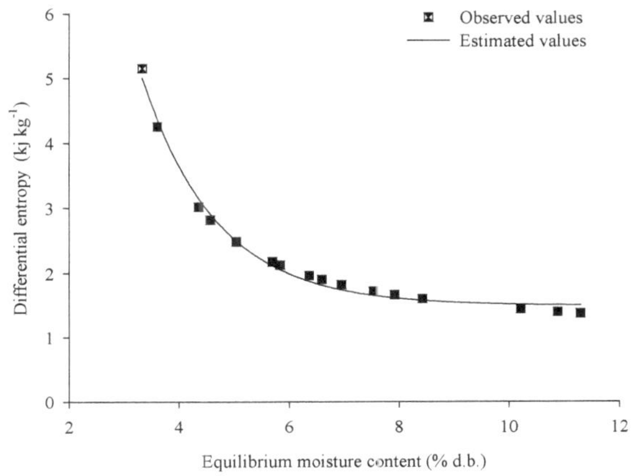

Figure 4. Estimated and experimental values of the differential entropy of the desorption of forage turnip seeds. **Significant at 1% by t test.

Figure 5 shows the values of differential enthalpy as a function of the values of the differential entropy of sorption for the desorption process of forage turnip seeds, as determined by Equation 8 for each equilibrium moisture content. We considered that for a given value of the moisture content, the differential enthalpy and entropy do not vary with temperature (AGUERRE et al., 1986).

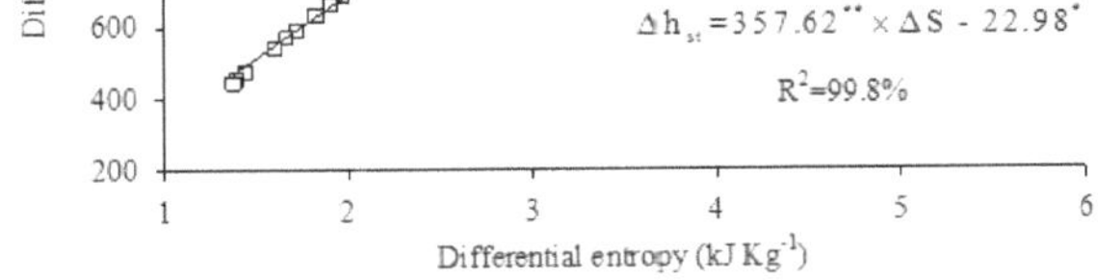

Figure 5. Enthalpy – entropy relationship for the process of water desorption from forage turnip seeds. **Significant at 1% by t test; *Significant at 5% by t test.

A linear relationship was shown for the desorption process, with a coefficient of determination (R^2) greater than 99%. As there is a high level of linearity between the values of differential enthalpy (Δh_{st}) and entropy (ΔS) of sorption, the enthalpy-entropy compensation theory, or isokinetic theory, can be considered valid for the phenomenon of the desorption of water from forage turnip seeds.

To test the validity of the enthalpy-entropy compensation theory, the isokinetic temperature was compared with the harmonic mean (T_{hm}) from the temperature range studied and used to determine the desorption process of forage turnip seeds. Enthalpy-entropy linear compensation is valid if the process is controlled by enthalpy when $T_B > T_{hm}$ or the process is controlled by entropy when $T_B < T_{hm}$ (GABAS et al., 2005).

The isokinetic temperature for the desorption of forage turnip seeds was 357.62±12.39 K. The calculated harmonic mean was 304.67 K, which is significantly different from the isokinetic temperature values described, confirming the phenomenon of the enthalpy-entropy compensation for the desorption process of forage turnip seeds. According to Liu and Guo (2001), the isokinetic temperature is that for which the reactions in series in the product should occur simultaneously, i.e., when it is in equilibrium.

Therefore, the mechanism of water desorption from forage turnip seeds can be controlled by enthalpy. These results are in agreement with various researchers who have successfully applied the isokinetic theory to the sorption of different products (CLADERA-OLIVERA et al., 2008; GONELI et al., 2010; MCMINN et al., 2005; MOREIRA et al., 2008; TELIS-ROMERO et al., 2005; TUNC; DUMAN, 2007).

Conclusion

With an increased equilibrium moisture content, there is an increase in the energy required for water evaporation in forage turnip seeds and an increase in differential entropy.The values of integral isosteric heat of desorption in the moisture content range of 3.33 to 11.30 (% d.b.) varied from 4,222.70 to 2,870.34 kJ kg^{-1}.The Gibbs free energy tended to increase with the elevation of the equilibrium moisture content, and its values were positive, demonstrating non-spontaneity in the desorption process.The enthalpy-entropy compensation theory, can be satisfactorily applied to the sorption phenomenon, and the process of water desorption seeds is controlled by enthalpy.

Acknowledgements

We thank the National Counsel of Technological and Scientific Development (Conselho Nacional de Desenvolvimento Científico e Tecnológico – CNPq), the Brazilian Federal Agency for the Support and Evaluation of Graduate Education (Coordenação de Aperfeiçoamento de Pessoal de Nível Superior – Capes) and IF Goiano - Câmpus Rio Verde for the financial support.

References

AGUERRE, R. J.; SUAREZ, C.; VIOLLAZ, P. E. Enthalpy-entropy compensation in sorption phenomena: application to the prediction of the effect of temperature on food isotherms. **Journal of Food Science**, v. 51, n. 6, p. 1547-1549, 1986.

AL-MUHTASEB, A. H.; MCMINN, W. A. M.; MAGEE, T. R. A. Water sorption isotherms of starch powders. Part 2: Thermodynamic characteristics. **Journal of Food Engineering**, v. 62, n. 2, p. 135-142, 2004.

ASCHERI, D. P. R.; MOURA, W. S.; ASCHERI, J. L. R.; JUNIOR, E. A. F. Propriedades termodinâmicas de adsorção de água do amido de rizomas do lírio-do-brejo (*Hedychium coronarium*). **Ciência e Tecnologia de Alimentos**, v. 29, n. 2, p. 454-462, 2009.

AVIARA, N. A.; AJIBOLA, O. O.; ONI, S. A. Sorption equilibrium and thermodynamic characteristics of soya bean. **Biosystems Engineering**, v. 87, n. 2, p. 179-190, 2004.

AVIARA, N. A.; AJIBOLA, O. O. Thermodynamics of moisture sorption in melon seed and cassava. **Journal of Food Engineering**, v. 55, n. 2, p. 107-113, 2002.

AVIARA, N. A.; AJIBOLA, O. O.; DAIRO, U. O. Thermodynamics of moisture sorption in sesame seed. **Biosystems Engineering**, v. 83, n. 4, p. 423-431, 2002.

AYALA-APONTE, A. Estimación de lãs isotermas de adsorción y del calor isostérico en harina de yuca. **Biotecnología en el Setor Agropecuario y Agroindustrial**, v. 9, n. 1, p. 88-96, 2011.

BERISTAIN, C. I.; GARCIA, H. S.; AZUARA, E. Enthalpy-entropy compensation in food vapor adsorption. **Journal of Food Engineering**, v. 30, n. 3-4, p. 405-415, 1996.

BRASIL. Ministério da Agricultura e Reforma Agrária. Secretaria Nacional de Defesa Agropecuária. **Regras para análise de sementes**. Brasília: Mapa/ACS, 2009.

BROOKER, D. B.; BAKKER-ARKEMA, F. W.; HALL, C. W. **Drying and storage of grains and oilseeds**. New York: The AVI Van Nostrand Reinhold, 1992.

CAMPOS, A. R. N.; SILVA, F. L. H.; GOMES, J. P.; OLIVEIRA, L. S. C.; OLIVEIRA, M. M. Isotermas de adsorção e calor isostérico da palma forrageira enriquecida proteicamente. **Revista Brasileira de Engenharia Agrícola e Ambiental**, v. 13, n. 6, p. 734-740, 2009.

CATELAM, K. T.; TRINDADE, C. S. F.; ROMERO, J. T. Adsorption isotherms and isosteric heat in passionfruit pulp dehydrated using a spray dryer and freeze dryer with additives and skim milk. **Ciência e Agrotecnologia**, v. 35, n. 6, p. 1196-1203, 2011.

CHEMKHIA, S.; ZAGROUBAA, F.; BELLAGIB, A. Thermodynamics of water sorption in clay. **Desalination**, v. 166, s/n., p. 393-399, 2004.

CLADERA-OLIVERA, F.; PETTERMANN, A.; NORENA, C.; WADA, K.; MARCZAK, L. Thermodynamic properties of moisture desorption of raw pinhão (*Araucaria angustifolia* seeds). **International Journal of Food Science and Technology**, v. 43, n. 5, p. 900-907, 2008.

CORRÊA, P. C.; GONELI, A. L. D.; JAREN C.; RIBEIRO, D. M.; RESENDE, O. Sorption isotherms and isosteric heat of peanut pods, kernels, and hulls. **Food Science and Technology International**, v. 13, n. 3, p. 231-238, 2007.

DOMINGUEZ, I. L.; AZUARA, E.; VERNON-CARTER, E. J.; BERISTAIN, C. I. Thermodynamic analysis of the effect of water activity on the stability of macadamia nut. **Journal of Food Engineering**, v. 81, n. 3, p. 566-571, 2007.

FASINA, O. O. Thermodynamic properties of sweet potato. **Journal of Food Engineering**, v. 75, n. 2, p. 149-155, 2006.

GABAS, A. L.; TELIS-ROMERO, J.; KOHAYAKAWA, M. N.; SILVEIRA JR., V.; PEDRO, M. A. M. Enthalpy-entropy compensation based on isotherms of mango. **Ciência e Tecnologia de Alimentos**, v. 25, n. 2, p. 297-303, 2005.

GHODAKE, H. M.; GOSWAMI, T. K.; CHAKRAVERTY, A. Moisture sorption and vaporization of withered leaves, black and green tea. **Journal of Food Engineering**, v. 78, n. 3, p. 827-935, 2007.

GONELI, A. L. D.; CORRÊA, P. C.; OLIVEIRA, G. H. H.; BOTELHO, F. M. Water desorption and thermodynamic properties of okra seeds. **Transactions of the ASABE**, v. 53, n. 1, p. 191-197, 2010.

HO, Y. S.; MCKAY, G. Sorption of dyes and copper ions onto biosorvents. **Process Biochemistry**, v. 28, n. 7, p. 1047-1061, 2003.

IGLESIAS, H.; CHIRIFE, J. Isosteric heats of water vapour sorption on dehydrated foods. Part II: hysteresis and heat of sorption comparison with BET theory. **Lebensmittel Wissenschaft and Technologie**, v. 9, p. 123-127, 1976.

KAYA, S.; KAHYAOGLU, T. Influence of dehulling and roasting process on the thermodynamics of moisture adsorption in sesame seed. **Journal of Food Engineering**, v. 76, n. 2, p. 139-147, 2006.

KRUG, R. R.; HUNTER, W. G.; GRIEGER, R. A. Enthalpy-entropy compensation.1- Some fundamental statistical problems associated with the analysis of Van't Hoffand Arrhenius data. **Journal of Physical Chemistry**, v. 80, n. 21, p. 2335-2341, 1976a.

KRUG, R. R.; HUNTER, W. G.; GRIEGER, R. A. Enthalpy-entropy compensation. 2- Separation of the chemical from the statistical effect. **Journal of Physical Chemistry**, v. 80, n. 21, p. 2341-2351, 1976b.

KUMAR, A. J.; SINGH, R. R. B.; PATIL, G. R.; PARE, A. A. Effect of temperature on moisture desorption isotherms of *kheer*. **Food Science and Technology**, v. 38, n. 3, p. 303-310, 2005.

LIU, L.; GUO, Q. X. Isokinetic relationship, isoequilibrium relationship, and enthalpy–entropy compensation. **Chemical Reviews**, v. 101, n. 3, p. 673-695, 2001.

MADAMBA, P. S.; DRISCOLL, R. H.; BUCKLE, K. A. The thin-layer drying characteristics of garlic slices. **Journal of Food Engineering**, v. 29, n. 1, p. 75-97, 1996.

MCMINN, W. A. M.; MAGEE, T. R. A. Thermodynamic properties of moisture sorption of potato. **Journal of Food Engineering**, v. 60, n. 2, p. 155-157, 2003.

MCMINN, W. A. M.; AL-MUHTASEB, A. H.; MAGEE, T. R. A. Enthalpy-entropy compesation in sorption phenomena of starch materials. **Journal of Food Engineering**, v. 38, n. 5, p. 505-510, 2005.

MOREIRA, R.; CHENLO, F.; TORRES, M. D.; VALLEJO, N. Thermodynamic analysis of experimental sorption isotherms of loquat and quince fruits. **Journal of Food Engineering**, v. 88, n. 4, p. 514-521, 2008.

NAYAK, L.; PANDEY, J. P. Free energy change and monolayer moisture content of paddy and milled ride. **Journal of the Institution of Engineers**, v. 80, p. 43-45, 2000.

OLIVEIRA, G. H. F.; CORRÊA, P. C.; ARAÚJO, E. F; VALENTE, D. S. M.; BOLTELHO, F. M. Desorption isotherms and thermodynamic properties of sweet corn cultivars (*Zea mays* L.). **International Journal of Food Science and Technology**, v. 45, n. 3, p. 546-554, 2010.

OLIVEIRA, G. H. H.; CORRÊA, P. C.; SANTOS, E. S. S.; TRETO, P. C.; DINIZ, M. D. M. S. Evaluation of thermodynamic properties using GAB model to describe the desorption process of cocoa beans. **International Journal of Food Science and Technology**, v. 46, n. 10, p. 2077-2084, 2011.

RESENDE, O.; CORRÊA, P. C.; GONELI, A. L. D.; MARTINAZZO, A. P.; RIBEIRO, R. M. Isotermas e calor isostérico de dessorção do arroz em casca. **Revista Brasileira de Armazenamento**, v. 31, n. 1, p. 86-94, 2006.

RIZVI, S. S. H. Thermodynamic properties of food in dehydration. In: RAO, M. A.; RIZVI, S. S. H. (Ed.). **Engineering properties of foods**. New York: Marcel Dekker, 1986. p. 133-214.

SILVA, A. S.; ALMEIDA, F. A. C.; ALVES, N. M. C.; MELO, K. S. Característica higroscópica e termodinâmica do coentro desidratado. **Revista Ciência Agronômica**, v. 41, n. 2, p. 237-244, 2010.

TELIS, V. R. N.; GABAS, A. L.; MENEGALLI, F. C.; TELIS-ROMERO, J. Water sorption thermodynamic properties applied to persimmon skin and pulp. **Thermochimica Acta**, v. 343, n. 1-2, p. 49-56, 2000.

TELIS-ROMERO, J.; KOHAYAKAWA, M. N.; SILVEIRA JR., V.; PEDRO, M. A. M.; GABAS, A. L.

Enthalpy–entropy compensation based on isotherms of mango. **Ciência e Tecnologia de Alimentos**, v. 25, n. 2, p. 297-303, 2005.

TOLABA, M. P.; PELTZER, M.; ENRIQUEZ, N.; POLLIO, M. L. Grain sorption equilibrium of quinoa grains. **Journal of Food Engineering**, v. 61, n. 3, p. 365-371, 2004.

TSAMI, E.; MARINOUS-KOURIS, D.; MAROULIS, Z. B. Water sorption isotherms of raisins, currants, figs, prunes, and apricots. **Journal of Food Science**, v. 55, n. 6, p. 1594-1597, 1990.

TUNC, S.; DUMAN, O. Thermodynamic properties and moisture adsorption isotherms of cottonseed protein isolate and different forms of cottonseeds samples. **Journal of Food Engineering**, v. 81, n. 1, p. 133-143, 2007.

WANG, N.; BRENNAN, J. G. Moisture sorption isotherm characteristics of potato at four temperatures. **Journal of Food Engineering**, v. 14, n. 4, p. 269-287, 1991.

YAZDANI, M.; AZIZI, P. M.; GHOBADI, P. Moisture sorption isotherms and isosteric heat for pistachio. **European Food Research and Technology**, v. 223, n. 5, p. 577-584, 2006.

10

Determination of the volumetric shrinkage in jatropha seeds during drying

Valdiney Cambuy Siqueira[1*], Osvaldo Resende[2] and Tarcísio Honório Chaves[2]

*[1]Programa de Pós-graduação em Engenharia Agrícola, Departamento de Engenharia, Universidade Federal de Lavras, Cx. postal 3037, 37200-000, Lavras, Minas Gerais, Brazil. [2]Instituto Federal de Educação, Ciência e Tecnologia Goiano, Rio Verde, Goiás, Brazil. *Author for correspondence. E-mail: vcambuy@yahoo.com*

ABSTRACT. The objective of this study was to determine the geometric diameter of volumetric shrinkage in Jatropha seeds subjected to drying under six air conditions and to fit different mathematical models to the experimental values. Seeds with an approximate initial moisture content of 0.61 kgw $kgdm^{-1}$ (kilogram of water per kilogram of dry matter) were dried in a convection oven at temperatures of 36, 45, 60, 75, 90, and 105°C until they reached a moisture content of 0.11 ± 0.006 kgw $kgdm^{-1}$. The experimental data were fitted to five mathematical models used to represent volumetric shrinkage in agricultural products. The models were analyzed using the coefficient of determination, chi-square, mean relative error, mean estimated error, and residual distribution. Geometric diameter data were subjected to regression analysis by adopting a 1% level of significance. It was concluded that moisture content and drying temperature influence volumetric shrinkage in Jatropha seeds, a phenomenon satisfactorily described by the Polynomial equation, where the geometric diameter linearly decreased with a reduction of the moisture content, regardless of the drying conditions.

Keywords: mathematical models, moisture content, Polynomial model.

Introduction

With the current oil crisis, research on alternative fuels has intensified, and biodiesel appears to be one possible immediate solution that is capable of being produced using vegetable oils and animal fats (GOLDFARB et al., 2010). Jatropha (*Jatropha curcas* L.) is among those plants with potential for biodiesel production. Indeed, Souza et al. (2009) have found an average concentration of 40.33 ± 1.91% lipids in dry Jatropha seeds.

In addition to its capacity for producing oil with all of the qualities necessary for biodiesel transformation, the geographical distribution of Jatropha is widespread due to its hardiness, resistance to long droughts, pests, and disease, and its adaptability to a wide range of climate conditions (SANTOS et al., 2009). Although Jatropha is a plant that has the ability to adapt to diverse conditions, cultivation practices are essential for healthy plant development and, consequently, to achieve higher productivity.

Among the steps after harvest, drying is of great importance, as it is directly related to the quality of the final product and may result in irreversible damage to the seeds. According to Resende et al. (2005), reducing the moisture content in seeds not only directly influences their physical properties, but also causes shrinking of the seed, a phenomenon known as volumetric shrinkage. Volumetric shrinkage can be very intensive, depending on the method and drying conditions (KROKIDA; MAROULIS, 1997), which affect the parameters of heat transfer and mass, and should be considered in the establishment of drying models (CORRÊA et al., 2006; RAMOS et al., 2003).

The volumetric shrinkage index determines the relationship between the seed volume, moisture content, and initial volume. This is of utmost importance during the drying process because it is possible to predict the behavior of the reduced volume occupied by the seed mass according to its moisture content reduction.

The development and improvement of equipment used for seed drying is also fundamentally important for drying simulation and to obtain theoretical information about the behavior of each product during water removal (RESENDE et al., 2010). For a simulation based on drying successively thin layers of the product, a mathematical model is used to satisfactorily represent water loss during the drying period (BERBERT et al., 1995; GINER; MASCHERONI, 2002).

According to Midilli et al. (2002), there are three types of models to represent thin layer drying, aimed at describing drying kinetics in agricultural products. These models include a theoretical model, which considers only the internal resistance and the transfer of heat and water between the product and the warm air, and semi-theoretical empirical models, which consider only the internal resistance, temperature, and relative humidity of the drying air.

Models used to represent drying in agricultural products have been developed, in most cases, by neglecting volumetric shrinkage in the product during the dehydration process (BROOKER et al., 1992). According to Ramos et al. (2005), changes in the product with regards to its volumetric shrinkage should also be included in models for the complete description, precision, and analysis of drying phenomena.

Given the prominence of Jatropha cultivation as an oilseed and the lack of theoretical information about the behavior of the seeds of this crop during the drying process, the goal of this study was to fit different mathematical models to experimental volumetric shrinkage data, identify the best model to represent the phenomenon, and determine the geometric diameter in Jatropha seeds submitted to drying under six air conditions.

Material and methods

The experiment was conducted at the Postharvest Laboratory of Plant Products at the Instituto Federal de Educação, Ciência e Tecnologia Goiano – *Campus* Rio Verde (IF Goiano – *Campus* Rio Verde).

Jatropha seeds with a moisture content of 0.61 kgw $kgdm^{-1}$ were used. The seeds were dried in a convection oven under six temperature conditions, 36, 45, 60, 75, 90, and 105°C, and at relative humidities of 31.7, 19.6, 9.4, 4.8, 2.6, and 1.5%, respectively, until reaching a moisture content of 0.11 ± 0.006 kgw $kgdm^{-1}$, determined at 105 ± 1°C for 24h in three replicates (BRASIL, 2009).

For the drying, 20 Jatropha seeds were individually enclosed in aluminum capsules with dimensions of 60.12 mm in diameter and 41.0 mm in height. The seeds were measured at three positions (length, width, and thickness) at 15-min. intervals with the aid of a digital caliper of 0.01-mm resolution. The moisture content reduction was monitored by weighing three trays with 300 g of the product under the same conditions of seed drying.

To evaluate the drying in Jatropha seeds, the rate of water reduction from the product was determined, according to the following expression described by Corrêa et al. (2001):

$$TRA = \left[(Ma_0 - Ma_i)/Ms \cdot (t_i - t_0)\right] \quad (1)$$

TRA: water reduction rate, kg kg^{-1} h^{-1};
Ma_0: total previous water mass, kg;
Ma_i: total current water mass, kg;
Ms: dry matter, kg;
t_0: total previous drying time, h;
t_i: total current drying time, h.

The volumetric shrinkage index of Jatropha seeds during the drying was determined by the relationship between the seed volume to each moisture content and the initial volume, as shown below:

$$\psi = V/V_o \quad (2)$$

ψ : volumetric shrinkage index, decimal;
V: volume for each moisture content, mm^3;
V_o: initial volume, mm^3.

The volume of each seed, in mm^3, was obtained throughout the drying process, according to the following expression proposed by Mohsenin (1986):

$$V_g = \left[(\pi \cdot a \cdot b \cdot c)/6\right] \quad (3)$$

V_g: seed volume, mm^3;
a: major axis of the seed, mm;
b: medial axis of the seed, mm;
c: minor axis of the seed, mm.

The characteristic dimensions of the orthogonal axes in Jatropha seeds were determined with the seed in its natural resting position, as illustrated in Figure 1.

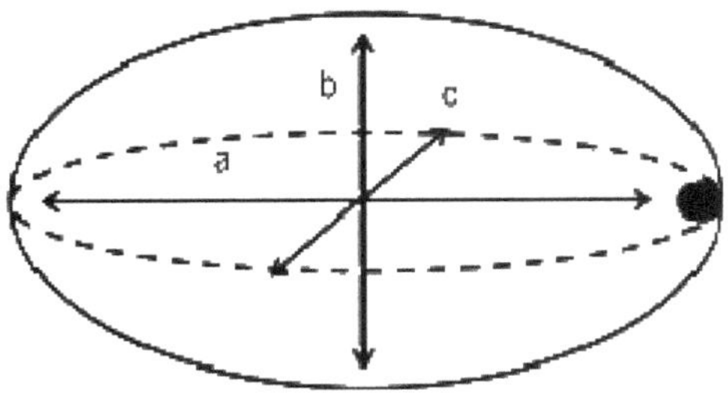

Figure 1. Schematic drawing of a Jatropha seed, considered spheroidal in shape, with its characteristic dimensions.

The geometric diameter of each seed was obtained throughout the drying process, according to the expression proposed by Mohsenin (1986), as follows:

$$D_g = (a \cdot b \cdot c)^{1/3} \tag{4}$$

D_g: geometric diameter, mm.

The experimental data of the unitary volumetric shrinkage index were fitted to the models described by the expressions listed in Table 1.

The mathematical models were fitted using nonlinear regression analysis by the Gauss-Newton method using a statistic software. The models were selected considering the magnitude of the coefficient of determination, the magnitude of the mean relative error, and the estimated error, chi-square, in addition to a verification of the behavior of residual distribution. Mean relative error values of less than 10% were considered as criteria for model selection, according to Mohapatra and Rao (2005).

The mean relative and estimated errors for each model were calculated conforming to the following expressions:

$$P = \left[(100/n) \sum \left(|Y - \hat{Y}| / Y\right)\right] \tag{10}$$

$$SE = \sqrt{\sum \left(Y - \hat{Y}\right)^2 / GLR} \tag{11}$$

$$\chi^2 = \left[\sum \left(Y - \hat{Y}\right)^2 / GLR\right] \tag{12}$$

P: mean relative error, %;
n: number of experimental observations;
$\hat{Y}$: the value calculated by the model;
Y: value experimentally observed;
SE: mean estimated error, decimal;
χ^2: chi-square.
GLR: degrees of freedom of the model, the number of observations minus the number of model parameters.

To assess the geometric diameter, the experiment was mounted in a factorial design, with factors such as different temperatures and moisture contents, the data were subjected to analysis of variance and regression adopting the 1% level of significance, with aid application program SISVAR 5.3 according to Ferreira (2008). Apart from this statistical analysis we used the mean relative error test to validate the model employed.

Results and discussion

Figure 2 shows the mean water removal rate in Jatropha seeds submitted to drying under six temperature conditions. The highest value was observed at a drying temperature of 105°C, with a magnitude of 0.1728 kg kg^{-1} h^{-1}. According to Resende et al. (2009), over the course of drying, the water reduction rate decreases because water is strongly bonded, requiring more energy for its evaporation. However, Figure 2 demonstrates that the water reduction rate increased at times, especially at temperatures of 36 and 45°C. This behavior was due to the greater distances between the reading intervals.

Table 1. Models used to simulate the volumetric shrinkage index in agricultural products.

Reference	Model	
Corrêa et al. (2004) - apud Corrêa et al. (2011)	$\psi = 1/\left[a_1 + b \cdot \exp(X)\right]$	(5)
Exponential - Exp.	$\psi = a_1 \cdot \exp(b \cdot X)$	(6)
Linear - Lin.	$\psi = a_1 + a_2 \cdot X$	(7)
Polynomial - Pol.	$\psi = a_1 + a_2 \cdot X + a_3 \cdot X^2$	(8)
Modified Bala and Woods (1984) - Mod. B. and W.	$\psi = 1 - a_1\{1 - \exp[-a_2(X_0 - X)]\}$	(9)

a_1, a_2, a_3 :parameters dependent on the product; X: moisture content of the product, kgw $kgdm^{-1}$; X_0: initial moisture content of the product, kgw $kgdm^{-1}$.

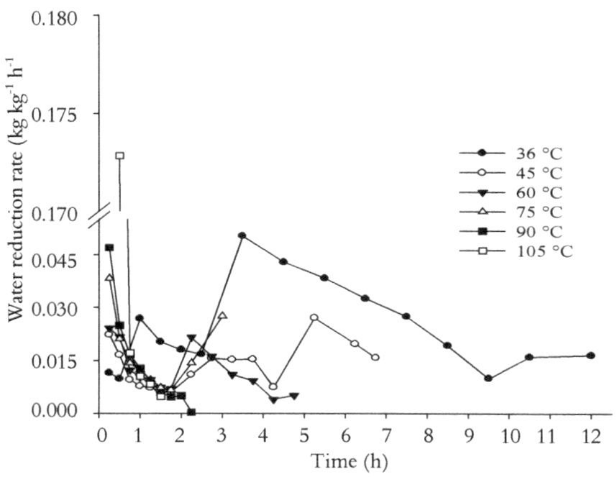

Figure 2. Water reduction rate (kg kg^{-1} h^{-1}) in Jatropha seeds subjected to drying under six air temperatures.

Figure 2 also shows that the period of the water reduction rate in Jatropha seeds was 1.5, 2.25, 3.0, 4.75, 6.75, and 12.0h at drying temperatures of 105, 90, 75, 60, 45, and 36°C, respectively, which corresponds to the time required for the seeds to reach a moisture content of 0.11 ± 0.006 kgw $kgdm^{-1}$. Therefore, increasing the temperature promoted a reduction of the drying time in Jatropha seeds, whereas higher temperatures also promoted a greater moisture gradient between the seed and drying air, which is an observation that has been made by many researchers of numerous agricultural products (AKPINAR et al., 2003; COSTA et al., 2011; LAHSASNI et al., 2004; MOHAPATRA; RAO, 2005; RESENDE et al., 2008; SIRISOMBOON; KITCHAIYA, 2009; SOUSA et al., 2011; ULLMANN et al., 2010).

Values for the mean estimated relative error (SE), mean relative error (P), and the coefficient of determination (R^2) for models fitted to the volumetric shrinkage index in Jatropha seeds subjected to drying under different temperature conditions are presented in Table 2. Regarding the coefficients of determination (R^2) relative to the adjusted mathematical models, note that only the polynomial and modified Bala and Woods models had values above 96%. Thus, the evaluation of other parameters, such as the mean estimated error, mean relative error, and residual distribution values, is necessary to choose the model that satisfactorily describes the volumetric shrinkage phenomenon.

As observed in the analyzed models, the mean estimated error (SE), which describes the standard deviation value for the estimate, showed little variation between the models and respective temperatures studied. However, the polynomial and modified Bala and Woods models showed the lowest SE values under all of the drying conditions. It is noteworthy that the lower the SE value was, the better the model fitted to the experimental data. Table 2 also shows that the models showed mean relative error values between 0.135 and 0.869%. According to Mohapatra and Rao (2005), mean relative error values below 10% indicate the adequate representation of the phenomenon. Thus, all of the models meet this criterion; however, as in the other evaluations, the polynomial and modified Bala and Woods models stand out. Corrêa et al. (2011) found that the modified Bala and Woods model best represented the volumetric shrinkage rate of red beans during drying.

Table 3 shows the residual distribution data for the models studied. The exponential and linear models showed a biased distribution under all of the conditions studied, and the polynomial model showed more drying conditions with random distribution, the desired behavior for the description of the volumetric shrinkage index in agricultural products.

Table 2. Coefficients of determination, mean relative error, and estimated error for the five models analyzed during volumetric shrinkage in Jatropha seeds under various temperature conditions (°C).

Temperature (°C)		Model				
		Cor.	Exp.	Lin.	Pol.	Mod. B. and W.
36	SE	0.005	0.005	0.005	0.005	0.006
	P	0.477	0.448	0.442	0.457	0.516
	R^2	98.4	98.4	98.3	98.5	97.9
45	SE	0.006	0.008	0.009	0.007	0.006
	P	0.553	0.675	0.702	0.583	0.535
	R^2	96.7	94.6	94.0	96.5	96.7
60	SE	0.004	0.004	0.004	0.004	0.004
	P	0.331	0.331	0.343	0.320	0.313
	R^2	98.9	98.9	98.7	99.1	99.0
75	SE	0.010	0.006	0.006	0.004	0.006
	P	0.845	0.584	0.524	0.347	0.528
	R^2	94.3	97.4	97.8	98.9	97.5
90	SE	0.010	0.006	0.006	0.005	0.006
	P	0.869	0.592	0.537	0.248	0.531
	R^2	95.1	97.7	98.1	98.7	97.9
105	SE	0.009	0.006	0.005	0.002	0.006
	P	0.665	0.455	0.411	0.135	0.391
	R^2	95.9	98.0	98.3	99.8	98.2

Illustrations of residual distribution (random for the polynomial model and biased for the model by Corrêa et al. (2004) apud Corrêa et al. (2011) during the modeling of volumetric shrinkage in Jatropha seeds subjected to drying at a temperature of 105°C are shown in Figure 3.

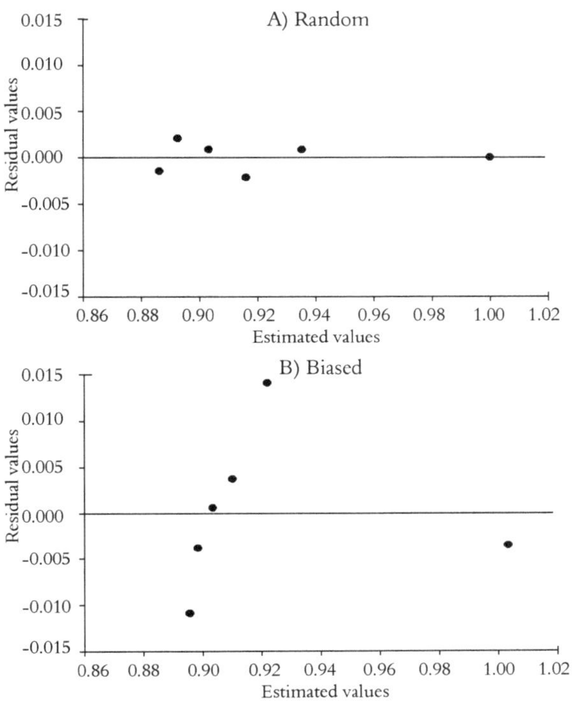

Figure 3. Illustration of the residual distribution during the modeling of volumetric shrinkage in Jatropha seeds: A) Random for the polynomial model; B) Biased for the model of Corrêa et al. (2004) apud Corrêa et al. (2011).

A model is considered random if the residual values found near the horizontal axis are close to zero and do not form defined figures, indicating no bias of the results. If a biased distribution is presented, the model is considered inadequate to represent the phenomenon in question (GONELI et al., 2011).

Table 4 presents the chi-square values obtained for the different models fitted to the experimental values of grain shrinkage of jatropha. It appears that the five models analyzed were found in the range of 99%. However, in general, the polynomial model was the one with the lowest chi-square. The lower the value of chi-square best fit of the model (AKPINAR et al., 2003; GÜNHAN et al., 2005; MIDILLI; KUCUK, 2003).

Data analysis indicates that the polynomial model is adequate for the mathematical description of the unitary volumetric shrinkage phenomenon in Jatropha seeds, demonstrating a high fitted coefficient of determination and reduced mean relative and estimated error values; therefore, the polynomial model was the model that showed the drying conditions with a random residual distribution. Afonso Júnior et al. (2003) have also observed that the polynomial model satisfactorily described volumetric shrinkage in coffee beans during drying.

Table 4 presents the coefficients of the polynomial model fitted to the experimental data of unitary volumetric shrinkage in Jatropha seeds subjected to drying at different temperatures.

As observed in Table 5, the coefficients of the polynomial model do not show a clear trend as a function of drying temperature. Thus, the unitary volumetric shrinkage in Jatropha seeds during drying occurred differently between the evaluated temperatures, as shown in Figure 4.

Table 3. Residual distribution for the models analyzed during volumetric shrinkage in Jatropha seeds under various temperature conditions.

Model	Temperature (°C)					
	36	45	60	75	90	105
Corrêa et al. (2004) apud Corrêa et al. (2011)	T	T	A	T	T	T
Exponential	T	T	T	T	T	T
Linear	T	T	T	T	T	T
Polynomial	T	T	A	A	T	A
Modified Bala and Woods (1984)	T	T	A	T	T	T

Table 4. Chi-square values calculated for the models used in the representation of grain shrinkage of jatropha.

Model	Temperature (°C)					
	36	45	60	75	90	105
Cor.	0.000031	0.000046	0.000019	0.000103	0.000103	0.000090
Exp.	0.000031	0.000074	0.000019	0.000047	0.000047	0.000043
Lin.	0.000034	0.000082	0.000023	0.000038	0.000039	0.000035
Pol.	0.000031	0.000052	0.000017	0.000021	0.000030	0.000004
Mod. B. and W.	0.000041	0.000045	0.000018	0.000044	0.000043	0.000039

Table 5. Coefficients of the polynomial model fitted to volumetric shrinkage in Jatropha seeds under different drying conditions.

Coefficients	Temperature (°C)					
	36	45	60	75	90	105
a_1	0.837**	0.852**	0.838**	0.831**	0.831**	0.848**
a_2	0.188**	0.066ns	0.185**	0.378**	0.356**	0.386**
a_3	0.096ns	0.252**	0.110*	-0.175*	-0.138ns	-0.226*

nsNot significant, *significant at 5% and **significant at 1% probability by the F test.

The unitary volumetric shrinkage values, both experimental and estimated by the polynomial model, are shown in Figure 4. The index of volumetric shrinkage was 0.147, 0.151, 0.145, 0.132, 0.140, and 0.115 at temperatures of 36, 45, 60, 75, 90, and 105°C, respectively. Note that the volumetric shrinkage was lower at higher temperatures (105, 90, and 75°C), especially in the seeds subjected to drying at 105°C. At this temperature, there was a high water reduction rate at the onset of drying, resulting in hardening of the integument that impaired seed shrinkage.

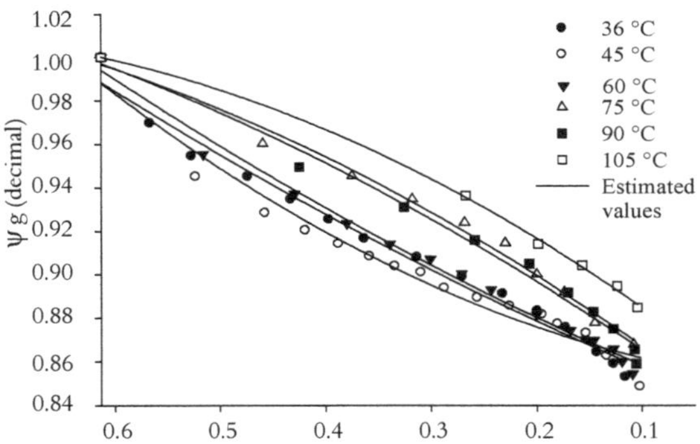

Figure 4. Unitary volumetric shrinkage values in Jatropha seeds, both experimental and estimated by the polynomial model, due to moisture content reduction.

Figure 5 shows the geometric diameter values in Jatropha seeds subjected to drying under different temperature conditions. It was observed that, with drying, the geometric diameter decreased from 12.26, 12.35, 12.27, 12.16, 12.31, and 12.20 mm to 11.77, 1.74, 11.70, 11.54, 11.80, and 11.56 mm, at temperatures of 105, 90, 75, 60, 45, and 36°C, respectively. Thus, different drying conditions did not significantly promote changes in the reduction of the geometric diameter. Sirisomboon and Kitchaiya (2009) have found that geometric diameters in Jatropha seeds were 9.92, 9.37, and 9.97 mm, when subjected to drying at temperatures of 80, 60, and 40°C, respectively. The differences in the geometric diameter values could be related to the seed variety and handling technique adopted during periods of pre- and/or post-harvest.

Table 6 shows the linear models fitted to the experimental data of the geometric diameter in Jatropha seeds as a function of the moisture content. It was observed that the models showed a high degree of significance according to the F test and low values of mean relative error. Thus, they satisfactorily described the behavior of the variable analyzed. These results agree with those found by Razavi et al. (2007), working with pistachios and other nuts.

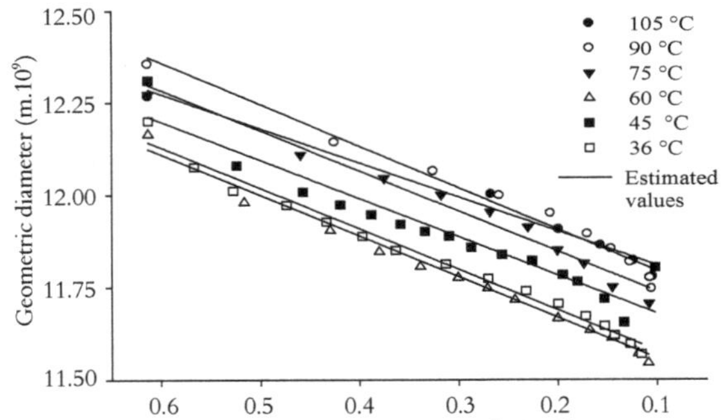

Figure 5. Geometric diameter in Jatropha seeds subjected to various drying conditions.

Table 6. Equations fitted to the geometric diameter values of Jatropha seeds as a function of moisture content.

Temperatura (°C)	Modelo	R^2 (%)	P (%)
105	$D_g = 0.927Ta + 11.71$	98.0**	0.15
90	$D_g = 1.130Ta + 11.68$	97.7**	0.19
75	$D_g = 1.091Ta + 11.62$	97.4**	0.19
60	$D_g = 1.104Ta + 11.44$	99.1**	0.11
45	$D_g = 1.039Ta + 11.57$	94.6**	0.27
36	$D_g = 1.100Ta + 11.46$	98.4**	0.15

**Significant at 1% probability by the F test.

Conclusion

Based on the results obtained, it can be concluded that the moisture content and drying temperature influence volumetric shrinkage in Jatropha seeds. The phenomenon satisfactorily described by the polynomial equation, where the geometric diameter linearly decreases with the reduction of moisture content, regardless of the drying conditions.

Acknowledgements

The authos acknowledge the Conselho Nacional de Desenvolvimento Científico e Tecnológico (CNPq), Coordenação de Aperfeiçoamento de Pessoal de Nível Superior (CAPES) and the Instituto Federal de Educação, Ciência e Tecnologia Goiano (IFGoiano – *Campus* Rio Verde) for the support on this study.

References

AFONSO JÚNIOR, P. C.; CORRÊA, P. C.; PINTO, F. A. C.; SAMPAIO, C. P. Shrinkage evaluation of five different varieties of coffee berries during the drying process. **Biosystems Engineering**, v. 86, n. 4, p. 481-485, 2003.

AKPINAR, E. K.; BICER, Y.; YILDIZ, C. Thin layer drying of red pepper. **Journal of Food Engineering**, v. 59, n. 1, p. 99-104, 2003.

BALA, B. K.; WOODS, J. L. Simulation of deep bed malt drying. **Journal of Agricultural Engineering Research**, v. 30, n. 3, p. 235-244, 1984.

BERBERT, P. A.; QUEIROZ, D. M.; SILVA, J. S.; PINHEIRO FILHO, J. B. Simulation of coffee drying in a fixed bed with periodic airflow reversal. **Journal of Agricultural Engineering Research**, v. 60, n. 3, p. 167-173, 1995.

BRASIL. Ministério da Agricultura Pecuária e Abastecimento. Secretaria de defesa Agropecuária. **Regras para análise de sementes**. Brasília: Mapa/ACS, 2009.

BROOKER, D. B.; BAKKER-ARKEMA, F. W.; HALL, C. W. **Drying and storage of grains and oilseeds**. Westport: The AVI Publishing Company, 1992.

CORRÊA, P. C.; MACHADO, P. F.; ANDRADE, E. T. Cinética de secagem e qualidade de grãos de milho-pipoca. **Ciência e Agrotecnologia**, v. 25, n. 1, p. 134-142, 2001.

CORRÊA, P. C.; RESENDE, O.; GARIN, S. A.; JAREN, C.; OLIVEIRA, G. H. H. Mathematical models to describe the volumetric shrinkage rate of red beans during drying. **Engenharia Agrícola**, v. 31, n. 4, p. 716-726, 2011.

CORRÊA, P. C.; RESENDE, O.; GONELI, A. L.; BOTELHO, F. M.; NOGUEIRA, B. L. Determinação do coeficiente de difusão liquida dos grãos de feijão. **Revista Brasileira de Produtos Agroindustriais**, v. 8, n. 2, p. 117-126, 2006.

COSTA, L. M.; RESENDE, O.; SOUSA, K. A.; GONÇALVES, D. N. Coeficiente de difusão efetivo e modelagem matemática da secagem de sementes de crambe. **Revista Brasileira de Engenharia Agrícola e Ambiental**, v. 15, n. 10, p. 1089-1096, 2011.

FERREIRA, D. F. SISVAR: um programa para análises estatísticas e ensino de estatística. **Revista Symposium**, v. 6, n. 2, p. 36-41, 2008.

GINER, S. A.; MASCHERONI, R. H. Diffusive drying kinetics in wheat, part 2: applying the simplified analytical solution to experimental data. **Postharvest Technology**, v. 81, n. 1, p. 85-97, 2002.

GOLDFARB, M.; DUARTE, M. E. M.; MATA, M. E. R. M. Armazenamento criogênico de sementes de pinhão manso (*Jatropha curcas* L.) Euphorbiaceae. **Revista Biotemas**, v. 23, n. 1, p. 27-33, 2010.

GONELI, A. L. D.; CORRÊA, P. C.; MAGALHÃES, F. E. A.; BAPTESTINI, F. M. Contração volumétrica e forma dos frutos de mamona durante a secagem. **Acta Scientiarum. Agronomy**, v. 33, n. 1, p. 1-8, 2011.

GÜNHAN, T.; DEMIR, V.; HANCIOGLU, E.; HEPBASLI, A. Mathematical modelling of drying of bay leaves. **Energy Conversion and Management**, v. 46, n. 11-12, p. 1667-1679, 2005.

KROKIDA, M. K.; MAROULIS, Z. B. Effect of drying method on shrinkage and porosity. **Drying Technology**, v. 15, n. 10, p. 2441-2458, 1997.

LAHSASNI, S.; KOUHILA, M.; MAHROUZ, M.; JAOUHARI, J. T. Drying kinetcs of prickly pear fruit (*Opuntia ficus indica*). **Journal of Food Engineering**, v. 61, n. 2, p. 173-179, 2004.

MIDILLI, A.; KUCUK, H. Mathematical modeling of thin layer drying of pistachio by using solar energy. **Energy Conversion and Management**, v. 44, n. 7, p. 1111-1122, 2003.

MIDILLI, A.; KUCUK, H.; YAPAR, Z. A. A new model for single-layer drying. **Drying Technology**, v. 20, n. 7, p. 1503-1513, 2002.

MOHAPATRA, D.; RAO, P. S. A thin layer drying model of parboiled wheat. **Journal of Food Engineering**, v. 66, n. 4, p. 513-518, 2005.

MOHSENIN, N. N. **Physical properties of plant and animal materials**. New York: Gordon and Breach Publishers, 1986.

RAMOS, I. N.; BRANDÃO, T. R. S.; SILVA, C. L. M. Structural changes during air drying of fruits and vegetables. **Food Science and Technology International**, v. 9, n. 3, p. 201-206, 2003.

RAMOS, I. N.; BRANDÃO, T. R. S.; SILVA, C. L. M. Integrated approach on solar drying, pilot convective drying and microstructural changes. **Journal of Food Engineering**, v. 67, n. 1, p. 195-203, 2005.

RAZAVI, S. M. A.; EMADZADEH, B.; RAFE, A.; AMINI, A. M. The physical properties of pistachio nut and kernel as a function of moisture content and variety: part I. Geometrical properties. **Journal of Food Engineering**, v. 81, n. 1, p. 209-217, 2007.

RESENDE, O.; ARCANJO, R. V.; SIQUEIRA, V. C.; RODRIGUES, S. Modelagem matemática para a secagem de clones de café (*Coffea canephora* Pierre) em terreiro de concreto. **Acta Scientiarum. Agronomy**, v. 31, n. 2, p. 189-196, 2009.

RESENDE, O.; CORRÊA, P. C.; GONELI, A. L. D.; CECON, P. R. Forma, tamanho e contração volumétrica do feijão (*Phaseolus vulgaris* L.) Durante a secagem. **Revista Brasileira de Produtos Agroindustriais**, v. 7, n. 1, p. 15-24, 2005.

RESENDE, O.; RODRIGUES, S.; SIQUEIRA, V. C.; ARCANJO, R. V. Cinética da secagem de clones de café (*Coffea canephora* Pierre) em terreiro de chão batido. **Acta Amazonica**, v. 40, n. 2, p. 247-256, 2010.

RESENDE, O.; CORRÊA, P. C.; GONELI, A. L. D.; BOTELHO, F. M.; RODRIGUES, S. Modelagem matemática do processo de secagem de duas variedades de feijão (*Phaseolus vulgaris* L.). **Revista Brasileira de Produtos Agroindustriais**, v. 10, n. 1, p. 17-26, 2008.

SANTOS, S. B.; MARTINS, M. A.; CARVALHO, F. M.; CARNEIRO, A. C. O. Determinação de algumas propriedades físicas dos grãos de pinhão manso (*Jatropha curcas L.*) In: DI LEO, N.; MONTICO, S.; NARDÓN, G. (Ed.). **Avances en ingeniería rural**: 2007 - 2009. Rosario: UNR Editora, 2009. p. 1067-1072.

SIRISOMBOON, P.; KITCHAIYA, P. Physical properties of Jatropha curcas L. kernels after heat treatments. **Biosystems Engineering**, v. 102, n. 2, p. 244-250, 2009.

SOUSA, K. A.; RESENDE, O.; CHAVES, T. H.; COSTA, L. M. Cinética de secagem do nabo forrageiro (*Raphanus sativus* L.). **Revista Ciência Agronômica**, v. 42, n. 4, p. 883-892, 2011.

SOUZA, A. D. V.; FÁVARO, S. P.; ÍTAVO, L. C. V.; ROSCOE, R. Caracterização química de sementes e tortas de

manso submetidas à secagem artificial. **Revista Ciência Agronômica**, v. 41, n. 3, p. 442-447, 2010.

pinhão-manso, nabo-forrageiro e crambe. **Pesquisa Agropecuária Brasileira**, v. 44, n. 10, p. 1328-1335, 2009.

ULLMANN, R.; RESENDE, O.; SALES, J. F.; CHAVES, T. H. Qualidade das sementes de pinhão

11

Comparative analysis of the performance of mixed terraces and level and graded terraces

Josiane Rosa Silva de Oliveira[1], **Fernando Falco Pruski**[1*], **José Márcio Alves da Silva**[1] **and Danilo Paulúcio da Silva**[2]

[1]*Departamento de Engenharia Agrícola, Universidade Federal de Viçosa, Av. P.H. Rolfs, s/n, 36570-000, Viçosa, Minas Gerais, Brazil.*
[2]*Departamento de Estudos Básicos e Instrumentais, Universidade Estadual do Sudoeste da Bahia, Rod. BR-415, Km 3, s/n, 45700-000, Itapetinga, Bahia, Brazil. *Author for correspondence. E-mail: ffpruski@ufv.br*

ABSTRACT. The terracing of agricultural land is the most widespread mechanical practice used among farmers in the control of water erosion. The objective of this paper was to carry out a comparative analysis between mixed terraces and level and graded terraces. The dimensioning of level terraces was carried out based on the surface runoff volume, while the graded terrace dimensions were based on the maximum runoff flow rate. For the mixed terraces, the dimensioning was carried by considering two surface runoff hydrographs, one for the determination of the terrace channel capacity and another to estimate the flow rate at the extremity of the terrace channel. Therefore, in order to contain and transport the excess volume, an additional value was calculated and added to the depth of the channel. A case study was performed for the Uberaba, Minas Gerais State (Brazil) rainfall conditions, considering events with return periods of 10, 30, and 50 years. The obtained results provided quantitative evidence that mixed terraces have a lower height than level terraces and a higher level than the graded terraces, resulting in direct consequences for the soil movement for the terrace construction.

Keywords: terraces, soil and water conservation, control of water erosion.

Introduction

Water erosion is the most important form of soil degradation. Besides leading to a reduction of soil fertility and declines in crop yields in farmlands, it causes strong financial and environmental impacts (BERTOL et al., 2007; ZUAZO et al., 2011).

Soil conservation Beneficial Management Practices (BMPs) play an important role in the control of water erosion (YANG et al., 2009; ZHOU et al., 2009). These practices are divided into three groups: fertilizer and soil amendments, which adjust the cropping system to maintain or improve soil fertility and also maintain its surface with greater coverage; vegetative, related to the use of vegetation to protect the soil against the direct raindrop action; and mechanical, based on the use of artificial structures in order to intercept and conduct surface runoff at a nonerosive velocity to stable outlets (PRUSKI, 2009).

Terracing, an agricultural technique for collecting water and reducing soil erosion (LIU et al., 2011; ZUAZO et al., 2005), is one of the best known and most widely used mechanical practices

(GRIEBELER et al., 2005). Terracing consists of a set of channels adequately spaced and placed so as to divide the long slope into short slopes (FAO, 2000; ZHANG et al., 2008), whose objective is to retain and infiltrate or intercept and transport, with a controlled flow velocity, the surface runoff originating from rains that exceed the infiltration capacity (AL ALI et al., 2008). A typical terrace unit consists of a terrace riser, a compacted central drain and a terrace bed that slopes backwards toward the central drain (VAN DIJK; BRUIJNZEEL, 2004). Depending on their function, terraces are classified as either level or graded.

Level terraces are built so that the channel is leveled, and the extremities are blocked so that the water coming from the surface runoff will be retained and infiltrated in the channel (FAO, 2000). Although this type of terrace is the most beneficial with respect to soil and water conservation, there are some risks associated with it, such as overtopping or, in the worst cases, break through. These are risks that can lead to the destruction of the downstream terraces, causing damage to the cultivated area and to the surrounding rural communities.

An alternative to reduce these risks is the construction of graded terraces, which are terraces constructed with the channel on a slight slope, accumulating the excess water and directing it to waterways that will safely transport water down slopes. However, these terraces present little effect with regard to water conservation and present a higher likelihood for erosion to occur in the waterway. Ehigiator and Anyata (2011), who studied the effects of land-clearing techniques and tillage systems on runoff and soil erosion, observed a high rate of erosion, even when graded-channel terraces were constructed to minimize soil erosion.

With this in mind, the use of mixed terraces, constructed such that the transport of the water to external areas only starts at the moment in which the water reaches a determined level, allows the advantages related to level terracing without taking on any of the risks associated with it. Up to a certain volume of runoff, mixed terraces act as level terraces, ensuring the accumulation and infiltration of water into the soil. After this volume is reached, the terraces begin working as graded terraces, promoting the transport of surplus runoff to outside areas. For events with little surface runoff, the mixed terraces ensure the infiltration of the total volume flow, and only remove the flow coming from extreme events when the retention of the total volume could put the hydraulic structures at risk. However, there are some limitations associated with mixed terracing, such as the difficulty of their design and location. With the aim of spreading the use of this type of terrace, a method based on physical concepts and mathematical techniques was developed for the design of mixed terracing systems (PRUSKI, 2009). Thus, the objective of this paper was to carry out a comparative analysis of the performances of mixed terraces and level and graded terraces.

Material and methods

The dimensioning of mixed terracing systems was based on the design of a surface runoff hydrograph using the principle that the flow increases until the time in which the water coming from the most distant point reaches the considered section, and after that, the flow decreases with time (PRUSKI, 2009). Two hydrograph limbs are identified (Figure 1): rising limb, where there is a growth of the flow rate over time because of the increased area of contribution to the runoff up until the considered section, and falling limb, where the flow rate decreases with time, starting at the time that water originating from the most remote cell reaches the considered section.

The hydrograph shown in the Figure 1 is also essential for the level and graded terracing systems. The necessary surface runoff volume for the level terracing systems design corresponds to the area under the hydrograph, while the maximum surface runoff flow rate, necessary for the graded terracing systems design, is equal to the hydrograph peak.

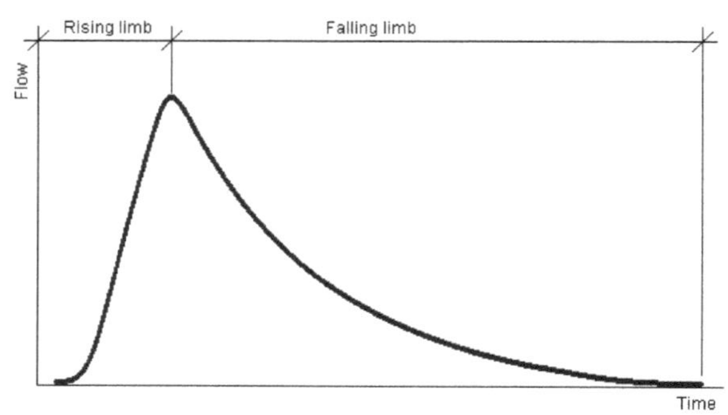

Figure 1. Graphic representation of a runoff hydrograph, showing the rising limb and the falling limb.

To obtain the hydrograph for any position of a slope field, terrace channels or surface drains, the slope field was divided using a matrix system that permits the analysis of the runoff at any position (Figure 2a). This is performed for two conditions: flow over the slope length and flow concentrated in the channel (Figure 2b).

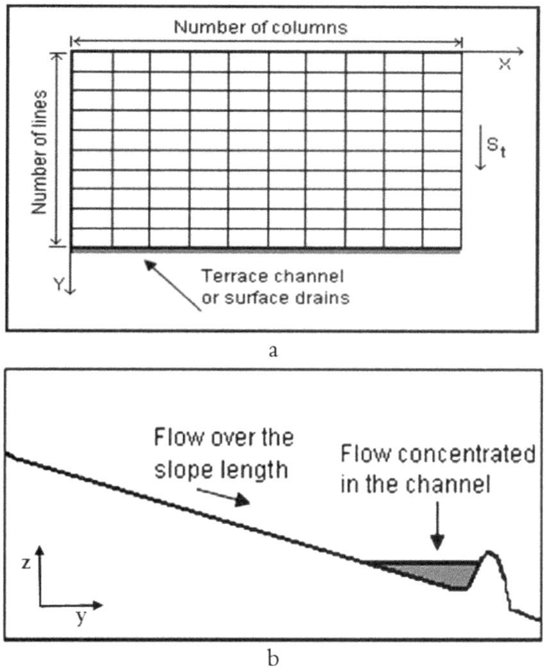

Figure 2. Schematic representation of the division of the slope in a matrix system (a) and flow over the slope length and concentrated flow in the channel (b).

The equations of continuity and momentum for gradually varied unsteady flow are often referred to as the Saint-Venant equations (SINGH, 1996). The kinematic wave model is one of the ways to apply the Saint-Venant equations, and, according to Chua et al. (2008), Naik et al. (2009) and Chua and Wong (2011), the model is given by:

$$\frac{\partial h}{\partial t}+\frac{\partial q}{\partial x}=i_i-\text{Tie} \quad (1)$$

$$S_f=S_0 \quad (2)$$

where:

h is the runoff depth (L), t is time (T), q is the flow per unit of slope width in the hill slope ($L^2\ T^{-1}$), x is the distance (L), i_i is the instantaneous rainfall intensity ($L\ T^{-1}$), Tie is the stable rate of infiltration ($L\ T^{-1}$), S_f is the slope of the energy line ($L\ L^{-1}$) and S_0 is the slope of the soil surface ($L\ L^{-1}$).

The kinetic wave model, considering $S_f = S_0$, was taken over the average cross-sectional area of the runoff on the hill. Using equations typically used for uniform conditions, the relationship between the flow and depth of the runoff was obtained, expressed by the following equation:

$$q=\alpha h^{\beta} \quad (3)$$

The α and β parameters were obtained using Manning's formula and expressed as:

$$\alpha=\frac{\sqrt{S_0}}{n} \text{ and } \beta=\frac{5}{3}, \quad (4)$$

where:

n is the roughness coefficient ($T\ L^{-1/3}$).

The value of the runoff depth was obtained by Equation 1 and solved using the method of finite differences according to the algorithm proposed by Bras (1990). This value was transformed into the flow value by Equation 3.

The instantaneous rainfall intensity is obtained from the rainfall intensity-duration-frequency equation, expressed by (PRUSKI et al., 1997):

$$i_i=i_m\left(1-\frac{c\,t}{t+b}\right) \quad (5)$$

$$i_m=\frac{K\,T^a}{(t+b)^c} \quad (6)$$

where:

i_m is the average maximum rainfall intensity ($L\ T^{-1}$), t is the rainfall duration (T), T is the return period (T), and K, a, b and c are parameters for a given geographic location.

According to Silva el al. (2009), Rai et al. (2010) and Chua and Wong (2010), for open channel flow, the kinematic wave model is given by:

$$\frac{\partial A}{\partial t}+\frac{\partial Q}{\partial x}=q \quad (7)$$

$$S_f=S_c \quad (8)$$

where:

A is the cross-sectional area of the terrace (L^2), Q is the total flow in the channel ($L^3\ T^{-1}$) and S_c is the slope of the canal bottom ($L\ L^{-1}$).

For the dimensioning of the mixed terracing system, it was assumed that until the full capacity of the channel was reached, all the water volume would be retained. After the capacity is surpassed, the flow in its extremity will then begin.

The channel capacity is calculated using an event with a given return period ($T_{storage}$) as a reference. In considering events with return periods greater than $T_{storage}$, two hydrographs are designed, one for $T_{storage}$ and another for the return period determined by the designer (T_{design}). The two hydrographs are superimposed (Figure 3), and the difference between the water volumes corresponds to the excess amount ($Q_{-Tstorage}$) that must be conducted to the extremities of the channel (PRUSKI, 2009). The vertical line

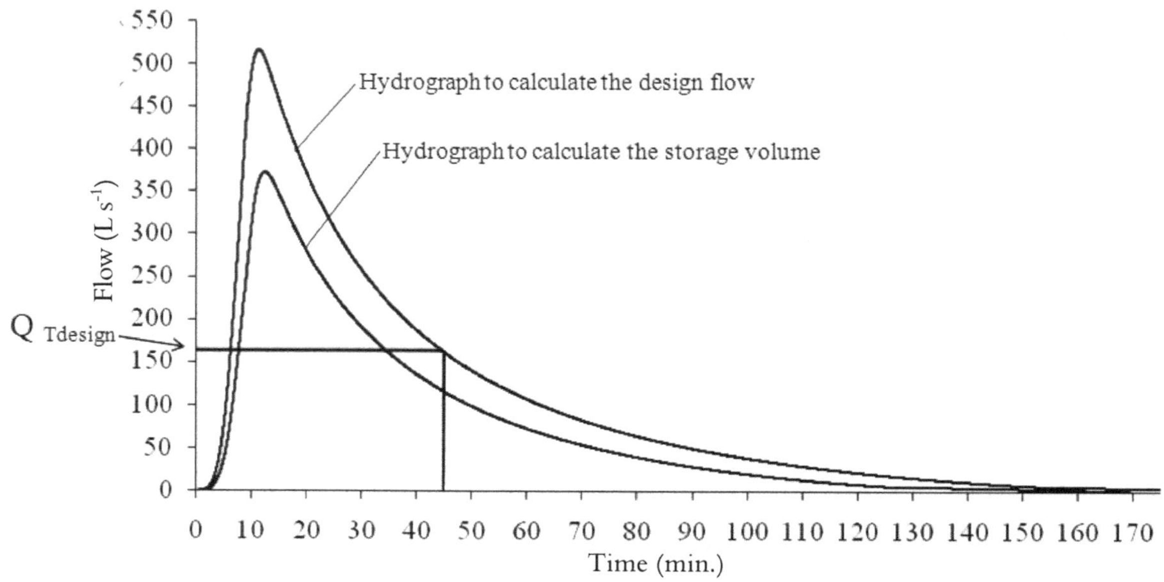

Figure 3. Graphic representation of runoff hydrographs for a return period equal to or greater than 10 years.

characterizes the time for which the flow volume, considering the hydrograph obtained with T_{design}, is equal to the total flow volume for the return period of $T_{storage}$ years.

To evacuate water, the mixed terraces have no gradient. They are built level with the lateral ends closed by a wall of height equal to the height (H) (Figure 4), determined according to the channel capacity to retain the amount corresponding to $T_{storage}$. When the runoff volume exceeds the channel storage capacity associated with the height H, the excess water volume should be drained to the outlets at one or both ends.

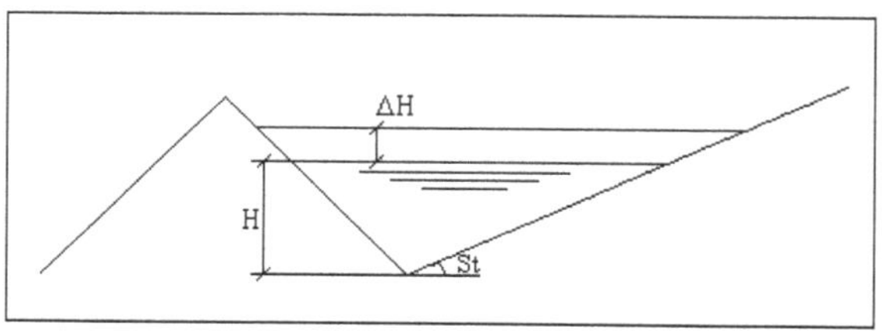

Figure 4. Representation of the cross-sectional area of a triangular terrace.

To contain and transport the excess water volume to the outlets, an increase in the depth of the channel (ΔH) is calculated using the Manning equation, which, for this condition, is solved using the Newton-Raphson method and expressed by:

$$\frac{Q_{max}\, n}{\sqrt{S_c}} = \frac{\Delta A^{5/3}}{\Delta P^{2/3}}, \quad (9)$$

where:

Q_{max} is the maximum flow rate to be transported by the channel ($L^3\ T^{-1}$), ΔA is the increment of the area necessary to transport Q (L^2), and ΔP is the increment of the wetted perimeter when transporting Q (L).

The hydraulic gradient is calculated using the following equation:

$$S_c = \frac{\Delta H}{L}, \text{ to one outlet} \quad (10)$$

$$S_c = \frac{\Delta H}{L/2}, \text{ to two outlets} \quad (11)$$

where:

L is the length of the channel (L).

The design of the waterway can be performed using Manning's formula and the maximum flow rates to be transported at the terrace channel outlets, considering a segment obtained by the division of the waterway into sections according to each terrace outlet.

To compare the mixed terracing systems with the level and graded terracing systems, a case study was performed for the Uberaba (Minas Gerais State, Brazil) rainfall conditions considering $T_{storage}$ equal to 10 years and T_{design} equal to 30 and 50 years, a noncropped area with a 50 m slope length, 7% slope, no tillage and a roughness coefficient equal to 0.120. A channel with a triangular shape was assumed; this format is the most typical, with grass and some weeds added as a protective cover, with a roughness coefficient equal to 0.03, a gradient of the front slope of 66.7% and a 500 m length with two outlets.

The case study was performed considering a stable rate of infiltration in the soil equal to 15 and 30 mm h^{-1} and a slope of the canal bottom of 0.002 and 0.003 m m^{-1}.

The design of the surface runoff hydrographs was obtained using the software Hidrograma 2.1 (http://www.ufv.br/dea/gprh/softwares.htm), considering events with return periods of 10, 30 and 50 years.

Results and discussion

The surface runoff hydrographs obtained for return periods equal to 10, 30 and 50 years for one of the corresponding conditions of the case study (Tie = 15 mm h^{-1} and S_c= 0.002 m m^{-1}) are presented in Figure 5. The vertical lines represent the time for which the water volume, considering the return periods of 30 and 50 years, is equal to the total water volume corresponding to the 10-year return period.

For level terraces, an increase in the stored volume (up to 48%) and terrace depth (21.6%) occurred when the return period was increased from 10 to 50 years. For graded terraces, a reduction in water conservation efficiency is perceived because, whereas level terraces present a storage capacity for all the surface runoff, graded terraces are the ones that transport the runoff. In fact, the implementation, management and mechanization costs for graded terraces seem as though they would be minimized because of the shorter height in relation to level systems. However, with these systems, it is also necessary to consider the cost of the waterway for the runoff to be transported by the terraces.

Considering that a mixed system, as its name already implies, provides an intermediary solution between the two systems, Table 1 shows the effect of employing this system on storage capacity, terrace height, and on the volume and flow to be drained. With respect to the efficiency of water storage, which is zero for graded terraces, mixed terracing represents 84.2, 76.4 and 67.6% of that corresponding to level terraces for return periods of 20, 30, and 50 years, respectively. With respect to terrace depth, it was shown that with the mixed system, intermediate values between those observed for level and graded systems were obtained, as was expected, representing an alternative for the reduction of implementation and maintenance costs of the terraces.

For extreme events with return periods greater than 10 years and with greater magnitudes of surface runoff, the excess volume to be transported by the channel to the outlet was significantly less for mixed terracing, corresponding to 15.8, 23.6 and 32.4% of the maximum flows that would be transported in the waterways of the graded terraces for return periods of 20, 30, and 50 years, respectively.

With increasing Tie, a smaller amount of water will be transformed into runoff (volume or flow), reducing the required capacity for the terrace channel (level, graded and mixed).

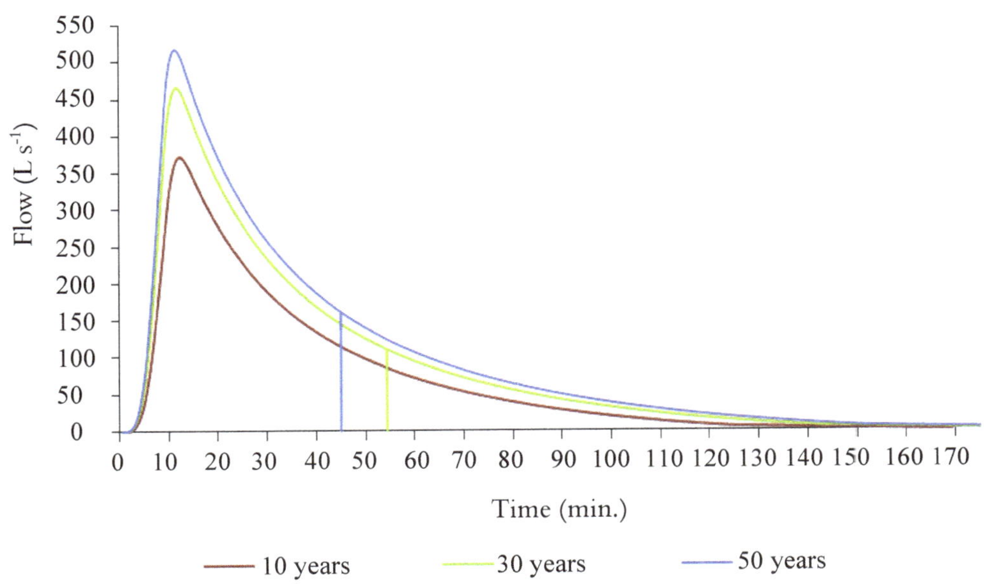

Figure 5. Surface runoff hydrographs for return periods equal to 10, 30, and 50 years for one condition (Tie = 15 mm h^{-1} and S_c = 0.002 mm^{-1}) of the conducted case study. The vertical lines represent the time for which the water volume, considering the return period of 30 and 50 years, is equal to the total water volume corresponding to the 10-year return period.

Comparing the results obtained for Tie values equal to 15 and 30 mm h^{-1}, it is evident that for level terraces, the channel height associated with 30 mm h^{-1} is approximately 67% of that corresponding to 15 mm h^{-1}, while for graded and mixed terraces, this ratio is 82 and 80%.

Regarding the slopes of the channel bottom, they only interfere with the behavior of graded terraces, and for these, the increase in slope from 0.002 to 0.003 mm h^{-1} is approximately 93%.

Similar behavior was observed for the other conditions of this case study—the values of the volume to be retained in the channel (V_s), the channel depth (H), the maximum flow (Q_{max}) and the volume to be transported in the channel (V_d) for each return period (T) and for each terracing system are presented in Table 1.

Table 1. Results of the hydrograph analysis.

S_c ($m\ m^{-1}$)	Tie ($mm\ h^{-1}$)	Terrace system	T (years)	V_s (m^3)	H (m)	Q_{max} ($m^3\ s^{-1}$)	V_d (m^3)
0.002	15	Level Terraces	10	730.8	0.75	-	-
			30	954.2	0.85	-	-
			50	1076.8	0.90	-	-
		Graded Terraces	10	-	0.40	0.372	730.8
			30	-	0.44	0.466	954.2
			50	-	0.45	0.516	1076.8
		Mixed Terraces	10	730.8	0.75	-	-
			30	730.8	0.78	0.111	223.4
			50	730.8	0.78	0.162	346.0
	30	Level Terraces	10	471.1	0.60	-	-
			30	649.4	0.70	-	-
			50	748.6	0.75	-	-
		Graded Terraces	10	-	0.38	0.319	471.1
			30	-	0.42	0.412	649.4
			50	-	0.44	0.463	748.6
		Mixed Terraces	10	471.1	0.60	-	-
			30	471.1	0.65	0.151	178.3
			50	471.1	0.66	0.213	277.5
0.003	15	Level Terraces	10	730.4	0.75	-	-
			30	953.8	0.85	-	-
			50	1076.4	0.90	-	-
		Graded Terraces	10	-	0.37	0.379	730.4
			30	-	0.41	0.474	953.8
			50	-	0.42	0.525	1076.4
		Mixed Terraces	10	730.4	0.75	-	-
			30	730.4	0.78	0.112	223.4
			50	730.4	0.78	0.163	346.0
	30	Level Terraces	10	471.4	0.60	-	-
			30	649.7	0.70	-	-
			50	748.8	0.75	-	-
		Graded Terraces	10	-	0.35	0.325	471.4
			30	-	0.39	0.420	649.7
			50	-	0.41	0.471	748.8
		Mixed Terraces	10	471.4	0.60	-	-
			30	471.4	0.65	0.152	178.3
			50	471.4	0.66	0.214	277.4

Conclusion

The comparison of the mixed terrace performance with level and graded terraces yielded quantitative evidence that led to the following observations:

- Mixed terraces exhibit a lower efficiency in surface water storage ability than level terraces and greater efficiency than gradient terraces, which has direct consequences on water loss; and

- Mixed terraces exhibit a lower height than level terraces and a higher height than gradient terraces, which has direct consequences on the volume of soil movement.

References

AL ALI, Y.; TOUMA, J.; ZANTE, P.; NASRI, S.; ALBERGEL, J. Water and sediment balances of a contour bench terracing system in a semi-arid cultivated zone (El Gouazine, central Tunisia). **Hydrological Sciences**, v. 53, n. 4, p. 883-892, 2008.

BERTOL, I.; COGO, N. P.; SCHICK, J.; GUDAGNIN, J. C.; AMARAL, A. J. Aspectos financeiros relacionados às perdas de nutrientes por erosão hídrica em diferentes sistemas de manejo do solo. **Revista Brasileira de Ciência do Solo**, v. 31, n. 1, p. 133-142, 2007.

BRAS, R. L. **Hydrology**: an introduction to hydrologic science. Massachusets: Addison - Wesley Publlishing Company, 1990.

CHUA, L. H. C.; WONG, T. S. W.; SRIRAMULA, L. K. Comparison between kinematic wave and artificial neural network models in event-based runoff simulation for an overland plane. **Journal of Hydrology**, v. 357, n. 3-4, p. 337-348, 2008.

CHUA, L. H. C.; WONG, T. S. W. Improving event-based rainfall–runoff modeling using a combined artificial neural network–kinematic wave approach. **Journal of Hydrology**, v. 390, n. 1-2, p. 92-107, 2010.

CHUA, L. H. C.; WONG, T. S. W. Runoff forecasting for an asphalt plane by Artificial Neural Networks and comparisons with kinematic wave and autoregressive moving average models. **Journal of Hydrology**, v. 397, n. 3-4, p. 191-201, 2011.

EHIGIATOR, O. A.; ANYATA, B. U. Effects of land clearing techniques and tillage systems on runoff and soil erosion in a tropical rain forest in Nigeria. **Journal of Environmental Management**, v. 92, n. 11, p. 2875-2880, 2011.

FAO-Food and Agriculture Organization. **Manual on integrated soil management and conservation practices**. Rome: FAO, 2000.

GRIEBELER, N. P.; PRUSKI, F. F.; TEIXEIRA, A. F.; SILVA, D. D. Modelo para o dimensionamento e a locação de sistemas de terraceamento em nível. **Engenharia Agrícola**, v. 25, n. 3, p. 696-704, 2005.

LIU, X.; HE, B.; LI, Z.; ZHANG, J.; WANG, L.; WANG, Z. Influence of land terracing on agricultural and ecological environment in the loess plateau regions of China. **Environmental Earth Sciences**, v. 62, n. 4, p. 797-807, 2011.

NAIK, M. G.; RAO, E. P.; ELDHO, T. I. A kinematic wave based watershed model for soil erosion and sediment yield. **Catena**, v. 77, n. 3, p. 256-265, 2009.

PRUSKI, F. F.; FERREIRA, P. A.; RAMOS, M. M.; CECON, P. R. Model to design level terraces. **Journal of Irrigation and Drainage Engineering**, v. 123, n. 1, p. 8-12, 1997.

PRUSKI, F. F. **Conservação de solo e água**: práticas mecânicas para o controle da erosão hídrica. 2. ed. Viçosa: UFV, 2009.

RAI, R. K.; UPADHYAY, A.; SINGH, V. P. Effect of variable roughness on runoff. **Journal of Hydrology**, v. 382, n. 1-2, p. 115-127, 2010.

SILVA, J. M. A.; PRUSKI, F. F.; RODRIGUES, L. N.; CECÍLIO, R. A. Modelo para a obtenção do hidrograma de escoamento superficial em bacias hidrográficas. 1. Desenvolvimento e análise de sensibilidade. **Revista Brasileira de Ciências Agrárias**, v. 4, n. 1, p. 78-84, 2009.

SINGH, V. P. **Kinematic wave modeling in water resources**: surface water hydrology. New York: Wiley, 1996.

VAN DIJK, A. I. J. M.; BRUIJNZEEL, L. A. Runoff and soil loss from bench terraces. 1. An event-based model of rainfall infiltration and surface runoff. **European Journal of Soil Science**, v. 55, n. 2, p. 299-316, 2004.

YANG, Q.; MENG, F.-R.; ZHAO, Z.; CHOW, T. L.; BENOY, G.; REES, H. W.; BOURQUE, C. P. A. Assessing the impacts of flow diversion terraces on stream water and sediment yields at a watershed level using SWAT model. **Agriculture, Ecosystems and Environment**, v. 132, n. 1-2, p. 23-31, 2009.

ZHANG, J. H.; SU, Z. A.; LIU, G. C. Effects of terracing and agroforestry on soil and water loss in hilly areas of the Sichuan basin, China. **Journal of Mountain Science**, v. 5, n. 3, p. 241-248, 2008.

ZHOU, X.; AL-KAISI, M.; HELMERS, M. J. Cost effectiveness of conservation practices in controlling water erosion in Iowa. **Soil and Tillage Research**, v. 106, n. 1, p. 71-78, 2009.

ZUAZO, V. H. D.; RUIZ, J. A.; RAYA, A. M.; TARIFA, D. F. Impact of erosion in the taluses of subtropical orchard terraces. **Agriculture, Ecosystems and Environment**, v. 107, n. 2-3, p. 199-210, 2005.

ZUAZO, V. H. D.; PLEGUEZUELO, C. R. R.; PEINADO, F. J. M.; GRAAFF, J.; MARTÍNEZ, J. R. F.; FLANAGAN, D. C. Environmental impact of introducing plant covers in the taluses of terraces: Implications for mitigating agricultural soil erosion and runoff. **Catena**, v. 84, n. 1-2, p. 79-88, 2011.

12

Construction and calibration of a bar weighing lysimeter

Daiane de Cinque Mariano[1], Rogério Teixeira de Faria[2], Paulo Sérgio Lourenço de Freitas[3*], Bruno Patias Lena[3] and Andre Luiz Johann[4]

[1]*Departamento de Agronomia, Universidade do Estado de Mato Grosso, Tangará da Serra, Mato Grosso, Brazil.* [2]*Departamento de Engenharia Rural, Faculdade de Ciências Agrárias e Veterinárias, Universidade Estadual Paulista, Jaboticabal, São Paulo, Brazil.* [3]*Departamento de Agronomia, Universidade Estadual de Maringá, Av. Colombo, 5790, 87020-900, Maringá, Paraná, Brazil.* [4]*Departamento de Engenharia de Biossistemas,* Escola Superior de Agricultura *"Luiz de Queiroz", Universidade Estadual Paulista, Piracicaba, São Paulo, Brazil. *Author for correspondence. E-mail: pslfreitas@uem.br*

ABSTRACT. Generally, standard lysimeters include a lever-load cell scale, which has become expensive compared to new electronic equipment available for experimental purposes. Currently, lysimeters containing weighing bars with up to 4 load cells are economically feasible. This goal of this study was to present the construction and calibration of four weighing lysimeters built using two rectangular tanks, with a surface area of 1.35 m^2 (1 m by 1.35 m) and a depth of 1.5 m, loaded on two weighing bars. Each bar had two high accuracy loading cells connected to a junction box and a datalogger by a coaxial cable. Calibration of each lysimeter was performed after field installation. The results showed a linear response ($R^2 > 0.9999$) of electric signal (mV) to mass increase or decrease (kg) with minimum hysteresis. The mean absolute error of the mass estimate (0.0245 to 0.2844 equivalent-mm) and a Willmott index between the observed and estimated data (d = 1.0000) indicate the high precision of the new equipment. The results allow concluding that the lysimeters in this study are appropriate for estimating evapotranspiration and other components of soil water balance.

Keywords: evapotranspiration, water balance, soil water.

Introduction

The complexity and difficulty to determine ET (evapotranspiration) in the field have been surpassed by technological advances in the electronic data collection systems that directly or indirectly monitor the different processes involved. Direct measures provide greater reliability and are typically performed by lysimeters; in addition, soil water balance methods are occasionally used (SILVA et al., 2009).

The use of weighing lysimeters in studies on crop water requirements, which can be used to determine ET or calibrate agrometeorological models, is essential for obtaining reliable results (CAMPECHE et al., 2011; SILVA et al., 1999). According to Silva et al. (2005), when properly installed, lysimeters allow accurate measurements of ET, especially if filled with soil similar to the soil layers of the external area.

The weighing lysimeter is fabricated by a box with impermeable walls that is filled with soil and placed in a field with either a bare surface or vegetation cover. A load cell measures the weight variation (VIANA et al., 2003).

The lysimetry technique has been used to accurately measure rainfall, evaporation and drainage (KIRKHAM et al., 1984; MENDONÇA et al., 2003) through the mass variation of a soil block due to the inflow or outflow of water using a load cell that monitors the water dynamics in the cultivated or bare soil (CAMPECHE et al., 2011).

Although precise, evapotranspiration (ET) measurements obtained using the lysimetry technique can also be affected by environmental and design factors such as advection, lysimeter size, soil moisture regime, wall thickness and distance, wall edge height, and differences in the height and density of vegetation inside and outside the lysimeter (CAMPECHE et al., 2011; MIRANDA et al., 1999).

The calibration of lysimeters is important because poor calibrations lead to inconsistent interpretations of ET values, especially when taken over short periods of time (CAMPECHE et al., 2011). According to Faria et al. (2006), the calibration aims to establish a relationship between the electrical output signal of the load cell and mass of the system and to check the linearity and hysteresis of the weighing system.

Campeche et al. (2011), Carvalho et al. (2007) and Faria et al. (2006) have reported high precision lysimeter measurements for standard weights added and removed from the system during calibration. In these studies, the deviations varied between 0.08 and 0.278 for 0.07 mm and 0.42 mm, respectively, with a linear response between the standard weights and readings of load cells; furthermore, the coefficients of determination were above 0.99.

In conventional lysimeters, the weighing system frequently includes a lever-load cell scale to expand the scale used (CAMPECHE et al., 2011; FARIA et al., 2006). In recent years, the cost of this type of scale has increased due to the rise in raw material and labor costs required for construction. Nevertheless, the electronics industry has offered increasingly inexpensive load cells, allowing the construction of new weighing systems to be less expensive than conventional lysimeters by minimizing the metal structure with the use of weighing bars.

The goal of this study is to present an analysis of weighing lysimeters, including their construction and calibration, to determine their ability to measure variations of soil water storage under field conditions.

Material and methods

Construction of lysimeters

Four weighing lysimeters were constructed and installed at excavated pits with rectangular walls in the Experimental Farm of IAPAR, Londrina, Paraná State. Each lysimeter consisted of double rectangular tanks with internal dimensions of 1.35 m^2 (1 m by 1.35 m) for the surface area and 1.5 m for the depth. Fiberglass and resin walls were strengthened with 2-mm steel beams placed at the edges and in the center (Figure 1). The external tank was constructed of walls fabricated of the same material and with surface dimensions of 1.1 m by 1.45 m and a depth of 1.6 m, and it was separated from the internal tank by 5 cm. The internal tank was supported on the smaller side by two 5-mm carbon steel U-shaped weighing bars that measured 0.15 m wide and 0.9 m long. Two load cells (Z6FC3/1T, HBM measurement, São Paulo, São Paulo State) were coupled on one side by steel screws 0.1 mm from the ends of each weighing bar and steel support brackets (Z6FC4 Z6/1T/ZEL, HBM measurement, São Paulo, São Paulo State).

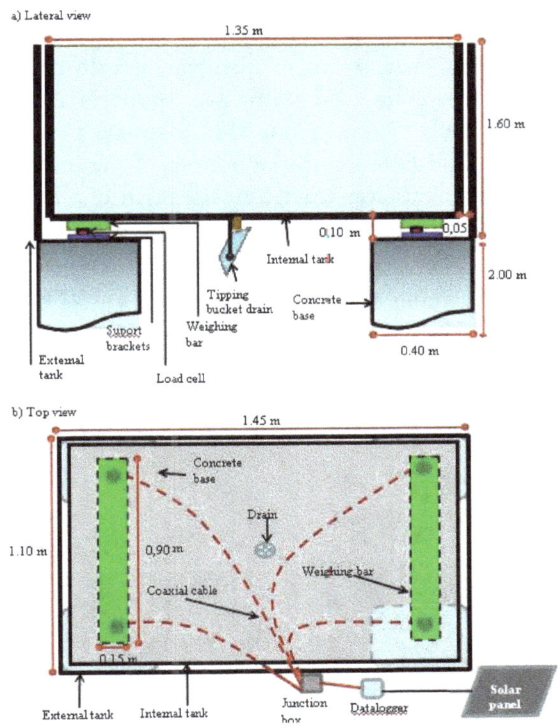

Figure 1. Lysimeter side (a) and top view (b).

At the bottom of the lysimeter pits, four vertical concrete bases of 0.4 x 0.4 x 2 m depth were installed to support the weighing system (Figure 1). The concrete bases supported the two weighing bars, which were leveled using 1 mm-steel sheet shims indicated by load measurements in each cell separately. The external tank was installed and the internal tank was filled as follows: a 5 cm-gravel layer, a 5 cm-sand layer, a geotextile mat (Bidim), Mexichem Brasil, São José dos Campos, São Paulo State, Brazil), and soil was placed at the bottom, which followed a series of five 0.30 m layers previously separated during the pit excavation, while maintaining a slight compression to reproduce the original soil profile.

Each lysimeter was provided with a drainage device consisting of a PVC pipe, 5 cm in length and 12 mm in diameter, to transport the water accumulated in the gravel layer on the bottom of the internal tank to a reservoir that was emptied when its maximum storage capacity of 1.23 mm was reached (Figure 2). This reservoir, called a tipping bucket drain, consisted of a PVC pipe of approximately 0.3 m length and 100 mm diameter with a PVC cap closed bottom and top opening beveled at 45°. The tipping bucket drain was supported by a shaft pierced into holes drilled parallel to the beveled side of the tipping bucket drain, which was placed approximately 1/3 of the distance above the base (Figure 2a). This shaft, built from a PVC rod, 1 cm in diameter and 18 cm in length, was connected with two supports placed approximately 7 cm from the base of a metal structure and fastened to the bottom of the internal tank, which was constructed in the form of a frame with rods measuring 20 cm long, 4 cm wide, and 1 cm thick. To ensure the reservoir was completely filled before tipping (Figure 2b and c), metal rings were glued on the bottom of the reservoir to counterbalance the moment of inertia.

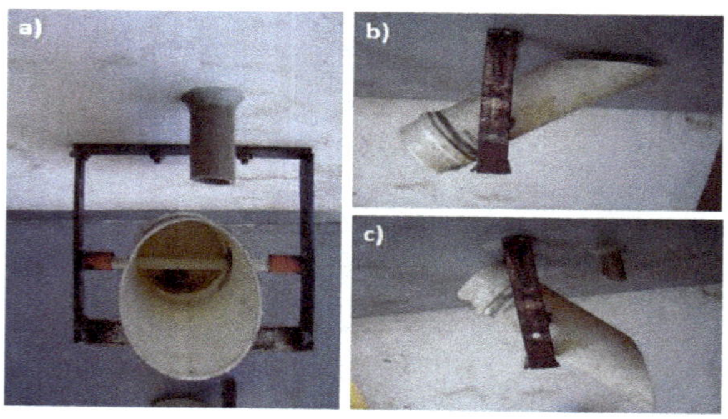

Figure 2. Tipping bucket drain: a) front view, b) side view before drainage and c) after drainage.

The four load cells of the pair of weighing bars at each lysimeter were connected by a coaxial cable with a junction box (4144, Alfa Instrumentos Eletrônicos LTDA, São Paulo, São Paulo State) installed on the soil surface to unify electrical signals resulting from the weighing system. A single coaxial cable was used to conduct electrical signals to the data acquisition system (Datalogger CR10X Campbell Scientific, Logan, USA), which was powered by a 12 V external battery supplied by a solar panel responsible for the power supply in case of outages. Data stored in the datalogger were transferred to a memory module and computer through the program PC208W (Campbell Scientific, Logan, USA).

Each load cell had a capacity of 1 t tensile, a protection degree of IP68 and a connection to the junction boxes by four wires; additionally, six wires were connected, offsetting the temperature effect in the cable between the junction box and the datalogger. According to the manufacturer (HBM measurement, São Paulo, São Paulo State), the sensitivity of the load cells is 2 +/- 0.05% mV. V^{-1} and the accuracy is 0.0180% of the full scale, which is warranted in the temperature range of -10 to +40°C.

The capacity of the reservoirs of the constructed tipping bucket drain was approximately 1.23 mm, with slight variation between devices (± 0.2 mm) due to the constructional features and counterweights used. Those differences were considered by computing the value for each lysimeter instead of the average.

Calibration of the lysimeters

First, the lysimeters' surfaces were covered with plastic to prevent evaporation. Wind effects, which cause oscillations in the lysimeters mass measurements, were minimized by placing windbreakers, made of plastic sheets approximately 1 m high, around each lysimeter. During the calibration period, the wind speed recorded by the IAPAR weather station located adjacent to the experiment varied between 70 and 250 km day, which was similar to or smaller than the usual mean values for the region (210 km day) (IAPAR, 2012).

Based on previous studies (CAMPECHE et al., 2011; FARIA et al., 2006), each lysimeter was calibrated separately. Mass readings were performed every three seconds in response to sequences of voltage measurements by adding and removing mass (sand) contained in sealed plastic bags previously determined by scales to be accurate to 0.01 g. After measuring the mass of the unloaded and covered lysimeter, two sandbags with a mass of 4.72 kg, corresponding to 7 mm, were placed in sequence on the 1.55 m^2 surface area of the lysimeters to reach 70 mm. Afterwards, sandbags with masses of 1.35 kg, corresponding to 1 mm, were added individually until the mass equivalent of 80 mm was reached. Finally, the complete load of 150 mm was achieved by adding ten loads of 7 mm. The unloading sequence from 150 mm was performed with the removal of sand bags from 7 mm to 80 mm and 1 mm to 70 mm; total discharge occurred with the removal of masses equivalent to 7 mm. The loading and unloading operations were repeated four times. Thus, 60 pairs of data per measurement sequence were acquired, with 240 pairs of data over four repetitions, which allowed for the calibration of each lysimeter.

The results were subjected to linear regression analyses that describe the precision of the systems and index of agreement (d), which determines the accuracy (WILLMOTT et al., 1985). The effectiveness of the lysimeter was confirmed by the behavior of the ET measured by the lysimeters and the ET estimated while conducting the cultures.

After calibration, the ability of the lysimeters to measure evapotranspiration was tested on a wheat crop cultivated on the lysimeters, from April to August.

Regression analysis was performed by the software SAEG (UFV, 1997) using data from four iterations. The statistical significance was evaluated by the F test.

Results and discussion

The values of the slope (a) of the regression equation between the weight and voltage for the four lysimeters ranged from 1271 to 1322 mm mV^{-1} (Table 1), which can be caused by the specific characteristics of the electrical strain gauges of each load cell (FARIA et al., 2006). In the current study, slope values were very close, indicating similar characteristics in weighing systems of the lysimeters. The values of the intercept (b) obtained from the regression varied between -2180 and -2015 (Table 1), resulting from the static weight (dead weight) of each set.

The calibration of the lysimeters presented a linear response, satisfactorily describing all cases and with coefficients of determination close to unity. The relationship between the weight and voltage, proven by the absolute error values of the linear regression estimation, ranged from 0.13 to 0.28 mm (Table 1). These results are within the range (0.08 - 0.42 mm) found by Faria et al. (2006), Carvalho et al. (2007) and Campeche et al. (2011). It is worth mentioning that resolutions of the lysimeters can vary depending on their design and construction and differ based on material, size and spatial arrangement for the assessment.

Table 1. Coefficients of determination (r^2) for linear regressions (y = ax + b) between weight (y, equivalent-mm), voltage (x in mV) and mean absolute error of weight estimation (equivalent-mm), which are obtained for the four lysimeters tested and are significant at $a \leq 0.01$ using the F test

Lysimeter	a	B	R^2	E
1	1322.40	-2179.70	1.00	0.28
2	1303.90	-2117.50	1.00	0.13
3	1270.80	-2021.50	1.00	0.15
4	1272.00	-2015.40	1.00	0.13

The graphic representation of the relationships between the weight variation and voltage of the four lysimeters analyzed (Figure 3) showed the absence of hysteresis, indicating a close relationship in the loading and unloading processes and voltage of the load cell; thus, results obtained from Miranda et al. (1999), Oliveira et al. (2008) and Fernandes et al. (2012) were confirmed. During the lysimeter calibration, Santos et al. (2008) also observed linear responses with negligible hysteresis of ET in the semi-arid Northeastern region.

The low values of the absolute error (Ae), mean square error (MSQ) and mean absolute error (MAE) resulting from the linear regression were satisfactory, which strengthened the accuracy of the evaluated equipment (Table 2). A Wilmott's index of agreement (d) value equal to unity indicated a good aggregation between the estimated and observed values, which resulted in perfect agreement. Santos et al. (2007) studied a diaphragm lysimeter and confirmed its capacity to determine the ETo of citrus seedlings with great accuracy; the coefficient of determination was 0.9975 and the index of agreement (d) was 0.99926.

Table 2. Absolute error (Ae), mean square error (MSQ), mean absolute error (MAE) and Wilmott index (d) resulting from the linear regression between the observed and estimated values.

Lysimeter	Ea	EMQ	EMA	D
1	0.0117	0.1289	0.2844	1.0000
2	0.0334	0.0268	0.0245	1.0000
3	0.0985	0.0424	0.0329	1.0000
4	0.0453	0.0232	0.0259	1.0000

Figure 4 illustrates the simple regression between the weight estimated by the model and weight observed in the experiment and the calculation of the regression parameters (intercept *a*, slope *b*, and correlation *r*). A high linearity of data was verified, and a close relationship occurred between the observed and measured values estimated in the four lysimeters, which gave the lysimeters the ability to adequately detect weight differences with high sensitivity in determining the water balance components in the soil-plant-atmosphere system.

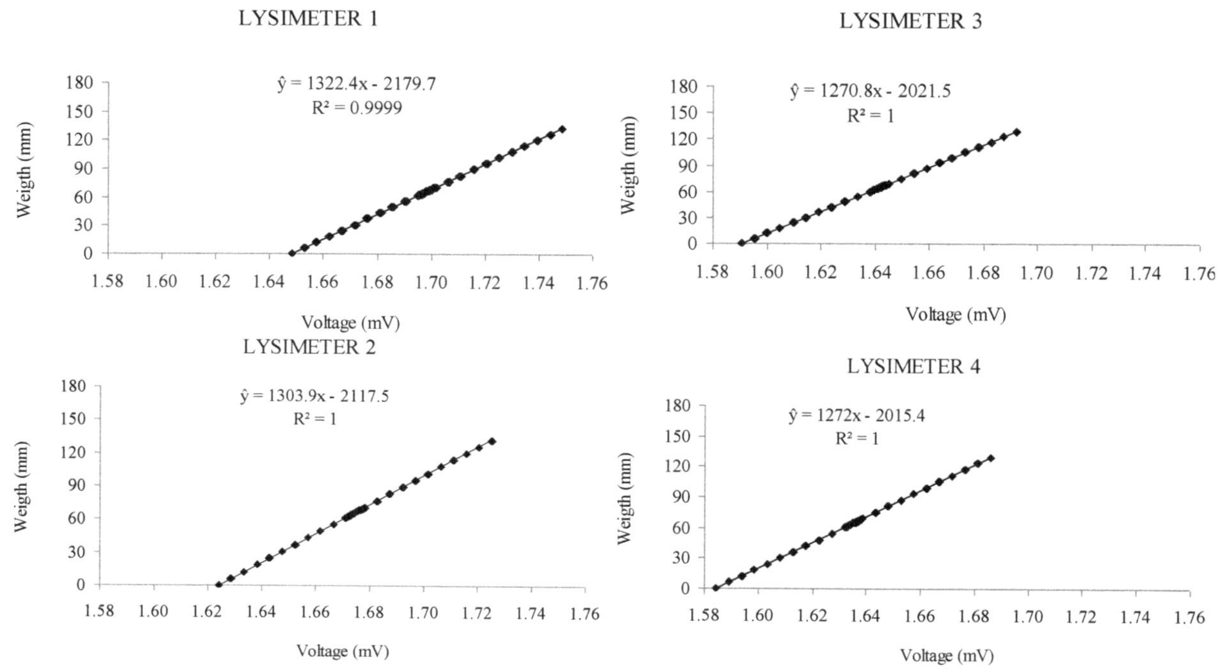

Figure 3. Weight and voltage relationship of the four lysimeters.

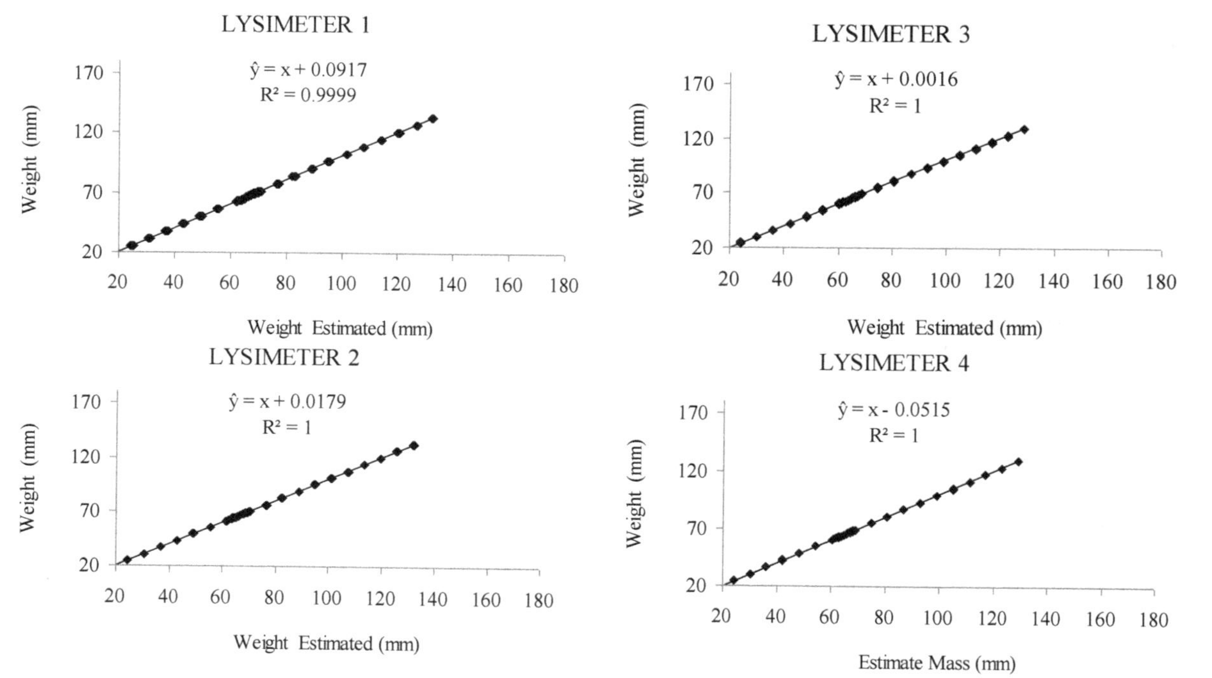

Figure 4. Estimated and observed weighing values as measured in the load cell during loading and unloading cycles and the respective slopes and coefficients of determination for the four lysimeters.

The hydrological component measurements obtained to prove the lysimeters' ability to monitor the water balance components under cultivation were taken over seven days, from May 6 to 12, 2011, when the wheat crop was in booting stage (Figure 5). Thus, the 30-mm irrigation performed on day 6 for approximately 17h resulted in increased water depth stored as approximately 29 mm. From the 8:00h of day 7, and over subsequent days, the stored quantity decreased due to the losses caused by evapotranspiration, demonstrating agreement between the weight variation in the periods of loading and unloading of soil water. Additionally, the equipment's sensitivity to oscillations in soil water storage can be demonstrated by the small peaks observed on days 9 and 12, close to 16h, which were caused by the weight increase/decrease due to the use of a neutron probe to monitor the soil water content (Figure 5). Pereira et al. (2002) observed adequate results of measured and estimated ETo as well as responses to variations of the main meteorological elements that influence the ETo.

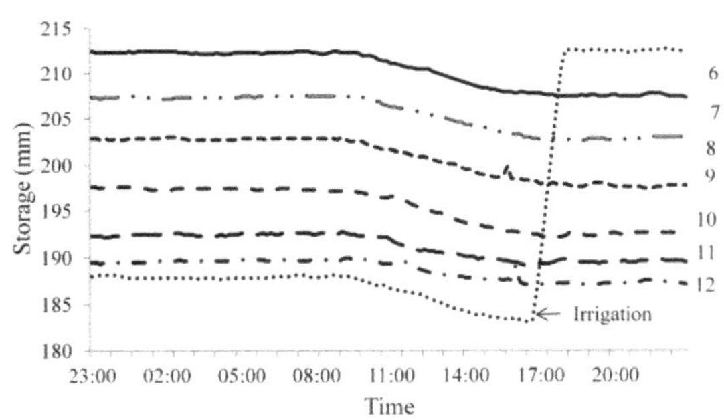

Figure 5. Soil water storage measured by a lysimeter cultivated with wheat from May 6 to 12, 2011.

To show the lysimeters ability of satisfactorily monitor the water balance components, Figure 6 presents hourly ETo and ETc of the wheat crop. ETo was calculated by the Penman-Monteith – FAO method using CLIMA software (FARIA et al., 2003) and ETc was measured in one of the built lysimeters during three days. In that period, there was neither rainfall nor drainage of the lysimeter; hence, the weight variation was caused only by losses of ETc, measured by the weight variation in the lysimeter. The values of ETc followed closely ETo through the three day period, with daily totals of 3.4 and 3.0 mm day, 3.5 and 3.2 mm day, and 3.1 and 3.0 mm day, respectively. Silva et al. (2003).

Likewise, Flumignan and Faria (2009) attained variable values of ET and Kc as a function of the irrigation method (sprinkle, drip irrigation, and non-irrigated), rainfall frequency, atmospheric demand and evolution of leaf area.

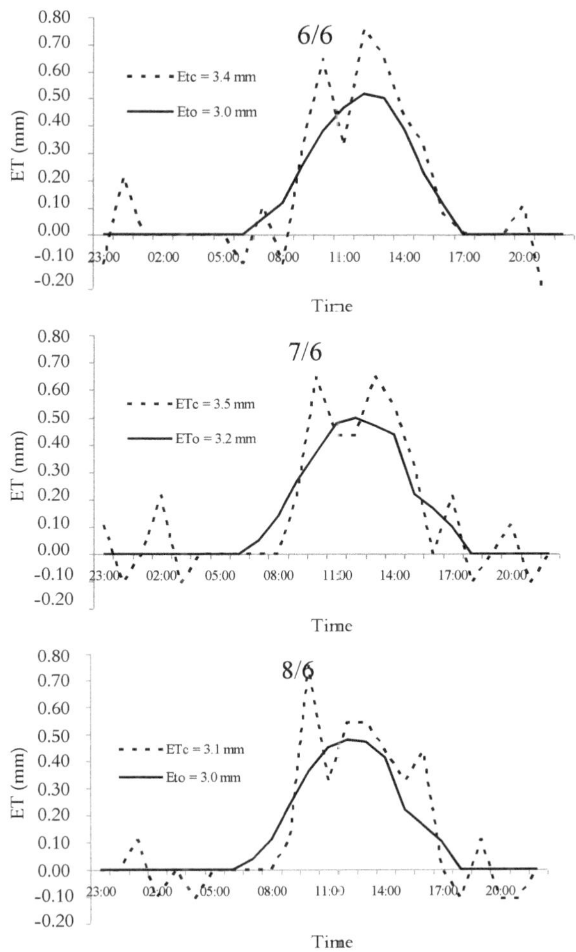

Figure 6. Wheat evapotranspiration (ETc) and reference evapotranspiration (ETo) measured by a lysimeter from Jun 6th to 8th, 2011.

To present the operating conditions of the drainage system, Figure 7a shows the variations of stored water in the lysimeter four days after rainfall (49 mm). Figure 7b shows the resulting drainage curves in the subsequent days. The reductions in the stored water depth were most likely due to the drainage, especially on the day after rainfall. The successive and gradual increases in values of accumulated drainage were due to the tipping of the amount stored in the drainage device installed at the bottom of the lysimeter. As observed in Figure 7b, the drained weight at each event was equal to 1.23 mm, which may have been due to the system.

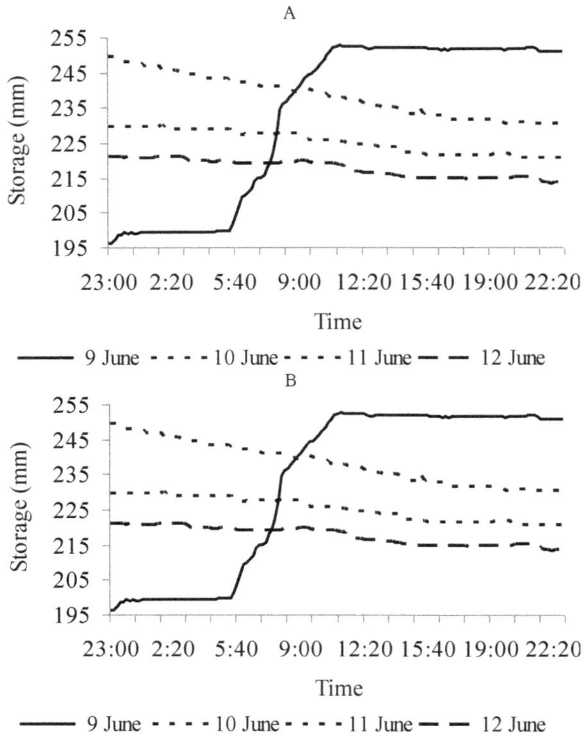

Figure 7. Soil water storage (a) and accumulated drainage (b) measured by a lysimeter cultivated with wheat from June 9 to 12, 2011.

Conclusion

The responses of the lysimeter load cells to applied weights during the calibration process were linear with coefficients of determination close to unity and sensitive to the weight variation equivalent to a water depth of 0.15 mm. Therefore, the weighing lysimeters with bars proved to be suitable for determining estimates of evapotranspiration and other water balance components.

References

CAMPECHE, L. F. M. S.; AGUIAR NETTO, A. O.; SOUSA, I. F.; FACCIOLI, G. G.; SILVA, V. P. R.; AZEVEDO, P. V. Lisímetro de pesagem de grande porte. Parte I: Desenvolvimento e calibração. **Revista Brasileira de Engenharia Agrícola e Ambiental**, v. 15, n. 5, p. 519-525, 2011.

CARVALHO, D. F.; SILVA, L. D. B.; GUERRA, J. G. M.; CRUZ, F. A.; SOUZA, A. P. Installation, calibration and operation of a weighing lysimeter. **Engenharia Agrícola**, v. 27, n. 2, p. 363-372, 2007.

FARIA, R. T.; CAMPECHE, F. S. M.; CHIBANA, E. Y. Construction and calibration of high precision lysimeters. **Revista Brasileira de Engenharia Agrícola e Ambiental**, v. 10, n. 1, p. 237-242, 2006.

FARIA, R. T.; CARAMORI, P. H.; CHIBANA, E. Y.; BRITO, L. R. S. CLIMA-computer program to organize and analyze meterorological data. **Engenharia Agrícola**, v. 23, n. 2, p. 372-387, 2003.

FERNANDES, C. N. V.; AZEVEDO, B. M.; VIANA, T. V. A.; BOMFIM, G. V.; MESQUITA, J. B. Lysimeter precision of load cell installation and calibration. **Irriga**, v. 17, n. 2, p. 251-263, 2012.

FLUMIGNAN, D. L.; FARIA, R. T. Evapotranspiration and crop coefficients of coffee trees during crop formation. **Bragantia**, v. 68, n. 1, p. 269-278, 2009.

IAPAR-Instituto Agronômico do Paraná. **Agrometeorologia**. Médias históricas. Available from: <http://www.iapar.br/arquivos/Image/monitoramento/Medias_Historicas/Londrina.htm>. Access on: Feb. 16, 2012.

KIRKHAM, R. R.; GEE, G. W.; JONES, T. L. Weighing lysimeters for long-term water balance investigations at remote sites. **Soil Science Society of America Journal**, v. 48, n. 5, p. 1203-1205, 1984.

MENDONÇA, J. C.; SOUSA, E. F.; BERNARDO, S.; DIAS, G. P.; GRIPPA, S. Comparison of estimation methods of reference crop evapotranspiration (ETo) for Northeren Region of Rio de Janeiro State, Brazil. **Revista Brasileira de Engenharia Agrícola e Ambiental**, v. 7, n. 2, p. 275-279, 2003.

MIRANDA, F. R.; YODER, R. E.; SOUSA, F. Installation and calibration of a weighing lysimeter in Paraipaba, CE. **Revista Brasileira de Engenharia Agrícola e Ambiental**, v. 3, n. 1, p. 107-110, 1999.

OLIVEIRA, L. M. M.; MONTENEGRO, S. M. G. L.; AZEVEDO, J. R. G.; SANTOS, F. X. Reference evapotranspiration in the experimental basin of Gameleira, PE, using lysimeter and indirect methods. **Revista Brasileira de Ciências Agrárias**, v. 3, n. 1, p. 58-67, 2008.

PEREIRA, A. R.; SANTIAGO, A. V.; MAGGIOTTO, S. R.; FOLEGATTI, M. V. Operational problems with a weighing lysimeter during a rainy season and dry windy days. **Revista Brasileira de Agrometeorologia**, v. 10, n. 1, p. 51-56, 2002.

SANTOS, F. X.; RODRIGUES, J. J. V.; MONTENEGRO, A. A. A.; MOURA, R. F. Low cost hydraulic weighing lysimeter performance in the semi-arid Northeast, Brazil. **Engenharia Agrícola**, v. 28, n. 1, p. 115-124, 2008.

SANTOS, R. A. S.; FOLEGATTI, M. V.; BOTREL, T. A.; ATARASSI, R. T.; ALVES JÚNIOR, J. Design, construction and evaluation of a weighing lysimeter with hydraulic load cell to determine water requirements of citrus nursery trees in greenhouse. **Irriga**, v. 12, n. 2, p. 202-215, 2007.

SILVA, F. C.; FOLEGATTI, M. V.; MAGIOTTO, S. R. Weighing lysimeter working analyse with load cell. **Revista Brasileira de Agrometeorologia**, v. 7, n. 1, p. 53-58, 1999.

SILVA, L. D. B.; FOLEGATTI, M. V.; VILLA NOVA, N. A. Evapotranspiração do capim Tanzânia obtida pelo método de razão de Bowen e lisímetro de pesagem. **Engenharia Agrícola**, v. 25, n. 3, p. 705-712, 2005.

SILVA, T. J. A.; MONTENEGRO, A. A.; RODRIGUES, J. J.; BONFIM, E. M. S. Aplicação de lisímetro de pesagem hidráulica na determinação da evapotranspiração de referência, em Petrolina-PE. **Engenharia Agrícola**, v. 23, n. 3, p. 511-520, 2003.

SILVA, V. P. R.; CAMPOS, J. H. B. C.; AZEVEDO, P. V. Water-use efficiency and evapotranspiration of mango orchard grown in northeastern region of Brazil. **Scientia Horticulturae**, v. 120, n. 4, p. 467-472, 2009.

VIANA, T. V. A.; FOLEGATTI, M. V.; AZEVEDO, B. M.; BOMFIM, G. V.; ELÓI, W. M. Evapotranspiration through the Bowen ratio system and a weighing lysimeter in greenhouse. **Irriga**, v. 8, n. 2, p. 113-119, 2003.

UFV-Universidade Federal de Viçosa. **SAEG** - Sistema de Análises Estatísticas e Genéticas. Viçosa: UFV, 1997.

WILLMOTT, C. J.; CKLESON, S. G.; DAVIS, R. E. Statistics for the Evaluation and comparison of models. **Journal of Geophysical Research**, v. 90, n. 5, p. 8995-9005, 1985.

13

The performance of the CROPGRO model for bean (*Phaseolus vulgaris* L.) yield simulation

Evandro Chaves de Oliveira[1*], José Maria Nogueira da Costa[2], Trazilbo José de Paula Júnior[3], Williams Pinto Marques Ferreira[4], Flávio Barbosa Justino[2] and Leonardo de Oliveira Neves[5]

*[1]Instituto Federal do Espírito Santo, BR-259, Km 70, Cx. Postal 256, 29709-910, Campus Itapina, Colatina, Espírito Santo, Brazil. [2]Departamento de Engenharia Agrícola, Centro de Ciências Agrárias, Universidade Federal de Viçosa, Viçosa, Minas Gerais, Brazil. [3]Empresa de Pesquisa Agropecuária de Minas Gerais, Zona da Mata, Viçosa, Minas Gerais, Brazil. [4]Empresa Brasileira de Pesquisa Agropecuária, Sete Lagoas, Minas Gerais, Brazil. [5]Instituto Federal Catarinense, Rio do Sul, Santa Catarina, Brazil. *Author for correspondence. E-mail: evandro.oliveira@ifes.edu.br*

ABSTRACT. The aim of this study was to evaluate the CROPGRO-Dry bean model for simulating dry bean yield. The model's genetic coefficients were calibrated based on the cultivars 'Pérola', 'Ouro Negro' and 'Ouro Vermelho' in Viçosa, State of Minas Gerais, Brazil. The coefficients were adjusted based on two experiments that were performed in 2003 with irrigated and nonirrigated water regimes. An additional experiment with irrigation was conducted in 2004. After calibration, the model simulated the bean yield for the period from 1975 through 2006. The simulations were based on daily data on maximum and minimum air temperatures, total precipitation and global solar radiation. The physical and hydric characteristics of the soil and crop management practices were also included. The results show that the crop model can correctly reproduce the observed yield. This finding may indicate that the model is a useful tool to evaluate the crop response to variability and changing climate.

Keywords: computer simulation, DSSAT, cultivars, *Phaseolus vulgaris*, yield, forecast date.

Introduction

The State of Minas Gerais is the second-highest producer of dry bean (*Phaseolus vulgaris* L.) in Brazil (IBGE, 2010). The most important groups of bean cultivated in Minas Gerais are "carioca" and "black" (VIEIRA et al., 2005). In the northern, southern and northwestern regions of the State, carioca cultivars are preferred. In the Zona da Mata region, some cultivars of the "red" group are also raised (RAMALHO et al., 1993). At some times of the year, the price of red beans may be twice the price of other beans (VIEIRA, 2005). The production of dry beans is one of the most important economic and social agricultural activities in Brazil. This activity employs approximately seven million people during the growing season and involves up to 295,000 growers (BORÉM; CARNEIRO, 2006).

The prediction of bean yield is an activity of great interest to the government, family farmers and agri-business companies. To improve the strategy of crop yield prediction, new methods may be adopted. These methods incorporate quantitative factors to improve the precision of estimation. Computational models to estimate crop yield allow the estimation of yield as a function of the soil management practices and climatic conditions that occur during the growing season. It should be emphasized that these models are usually composed of sub-modules that aim to represent different processes occurring in the soil-plant-atmosphere system to simulate crop growth, development and yield.

The last two decades have been marked by the development of more sophisticated crop models. These models include improved representations of crop phenology as a function of accumulated heat units (e.g., the CERES and EPIC models) and relatively detailed soil biogeochemical processes for carbon and nitrogen (e.g., Century, DNDC and EPIC). Recently, process-based models with different levels of complexity have been developed to estimate crop yield in different regions (BETTS, 2005; COSTA et al., 2009; HANSEN, 2005). The Decision Support System for Agrotechnology Transfer (DSSAT) includes a set of models of crop growth and has been widely used in recent years as a useful computational tool in the evaluation of management options associated with environmental conditions (HOOGENBOOM et al., 2003). The modular structure of the DSSAT includes an outstanding model known as CROPGRO. This generic model for legumes allows the simulation of the growth and development of several crops, such as the dry bean, cowpea, peanut and soybean crops.

Examples of the simulation efficiency of CROPGRO were presented by Ruíz-Nogueira et al. (2001), who used the model to evaluate soybean growth and yield under water limitation conditions for three cultivars in different climatic conditions of northeast Spain. Hartkamp et al. (2002) evaluated the performance of CROPGRO to predict phenology, growth, senescence and nitrogen accumulation in locations that represented different scenarios of environmental and agronomic management. The evaluation involved sensitivity analysis of various regimes of temperature and precipitation to identify regions suitable for the production of Mucuna aterrima as green manure. Jagtap and Jones (2002) tested the performance of CROPGRO in predicting soybean yield and observed that the model achieved a performance criterion of 95% in predicting interannual variations in yield.

In Brazil, Faria et al. (1997) used simulations with CROPGRO-Dry bean to evaluate different strategies for the irrigation management of beans in the State of Paraná. Meireles et al. (2003) calibrated the CROPGRO-Dry bean model to quantify decreases in the yield of beans sown on 36 sowing dates in Santo Antônio de Goiás, Goiás State. In a study conducted in Maringá, State of Paraná, Dallacort et al. (2005) used the model's simulation of the growth and development of the bean to determine the best sowing times for the cultivar IAPAR 57 from 1980 through 2000.

Sensitivity experiments with CROPGRO to evaluate the effect of meteorological variables on the growth, development and yield of beans may contribute to improved management practices and to a better understanding of the interactions between climate and plants (COSTA et al., 2009; JONES et al., 2003).

In this sense, this study may serve as a tool for small farmers, agro-industrial companies, extension institutions, technical assistance and research institutions, government agencies and development and funding institutions to help in planning agricultural activities with less risk of adverse weather effects.

In view of the importance of beans (*Phaseolus vulgaris* L.) for the State of Minas Gerais and Brazil and the need for a system of yield prediction, this study was conducted to evaluate the CROPGRO-Dry bean model as a crop-yield forecasting tool for the bean cultivars 'Pérola', 'Ouro Negro' and 'Ouro Vermelho' in Viçosa, Zona da Mata region (State of Minas Gerais, Brazil).

Material and methods

The simulations were performed with the CROPGRO-Dry bean model, included in the DSSAT system, version 4.0.2.0 (HOOGENBOOM et al., 2004), which simulates the growth, development and yield of beans (*Phaseolus vulgaris* L.), Figure 1. The construction of the required files for the simulation was described in detail by Boote et al. (1998a, 1998b and 2003). The model allows simulations of the specific conditions of each soil for the principal physical and physiological processes associated with the crop. These processes include photosynthesis, respiration, biomass accumulation and partitioning, phenology, soil water extraction, evapotranspiration and common bean growth and leaf area for both stems and roots as a function of daily climatological variables (rainfall, solar radiation and maximum and minimum temperatures).

Simulations were performed based on field experiments conducted in the Prof. Diogo Alves de Melo experimental area. This area is part of the Federal University of Viçosa (UFV), Viçosa, Minas Gerais State, located at latitude 20° 45'S, longitude 42° 51'W and an altitude of 690 m.

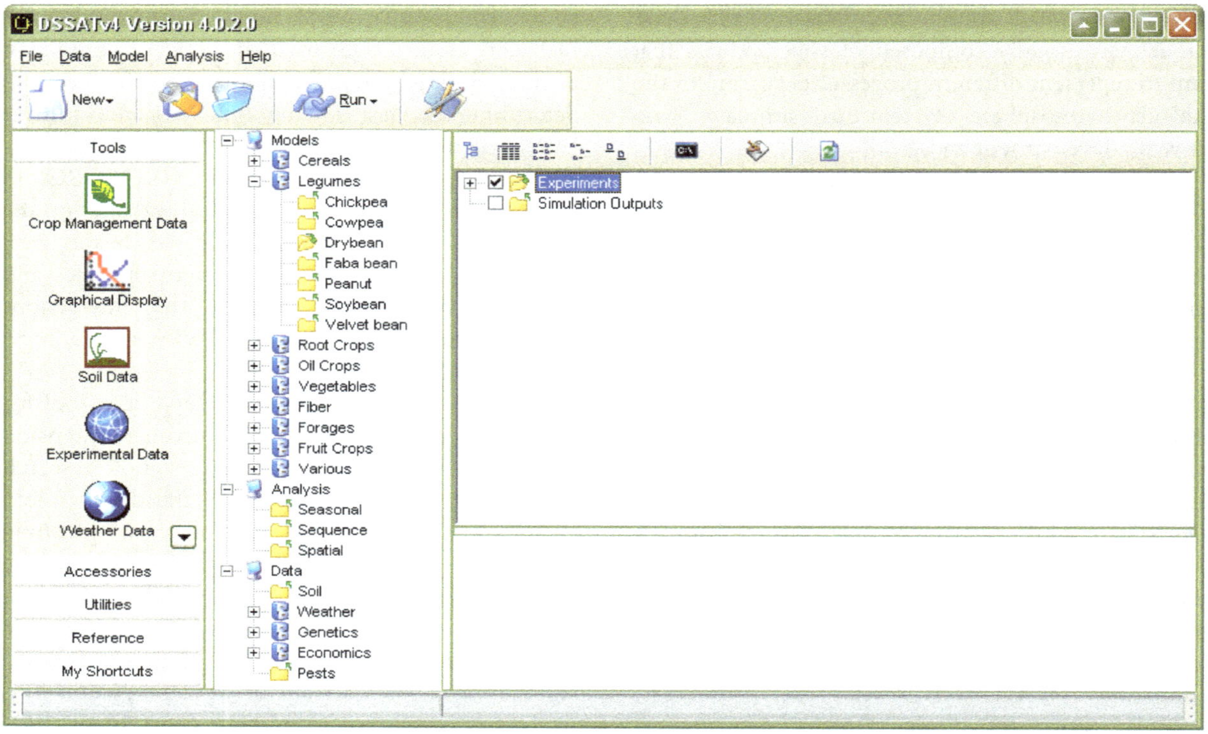

Figure 1. Window illustrating the atmosphere-soil-plan database system for the CROPGRO-Dry bean simulation in DSSAT.

Data used in the simulation

Daily weather data and edaphic and crop management information were used as input data for the model. The set of daily weather input variables were taken from 1975 through 2006 and included precipitation, maximum and minimum air temperatures and solar radiation. These data were provided by the 5th District of Meteorology (5 DISME) of the National Institute of Meteorology (INMET). The location evaluated was Viçosa, Minas Gerais State. The soil of the experimental area is classified as red-yellow podzolic Cambic. Its chemical and physical characteristics are shown in Table 1.

Table 1. Results of sample analysis and classification of the soil (0-40 cm depth) of the experimental area. The chemical and physical analyses were performed at the Laboratory of Soil Analysis (UFV).

pH in H_2O	H + Al	Al	Ca	Mg	K	P	OM	Soil fraction (%)			
	Cmol(+) dm^{-3}				μg dm^{-3}		dag kg^{-1}	Grit	Fine Sand	Silt	Clay
5,9	3.3	0	2.3	0.9	42	9	2.2	19	14	21	46

Texture Class – Clayey; Soil Classification – Red-Yellow Podzolic Cambic (LRd); Organic Matter (OM).

The bean cultivars 'Pérola' (carioca group), 'Ouro Negro' (black group) and 'Ouro Vermelho' (red group) were chosen owing to their economic importance. Fertilization was applied as recommended by Ribeiro et al. (1999). The control of weeds and pests was performed in accordance with the recommendations for the crop. The sowing dates were March 27, 2003 and March 20, 2004 for the experiments conducted with irrigation. November 20, 2003 was chosen for the experiment conducted during the rainy season and without irrigation.

The sowing was done at a 3-cm depth after the soil was prepared in a conventional manner. The experimental design was completely randomized, with three replications and plots consisting of four rows 0.5 m apart and 4 m long. The average values for adjustment of the genetic coefficients of the model were obtained from these plots.

Calibration and testing of CROPGRO-Dry bean

The calibration and testing of the CROPGRO-Dry bean model for the experimental conditions were performed by adjusting genetic coefficients that characterize the essential aspects of beans, as recommended by Hoogenboom et al. (2003) and Jones et al. (2003).

The sequence of the procedures for the calibration of the genetic coefficients was as follows: 1) the genetic coefficients were selected from a given genotype from those in the same maturity group or similarly adapted; 2) the model was run for location, cultivar or treatment; and 3) the values were assigned to specific genetic coefficients for growth, beginning with the parameters for phenology and followed by the growth parameters of the crop (HOOGENBOOM et al., 1992). This procedure was performed by a trial-and-error process, i.e., a process in which values were assigned to each factor by determining whether the model generated results close to those measured under field conditions.

The experiment conducted without irrigation in 2003 was used to perform the calibration of CROPGRO-Dry bean, whereas the experiments conducted in the rainy season (2003) and in 2004 were used as independent data for evaluation.

The observed dates of flowering and physiological maturity and yield were statistically compared to the simulated values with the coefficient of determination (r2), the index of agreement (d) (WILLMOTT et al., 1985) and the mean square error (RMSE) (LOAGUE; GREEN, 1991).

Simulation of yield

To evaluate the model as a tool for forecasting bean yield, simulations were performed using a 31-year time series of climate data for the area (1975-2006). The model was developed to simulate the bean yield for the November 20, 2003 sowing date (test year). In the first step, the daily weather data observed during the base year were used. In the second step, the same sowing date was used with a seasonal analysis of the data for all years of the series. In the third step, the seasonal analysis was performed with climate data from the test year for the initial phase of the cycle (first 20 days) and for the rest of the cycle with data for all years of the series. Finally, the historical data were replaced by weekly data that were observed for the year of interest to test the physiological maturity of the crop. The values of the crop yield obtained with simulations using the observed data for the full year were compared with the values obtained by substituting part of the climate data for the historical data. Thus, it was possible to generate a crop yield forecast and define the period of prediction.

Results and discussion

The seasonal temperature and rainfall patterns in the two experimental years were different (Table 2). In 2003, the average air temperature was 1.1 and 0.9°C above that observed in 2004 and in the historic series of 1975-2006, respectively. The total cumulative seasonal rainfall was 17% (2003) and 40% (2004) in relation to the historical average from 1975 through 2006. However, after the rainy season (November through February), irrigation was required from May through July. In 2003, 263 mm of water was applied, and 120 mm was applied in 2004.

Table 2. Mean air temperature and monthly and seasonal rainfall in 2003 and 2004. Viçosa, Minas Gerais State, Brazil.

Year	Nov	Dec	Jan	Feb	Mar	Apr	May	Jun	Jul	Seasonal
Temperature (°C)										Average
2003	22.5	24.6	24.3	25.1	24.7	22.5	20.3	19.2	17.4	22.8
2004	22.5	22.8	23.0	23.2	23.2	22.3	20.0	17.0	17.8	21.7
(1975-2006)	22.4	23.0	23.5	23.9	23.3	21.9	19.8	17.8	17.5	21.9
Precipitation (mm)										Total
2003	163.8	252.1	432.7	27.9	81.4	16.6	21.2	0.0	7.5	1023.5
2004	195.9	276.0	405.9	388.4	174.4	136.7	41.6	40.7	35.6	1725.8
(1975-2006)	215.9	239.2	237.9	187.9	141.6	57.4	38.2	15.7	16.3	1234.0
Irrigation (mm)										Total
2003	-	-	-	-	0.0	0.0	59.0	152.0	52.0	263.0
2004	-	-	-	-	0.0	0.0	38.0	61.0	21.0	120.0

Data not collected (-).

Test of the CROPGRO-Dry bean model

Table 3 presents the genetic coefficients of the three bean cultivars for the calibration of the CROPGRO-Dry bean model for the estimation of the growth and development of the plants in the soil conditions and climate of Viçosa, Minas Gerais State. Seven coefficients define the life cycle development characteristics, and eight coefficients define the growth characteristics (BOOTE et al., 2003). According to Didonet and Silva (2004), the Brazilian bean cultivars are generally insensitive to photoperiod (PPSEN = 0), i.e., their development and growth are controlled only by the air temperature, with thermal requirements in photothermal days (dft).

The three cultivars exhibited differences in the following genetic coefficients: the duration of the period between the emergence of the seedlings and the appearance of the first flower; the duration of the period between the appearance of the first flower and the first pod; the duration of the period between the appearance of the first flower and the onset of seed formation; and the duration of the period between the beginning of the formation of seeds and physiological maturity.

The CROPGRO-Dry bean model showed high sensitivity to variation in the genetic coefficients. The simulated yields varied under the same climate and soil conditions. Similarly, Dallacort et al. (2006) reported that a sensitivity analysis of the CROPGRO model showed a marked response owing to changes in the genetic coefficients of the soy cultivars CD 202, CD 204, CD 206 and CD 210 for the municipality of Palotina, State of Paraná.

Table 3. Genetic coefficients used for the calibration of the CROPGRO-Dry bean model for three cultivars.

Coefficients	Cultivars		
	Pérola	Ouro Negro	Ouro Vermelho
CSDL	12.17	12.17	12.17
PPSEN	0.0	0.0	0.0
	Development parameters		
EM-FL	35.5	32.0	34.0
FL-SH	6.0	5.0	6.0
FL-SD	11.0	9.0	10.0
SD-PM	17.0	15.5	16.0
FL-LF	18.0	18.0	18.0
LFMAX	1.0	1.0	1.0
	Growth parameters		
SLAVAR	270.0	270.0	270.0
SIZLF	100.0	100.0	100.0
XFRT	1.0	1.0	1.0
WTPSD	0.256	0.255	0.255
SFDUR	9.0	9.0	9.0
SDPDV	5.0	5.0	5.0
PODUR	6.5	6.5	6.5

CSDL – critical day length, above which reproductive development is not affected (h); PPSEN – response inclination for the development of the photophase over time (1 h^{-1}); EM-FL – period between plant emergence and the appearance of the first flower (R1) (photothermal days); FL-SH – period between the appearance of the first flower and the first pod (R3) (photothermal days); FL-SD – period between the appearance of the first flower and the start of seed formation (R5) (photothermal days); SD-PM – period between the start of seed formation and physiological maturity (R7) (photothermal days); FL-LF – period between the appearance of the first flower (R1) and the end of leaf expansion; LFMAX - maximum leaf photosynthesis rate at an optimal temperature of 30 °C; SLAVARN - specific leaf area under standard growth conditions (cm^2); SIZLF – maximum size of a completely expanded leaf (cm^2); XFRT – maximum fraction of the daily growth that is partitioned between the seed and the pod; WTPSD – maximum weight per seed (g); SFDUR – duration of the grain swelling period in the pods under standard growth conditions (photothermal days); SDPDV – mean seeds per pod under standard growth conditions (photothermal days); PODUR – time necessary for the cultivar to reach ideal pod conditions (photothermal days).

Estimated duration of the phenological stages of the bean

Considerable agreement was evident between the observed and simulated values of the number of days from sowing to flowering and from sowing to maturity for these three bean cultivars (Figure 2). The simulated number of days from sowing to flowering (SF) showed a high coefficient of determination between the simulated and observed values. In addition, the agreement index (d) showed values as high as 0.99. Moreover, the mean square error (RMSE) was 2.97%. These findings show that the CROPGRO-Dry bean model effectively simulated the phenology of the three bean cultivars. The same outcome was observed for the period from sowing to maturity (SM). As shown in Figure 2, the model accurately estimated the period from sowing to physiological maturity and showed high values for the coefficient of determination (r^2 = 0.99) and for the index of agreement (d = 0.99). In addition, the RMSE was low (2.47%). These results are consistent with previous studies that showed that the model appeared to simulate the phenology of the bean cultivars adequately (FARIA et al., 1997; MEIRELES et al., 2002).

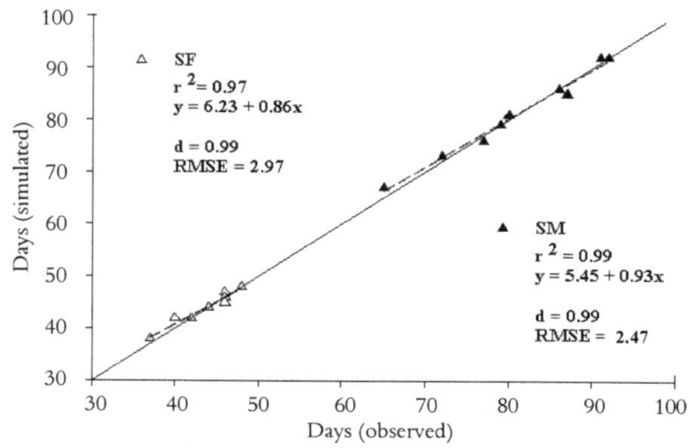

Figure 2. Relationship between the number of observed and simulated days from sowing to flowering (SF) and sowing to physiological maturity (SM) for three bean cultivars in trials conducted in 2003 and 2004 in Viçosa, Minas Gerais State, Brazil.

Estimated crop yield

Crop yield was satisfactorily simulated by the CROPGRO-Dry bean model for the three bean cultivars (Table 4). For 'Perola' and 'Ouro Negro', the observed and simulated values were very similar. The percentage of deviation of the estimate ranged from -5.58 to 6.24%, with rates of agreement that were (d) high (between 0.88 and 0.99) and values of the mean square error (RMSE) under 5%. The performance of the model for the cultivar 'Ouro Vermelho' was lower than that for the other two cultivars. The value of the RMSE was 12.63%, and the value of d was 0.91. These differences may be related to the spatial variability of the experimental data.

Table 4. Mean values of the simulated and observed grain yield with the CROPGRO-Dry bean model in the calibration phase for three bean cultivars. Viçosa, Minas Gerais State, Brazil.

Cultivars	Experiment	Simulated (kg ha^{-1})	Observed (kg ha^{-1})	PD[(1)] (%)	RMSE (%)	d
Perola	Rainy season 2003	1792	1898	-5.58		
	No irrigation 2003	2222	2125	4.56	4.85	0.88
	2004	2166	2074	4.44		
Ouro Negro	Rainy season 2003	1600	1545	3.56		
	No irrigation 2003	2249	2261	-0.53	3.88	0.99
	2004	1908	1796	6.24		
Ouro Vermelho	Rainy season 2003	1767	1603	10.23		
	No irrigation 2003	2558	2801	-8.68	12.63	0.91
	2004	2230	2653	-15.94		

(1) PD (%) = [(Estimated - Observed) / Observed] x 100.

In 2004, a delay in sowing (March 20, 2004) compared with the previous year (February 27, 2003) may have caused the flowering of the bean cultivars in 2004 to occur during a prolonged period of low air temperatures. It is probable that these low temperatures contributed to a decrease in yield from 2125 to 2074 kg ha^{-1} (2.4%) for 'Pérola', 2261 to 1796 kg ha^{-1} (20.6%) for 'Ouro Negro', and 2801 to 2653 kg ha^{-1} (5.3%) for 'Ouro Vermelho'. In the simulation, the CROPGRO-Dry bean exhibited the same trend in yield for the three cultivars. However, it showed reductions of up to 2.5 and 12.8% for 'Pérola' and 'Ouro Vermelho', respectively. The 'Ouro Negro' cultivar was slightly lower (15.1%) compared with the experimental data.

Forecast of the bean yield

Based on the meteorological data for 2003, a simulation of the bean yield from December 5 to February 20 was performed. Figure 3 shows the estimates of the yield for the three bean cultivars and the standard deviations of the estimates. We performed simulations based on weather data for the three cultivars for December 5th through February 20th. The simulation that used the beginning of the growing season as its initial data exhibited unrealistic values of the yield and high standard deviations. Because no irrigation was used, it is possible that this result was associated with the shortage of water.

In contrast, the simulation of the phenological stages for better-established cultivars showed predicted yields close to the observational data and standard deviations close to zero. The standard deviations of the estimates of yield for December 20th and January 20th for 'Pérola', 'Ouro Negro' and 'Ouro Vermelho' ranged from 494 to 222 kg ha^{-1}, 401 to 89 kg ha^{-1} and 411 to 131 kg ha^{-1}, respectively. The most reliable estimates of yields with CROPGRO-Dry bean could be made 30 days in advance of the harvest for the three cultivars. These results are promising because farmers and decision makers could have access to accurate yield forecasts prior to the final harvest. Similarly, the results obtained by Thornton et al. (1997) and Soler et al. (2007), using similar methodology, showed that the accuracy of the estimates of expected income increased toward the end of the growing season. Soler et al. (2007) reported that the CSM-CERES-Maize model can be a promising tool for yield forecasting for maize hybrids grown off season in Piracicaba, São Paulo State, Brazil. An accurate yield forecast was obtained approximately 45 days prior to harvest.

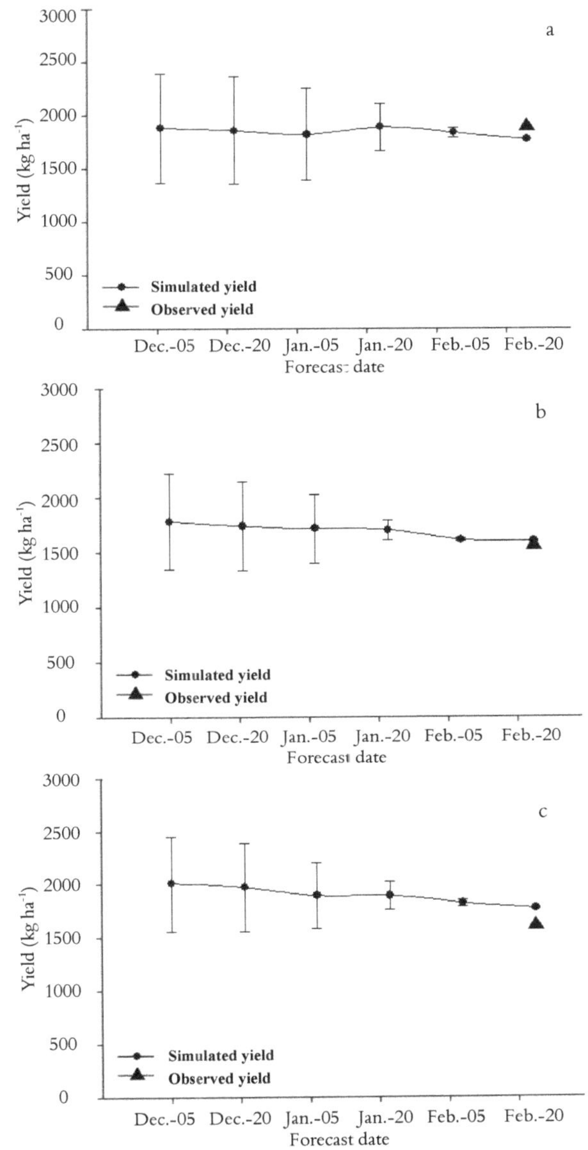

Figure 3. Yield and standard deviations for beans cultivated in the rainy season of 2003 for the 'Pérola' (a) 'Ouro Negro' (b) and 'Ouro Vermelho' (c) cultivars.

Conclusion

The CROPGRO-Dry bean model accurately simulated the phenological development and yield of the bean cultivars 'Pérola', 'Ouro Negro' and 'Ouro Vermelho' for the soil conditions and climate of Viçosa, Minas Gerais State, Brazil.

In addition, the model showed high sensitivity to genetic variation coefficients and exhibited differences between cultivars in the simulations of phenological development and yield and proved to be viable as a tool for forecasting the yield of beans cultivated in the rainy season. Good estimates of the yield were obtained 30 days in advance of the harvest for the bean cultivars 'Pérola', 'Ouro Negro' and 'Ouro Vermelho'.

Acknowledgements

We thank the National Council of Scientific and Technological Development (CNPq), the Foundation for Research Support of Minas Gerais (FAPEMIG) and the Coordination of Improvement of Higher Education Personnel (CAPES) for funding the project and for the scholarships granted.

References

BETTS, R. A. Integrated approaches to climate–crop modelling: needs and challenges. **Philosophical Transactions of the Royal Society B**, v. 360, n. 1463, p. 2049-2065, 2005.

BOOTE, K. J.; JONES, J. W.; HOOGENBOOM, G. Simulation of crop growth: CROPGRO model. In: PEART, R. M.; CURRY, R. B. (Ed.). **Agricultural systems modeling and simulation**. New York: M. Dekker, 1998a. p. 651-691.

BOOTE, K. J.; JONES, J. W.; HOOGENBOOM, G.; PICKERING, N. B. The CROPGRO model for grain legumes. In: TSUJI, G. Y.; HOOGENBOOM, G.; THORNTON, P. K. (Ed.). **Understanding options for agricultural production**. Dordrecht: Kluwer, 1998b. p. 99-128.

BOOTE, K. J.; JONES, J. W.; BATCHELOR, W. D.; NAFZIGER, E. D.; MYERS, O. Genetic coefficients in the CROPGRO-Soybean Model: links to field performance and genomics. **Agronomy Journal**, v. 95, n. 1, p. 32-51, 2003.

BORÉM, A.; CARNEIRO, J. E. S. A Cultura. In: VIEIRA, C.; PAULA JÚNIOR, T. J.; BORÉM, A. (Ed.). **Feijão**. 2. ed. Viçosa: UFV, 2006. p. 13-18.

COSTA, L. C.; JUSTINO, F.; OLIVEIRA, L. J. C.; SEDIYAMA, G.; FERREIRA, W. P. M.; LEMOS, C. F. Potential forcing of CO2, technology and climate changes in maize (*Zea mays*) and beans (*Phaseolus vulgaris*) yields in the southern part of Brazil. **Environmental Research Letters**, v. 4, n. 1, p. 14013-14023, 2009.

DALLACORT, R.; FREITAS, P. S. L.; FARIA, R. T.; GONÇALVES, A. C. A.; REZENDE, R.; BERTONHA, A. Utilização do modelo Cropgro-soybean na determinação de melhores épocas de semeadura da cultura da soja, na região de Palotina, Estado do Paraná. **Acta Scientiarum. Agronomy**, v. 28, n. 4, p. 583-589, 2006.

DALLACORT, R.; REZENDE, R.; FREITAS, P. S. L.; FARIA, R. T.; AZEVEDO, T. L. F.; JÚNIOR, J. B. T. Utilização do modelo Cropgro-drybean na determinação das melhores épocas de semeadura da cultura do feijão para a região de Maringá, Estado do Paraná, Brasil. **Acta Scientiarum. Agronomy**, v. 27, n. 2, p. 349-355, 2005.

DIDONET, A. D.; SILVA, S. C. Elementos climáticos e produtividade do feijoeiro. **Informe Agropecuário**, v. 25, n. 223, p. 13-19, 2004.

FARIA, R. T.; FOLEGATTI, M. V.; FRIZZONE, J. A.; SAAD, A. M. Determination of a long-term irrigation strategy for drybeans in Parana State-Brazil. **Scientia Agricola**, v. 54, special issue, p. 155-164, 1997.

HANSEN, J. W. Integrating seasonal climate prediction and agricultural models for insights into agricultural practice. **Philosophical Transactions of the Royal Society B**, v. 360, n. 1463, p. 2037-2047, 2005.

HARTKAMP, A. D.; HOOGENBOOM, G.; GILBERT, R. A.; TARAWALI, A. J.; GIJSMAN, A. J.; BOWEN, W.; WHITE, J. W. Adaptation of the CROPGRO growth model to velvet bean (*Mucuna prurients*) II. Culitvar evaluation and model testing. **Field Crops Research**, v. 78, n. 1, p. 27-40, 2002.

HOOGENBOOM, G.; JONES, J. W.; BOOTE, K. J. Modeling growth, development, and yield of grain legumes using SOYGRO, PNUTGRO, and BEANGRO: a review. **Transactions of the ASAE**, v. 35, n. 6, p. 2043-2055, 1992.

HOOGENBOOM, G.; JONES, J. W.; WILKENS, P. W.; HUNT, L. A.; PORTES, C. H.; BATCHELOR, W. D.; HUNT, L. A.; BOOTE, K. J.; SINGH, U.; UEHARA, G.; BOWEN, W. T.; GUSMAN, A. J.; DU TOIT, A. S.; WHITE, J. W.; TSUJI, G. Y. **Decision support system for agrotechnology transfer**: version 4.0. Honolulu: University of Hawaii, 2003. [CD-ROM]

HOOGENBOOM, G.; JONES, J. W.; WILKENS, P. W.; PORTER, C. H.; BATCHELOR, W. D.; HUNT, L. A.; BOOTE, K. J.; SINGH, U.; URYASEV, O.; BOWEN, W. T.; GIJSMAN, A. J.; DU TOIT, A.; WHITE, J. W. E.; TSUJI, G. Y. **Decision support system for agrotechnology transfer**: version 4.0. Honolulu: University of Hawaii, 2004. [CD-ROM]

IBGE-Instituto Brasileiro de Geografia e Estatística. **Levantamento Sistemático da Produção Agrícola**. Rio de Janeiro, v. 23, n. 9, p. 1-80, set. 2010. Available from: <http://www.ibge.gov.br/>. Access on: July 19, 2011.

JAGTAP, S. S.; JONES, J. W. Predicting regional soybean yield and production: Adaptation of the CROPGRO soybean simulation model and scaling up yield. **Agriculture, Ecossytems and Environment**, v. 93, n. 1-3, p. 78-85, 2002.

JONES, J. W.; HOOGENBOOM, G.; PORTER, C. H.; BOOTE, K. J.; BATCHEALOR, W. D.; HUNT, L. A.; WILKENS, P. W.; SINGH, U.; GIJSMAS, A. J.; RITCHIE, J. T. The DSSAT cropping systems model. **European Journal of Agronomy**, v. 18, n. 3-4, p. 235-265, 2003.

LOAGUE, K.; GREEN, R. E. Statistical and graphical methods for evaluating solute transport models: overview and application. **Journal of Contaminant Hydrology**, v. 7, n. 1-2, p. 51-73, 1991.

MEIRELES, E. J. L.; PEREIRA, A. P.; SENTELHAS, P. C.; STONE, L. F.; ZIMMERMANN, F. J. P. Calibration and test of the cropgro-dry bean model for edaphoclimatic conditions in the savannas of central Brazil. **Scientia Agricola**, v. 59, n. 4, p. 723-729, 2002.

MEIRELES, E. J. L.; PEREIRA, A. R.; SENTELHAS, P. C.; STONE, L. F.; ZIMMERMANN, F. J. P. Risco Climático de quebra de produtividade da cultura do feijoeiro em Santo Antônio de Goiás, GO. **Bragantia**, v. 62, n. 1, p. 163-171, 2003.

RAMALHO, M. A. P.; ABREU, A. F. B.; RIGHETTO, G. U. Interações de cultivares de feijão por épocas de semeadura em diferentes localidades do Estado de Minas Gerais. **Pesquisa Agropecuária Brasileira**, v. 28, n. 10, p. 1183-1189, 1993.

RIBEIRO, A. C.; GUIMARÃES, P. T. G.; ALVAREZ, V. H. **Recomendações para o uso de corretivos e fertilizantes em Minas Gerais**. Viçosa: Comissão de Fertilidade do Solo do Estado de Minas Gerais, 1999. (5° aproximação).

RUÍZ-NOGUEIRA, B.; BOOTE, K. J.; SAU, F. Calibration and use of CROPGRO-soybean model for improving soybean management under rainfed conditions. **Agricultural Systems**, v. 68, n. 2, p. 151-173, 2001.

SOLER, C. M. T.; SENTELHAS, P. C.; HOOGENBOOM, G. Application of the CSM-CERES-Maize model for planting date evaluation and yield forecasting for maize grown off-season in a subtropical environment. **European Journal of Agronomy**, v. 27, n. 2-4, p. 165-177, 2007.

THORNTON, P. K.; BOWEN, W. T.; RAVELO, A. C.; WILKENS, P. W.; FARMER, G.; BROCK, J.; BRINK, J. E. Estimating millet production for famine early warning: an application of crop simulation modeling using satellite and ground-based data in Burkina Faso. **Agricultural and Forest Meteorology**, v. 83, n. 1-2, p. 95-112, 1997.

VIEIRA, C. **Memórias de meio século de estudo sobre a cultura do feijão**. Viçosa: UFV, 2005.

VIEIRA, C.; BORÉM, A.; RAMALHO, M. A. P.; CARNEIRO, J. E. S. **Melhoramento de espécies cultivadas**. Viçosa: UFV, 2005.

WILLMOTT, C. J.; AKLESON, G. S.; DAVIS, R. E.; FEDDEMA, J. J.; KLINK, K. M.; LEGATES, D. R.; O'DONNELL, J.; ROWE, C. M. Statistic for the evaluation and comparison of models. **Jounal of Geophysical Research**, v. 90, n. 5, p. 8995-9005, 1985.

14

Thermodymnamic properties of crambe fruits

Daniel Emanuel Cabral de Oliveira[1*], Osvaldo Resende[2], Lílian Moreira Costa[2] and Hellismar Wakson da Silva[3]

*[1]Instituto Federal de Educação, Ciência e Tecnologia Goiano, Campus Iporá, Avenida Oeste, 350, Parque União, 76200-000, Iporá, Goiás, Brazil. [2]Instituto Federal de Educação, Ciência e Tecnologia Goiano, Rio Verde, Goiás, Brazil. [3]Universidade Federal de Lavras, Lavras, Minas Gerais, Brazil. *Author for correspondence. E-mail: daniel.oliveira@ifgoiano.edu.br*

ABSTRACT. This study aimed to determine and evaluate the thermodynamic properties of crambe fruit at different equilibrium moisture contents. The dynamic-gravimetric method was used to collect experimental data. Crambe fruits with an initial moisture content of approximately 26% db (dry basis) were used. The thin-layer desorption of the product was conducted at different controlled temperatures (25, 30, 35, 40, and 45°C) and at water activity levels ranging from 0.10 to 0.89 until the product reached the equilibrium moisture content at the specified air condition. In conclusion, the thermodynamic properties of crambe fruit are affected by temperature and moisture content. The isokinetic theory is valid for the desorption process, which is controlled by enthalpy, and the Gibbs free energy is positive at all of the tested temperatures, which indicates that water desorption is not a spontaneous process.

Keywords: *Crambe abyssinica*, enthalpy, entropy, Gibbs free energy.

Introduction

Nowadays concerns about climate change and sustainability along with the need for development and the increasing demand for energy resources have driven researchers to discover new fuel sources that can replace fossil fuels. Thus, the crop potential of crambe (*Crambe abyssinica*) fruit has been investigated in Brazil because of high concentrations of oil (44.10±1.46%) and proteins (21.3±1.86%), what make it an excellent alternative for biodiesel production (Souza, Fávaro, Ítavo, & Roscoe, 2009).

Following harvest, crambe fruit should bedried to reduce moisture content to suitable levels, and to enable preservation during storage. However, the existing relationship between the product and surrounding air must be understood to establish the best drying and storage conditions for agricultural products (Goneli, Corrêa, Oliveira, Gomes, & Botelho, 2010a; Corrêa, Botelho, Botelho, & Goneli, 2014), including crambe fruits.

The relationship between the product and the surrounding air may be best understood using sorption isotherms or hygroscopic equilibrium curves that depict the relationship between the hygroscopic equilibrium moisture content of a particular product and the relative humidity at a specific temperature (Resende, Corrêa, Goneli, & Ribeiro, 2006; Cladera-Olivera, Marczak, Noreña, & Pettermann, 2011).

Sorption isotherms are a valuable and widely used tool in preventing possible changes in food stability during storage, packaging development and the design and optimization of drying equipment (Cladera-Olivera, Pettermann, Noreña Wada, & Marczak, 2008).

In several agricultural product studies, researchers used sorption isotherms to determine certain thermodynamic properties to describe the water sorption of a material and the energy requirements involved in the drying processes (Kaya & Kahyaoglu, 2006; Corrêa, Oliveira, Botelho, Goneli, & Carvalho, 2010; Smaniotto, Resende, Oliveira, Sousa, & Campos, 2012; Oliveira, Resende, Campos, & Sousa, 2014a; Koua, Koffi, Gbaha, & Toure, 2014).

Additionally, thermodynamic properties are required and used in drying equipment projects; in the study of the adsorbed water properties; and in the evaluation of food microstructure and understanding of physical phenomena that occur on the surface of foods (Kaya & Kahyaoglu, 2006; Corrêa et al., 2010).

The latent heat of vaporization (Corrêa et al., 1998; Ojediran, Raji, & Owamah, 2013; Oliveira, Resende, Chaves, Sousa, & Smaniotto, 2014b), differential entropy and enthalpy (Rosa, Moraes, & Pinto, 2010; Cladera-Olivera et al., 2011), the theory of enthalpy-entropy compensation or isokinetic theory (Thys, Noreña, Marczak, Aires, & Cladera-Olivera, 2010; Oliveira et al., 2014a) and Gibbs free energy (Goneli et al., 2010a; Goneli, Corrêa, Oliveira, & Botelho, 2010b) are the most studied thermodynamic parameters. Thus, this study aimed to determine and analyze the thermodynamic properties of crambe fruit at different temperatures and equilibrium moisture contents.

Material and methods

This study was carried out at Postharvest Laboratory of Plant Products of the Instituto Federal de Educação, Ciência e Tecnologia Goiano - *Campus* Rio Verde, located in Rio Verde, Goiás State, Brazil. Crambe (*Crambe abyssinica*) fruit was harvested in experimental area of IF Goiano - Câmpus Rio Verde were used to perform the experiment executed in this study.

Crambe fruits with a moisture content of approximately 26 (% db) were hand-picked. The moisture contents were measured by gravimeter method using an oven at 105±3°C for 24 hours (Brasil, 2009).

The dynamic-gravimetric method was used to determine hygroscopic equilibrium moisture content of crambe fruits. A thin-layer desorption of the product was investigated at different controlled temperatures (25, 30, 35, 40, and 45°C) and water activity range from 0.10 to 0.89 until product reached equilibrium moisture content at a specific air conditions.

The environmental conditions required to perform hygroscopicity tests were achieved using an environmental chamber, which controls the temperature and relative humidity. Samples containing 10 g of product were wrapped in a permeable fabric (i.e., voile) to allow air circulation through the product; the samples were then placed inside of the device.

The Sigma Copace model, which showed the best experimental data fit, was used to calculate the thermodynamic properties of the crambe fruits (Costa, Resende, & Oliveira, 2015). Water activities were determined using the Equation 1:

$$Xe = Exp\{0.571297^{**} - (0.011807^{**} \cdot T) + [0.967917^{**} \cdot Exp(a_w)]\} \quad (1)$$

where: Xe^* = equilibrium moisture content, % db; a_w = water activity, decimal; and T = temperature, °C.

**Significant at 1% based on a Student's t-test.

Based on Clausius-Clapeyron, Othmer (1940) studies, it was established the Equation 2 to calculate partial vapor pressure contained in porous systems:

$$Ln(Pv) = \left(\frac{L}{L'}\right) \cdot Ln(Pvs) + C \quad (2)$$

where: Pvs = Saturation vapor pressure of free water at a specific equilibrium temperature (°C); Pv = vapor pressure of free water at a specific equilibrium temperature (°C); L = latent heat of vaporization of the product water (kJ kg^{-1}); L' = latent heat of vaporization of free water at the equilibrium temperature (kJ kg^{-1}); and C = constant of integration.

Based on the sorption isotherms of crambe fruits, the value of the L/L' ratio of Equation 3 was assessed at different equilibrium moisture contents, Xe (decimal), which adjusted the equation to the enthalpy of vaporization of water, as reported by Rodrigues - Arias (Brooker, Bakker-Arkema, & Hall, 1992). This equation also includes an additional coefficient to improve the estimation of L/L' ratio (Corrêa, Christ, Martins, & Mantovani, 1998):

$$\frac{L}{L'} - 1 = a \cdot Exp(-b \cdot Xe^m) \quad (3)$$

where: a, b and m = estimated coefficients.

The latent heat of vaporization of free water (kJ kg^{-1}) at the equilibrium temperature (°C) was calculated using the mean temperature (T) of the range tested in °C using the Equation 4:

$$L' = 2502.2 - 2.39 \cdot T \quad (4)$$

The saturation vapor pressure of free water, Pvs, was calculated using the Thétens equation 5 (Corrêa et al., 1998):

$$Pvs = 0.61078 \cdot 10^{((7.5 \cdot T)/(273.3+T))} \quad (5)$$

The vapor pressure (Pv) was determined according to Equation 6:

$$Pv = a_w \cdot Pvs \quad (6)$$

The combination of Equations 3 and 4 results in the Equation 7, which is used to estimate the latent heat of vaporization of the product (Corrêa et al., 1998):

$$L = (2502.2 - 2.39 \cdot T) \cdot \left[1 + a \cdot Exp\left(-b \cdot Xe^{m}\right)\right] \quad (7)$$

The differential sorption entropy was calculated using the Gibbs-Helmholtz equation 8 (Rizvi, 1995):

$$\Delta S = \frac{\Delta h_{st} - \Delta G}{T_a} \quad (8)$$

where: ΔS = differential sorption entropy (kJ kg^{-1} K^{-1}); Δh_{st} = differential enthalpy (kJ kg^{-1}); T_a = absolute temperature (K); and ΔG = Gibbs free energy (kJ kg^{-1}).

Gibbs free energy was assessed using the following Equation: 9

$$\Delta G = -R \cdot T_a \cdot Ln(a_w) \quad (9)$$

where: R = universal gas constant, 8.314 kJ $kmol^{-1}$ K^{-1}, which is 0.4619 kJ kg^{-1} K^{-1} for water vapor.

The effects of the changes in water sorption on Gibbs free energy are typically followed by changes in the values of the enthalpy and entropy. Thus, the following equation 10 results from substituting Equation 8 into Equation 9 and rearranging:

$$Ln(aw) = \frac{\Delta h_{st}}{R \cdot T_a} - \frac{\Delta S}{R} \quad (10)$$

The differential sorption entropy (ΔS) and differential enthalpy (Δ_{hst}) were estimated using Equation 11 and were correlated using the Equation below (Beristain, Garcia, & Azuara, 1996):

$$\Delta h_{st} = T_B (\Delta S) + \Delta G_B \quad (11)$$

where: T_B = isokinetic temperature (K); and ΔG_B: Gibbs free energy at isokinetic temperature (kJ kg^{-1}).

The isokinetic temperature is the temperature at which reactions in series occur at the same rate. The theory of enthalpy-entropy compensation is presumed to be valid for sorption because enthalpy and entropy are highly correlated (Beristain et al., 1996). To confirm the existence of compensation, the isokinetic temperature was compared to the harmonic mean temperatures used to assess the sorption isotherms based on Equation 12 (Krug, Hunter, & Grieger, 1976):

$$T_{hm} = \frac{n}{\sum\left(\frac{1}{T}\right)} \quad (12)$$

where: T_{hm} = harmonic mean temperature (K); and n = number of temperatures used.

Based on Krug et al. (1976), linear chemical compensation or enthalpy-entropy compensation theory will occur if the isokinetic temperature (T_B) is different from the harmonic mean temperature (T_{hm}). An approximate confidence interval of (1-α) 100% was established for the isokinetic temperature using the following equations 13-15 :

$$T_B = \hat{T} \pm t_{m-2,\alpha/2} \sqrt{Var(T_B)} \quad (13)$$

where:

$$\hat{T}_B = \frac{\sum\left(\Delta h_{st} - \overline{\Delta h_{st}}\right)\left(\Delta S - \overline{\Delta S}\right)}{\sum\left(\Delta S - \overline{\Delta S}\right)^2} \quad (14)$$

and:

$$Var(T_B) = \frac{\sum\left(\Delta h_{st} - \overline{\Delta G_B} - \hat{T}_B \Delta S\right)^2}{(m-2)\sum\left(\Delta S - \overline{\Delta S}\right)^2} \quad (15)$$

where: m = number of data pairs of enthalpy and entropy; $\overline{\Delta h_{st}}$ = mean enthalpy, kJ kg^{-1}; $\overline{\Delta S}$ = mean entropy, kJ kg^{-1}.

If the harmonic mean temperature T_{hm} falls within the calculated range of the isokinetic temperature T_B, the ratio between the values of differential sorption entropy and enthalpy only describes the experimental errors and not the existence of chemical and physical factors that react based on the theory of compensation (Beristain et al., 1996). A 99% confidence interval was used for T_B in the experimental range.

Results and discussion

Table 1 shows the water activity data estimated using the Sigma Copace model (Equation 1) at 25,

30, 35, 40, and 45°C and at equilibrium moisture contents range from 3.9 to 10.9 (% db).

Table 1. Water activity values (decimal) estimated using the Sigma Copace model based on the temperature and equilibrium moisture content (Xe) of crambe (*Crambe abyssinica*) fruits.

Xe (% db)	Temperature (°C)				
	25	30	35	40	45
3.9	0.1036	0.1572	0.2080	0.2563	0.3025
4.0	0.1293	0.1815	0.2312	0.2784	0.3236
4.1	0.1661	0.2165	0.2644	0.3102	0.3539
4.3	0.1919	0.2410	0.2878	0.3326	0.3754
4.6	0.2628	0.3086	0.3524	0.3944	0.4347
4.8	0.2964	0.3407	0.3832	0.4239	0.4631
4.9	0.3037	0.3477	0.3899	0.4303	0.4692
5.5	0.3931	0.4335	0.4722	0.5096	0.5456
6.1	0.4626	0.5003	0.5366	0.5717	0.6055
6.2	0.4714	0.5088	0.5448	0.5795	0.6131
6.3	0.4745	0.5118	0.5477	0.5823	0.6158
7.0	0.5468	0.5815	0.6151	0.6475	0.6789
7.4	0.5775	0.6112	0.6438	0.6753	0.7059
7.7	0.6003	0.6332	0.6651	0.6960	0.7259
7.8	0.6026	0.6355	0.6673	0.6981	0.7280
8.0	0.6252	0.6573	0.6884	0.7186	0.7479
8.4	0.6473	0.6788	0.7092	0.7388	0.7675
9.7	0.7260	0.7551	0.7833	0.8108	0.8376
10.9	0.7789	0.8065	0.8333	0.8595	0.8850

When studying the thermodynamic properties of corn kernels and oil radish seeds, respectively, Oliveira, Resende, Smaniotto, Sousa, and Campos (2013) and Sousa, Resende, Goneli, Smaniotto, and Oliveira (2015) observed a decrease in equilibrium moisture content at the same water activity with temperature increase. Increases in water activity and temperature are related considering the same moisture content. These results are similar to those found for crambe fruit.

Table 2 outlines values of L/L' ratios for different moisture contents. Decreases in moisture content lead to increases in L/L' ratios of crambe fruits.

Table 2. L/L' ratios for different moisture contents of crambe (*Crambe abyssinica*) fruit.

Moisture content (% db)	L/L' ratio	Moisture content (% db)	L/L' ratio
3.9	1.9544	6.3	1.2353
4.0	1.8199	7.0	1.1954
4.1	1.6786	7.4	1.1812
4.3	1.6027	7.7	1.1716
4.6	1.4533	7.8	1.1707
4.8	1.4022	8.0	1.1619
4.9	1.3923	8.4	1.1538
5.5	1.2957	9.7	1.1292
6.1	1.243	10.9	1.1154
6.2	1.2373		

The L/L' ratio noticeably increases as the moisture content decreases, reaching values near 1.0 at high moisture contents. Corrêa et al. (1998) and Ojediran et al. (2013) assessed similar performances in their studies of popcorn seeds and cottonseeds, respectively.

Table 3 shows "a", "b" and "m" parameters used to estimate the ratio between the latent heat of vaporization of water in agricultural products and the latent heat of free water through non-linear regression.

Table 3. Parameters "a", "b" and "m" used to calculate the ratio between the latent heat of vaporization of water in agricultural products and the latent heat of free water (L/L').

A	B	m	R^2 (%)
0.102989**	-39.40607**	-2.13472**	99.87

**Significant at 1% based on Student's t-test.

The following Equation (16) to calculate the latent heat of vaporization of water for crambe fruit results from changing the values of "a", "b" and "m" in the equation proposed by Corrêa et al. (1998):

$$L = (2502.2 - 2.39 \cdot T) \cdot \left[1 + 0.102989 \cdot \exp\left(39.40607 \cdot Xe^{-2.13472}\right)\right] \quad (16)$$

Figure 1 shows the curves of the latent heat of vaporization of water of crambe fruits at 25, 30, 35, 40, and 45°C.

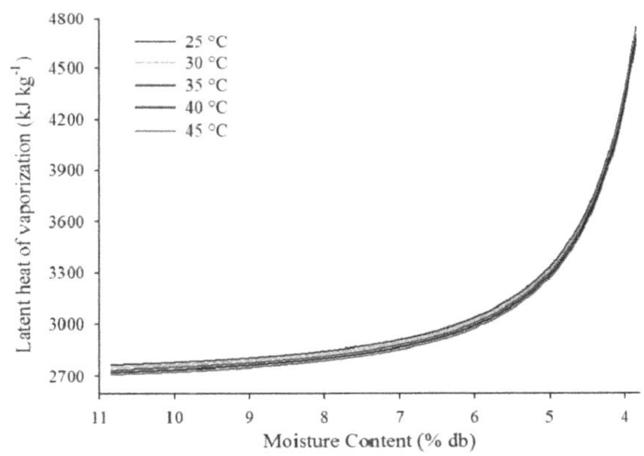

Figure 1. Experimental and estimated values of the latent heat of vaporization of water as a function of the equilibrium moisture content of crambe fruits (*Crambe abyssinica*).

The latent heat of vaporization is indirectly proportional to the moisture content of crambe fruits, and the increase in temperature at the same moisture content reduced the latent heat of vaporization.

Based on Wang and Brennan (1991), the latent heat of vaporization is indicative of the intermolecular forces of attraction between the water vapor adsorption sites of the product. Thus, the analysis of Figure 1 shows the existence of water molecules in a free state when crambe fruits have high moisture contents because the latent heat of vaporization of water in the product is near the latent heat of vaporization of pure water (2,418.55 kJ kg^{-1}).

The latent heat of vaporization of crambe fruits ranged from 4,733.47 to 2,708.87 kJ kg^{-1}, with moisture contents ranging from 3.9 to 10.9 (% db). Oliveira et al. (2014b) reported lower latent heats of vaporization, which ranged from 2,762.92 to 2,495.56 kJ kg^{-1}, for physic nut seeds, which had moisture contents ranging from 5.61 to 13.42 (% db). Smaniotto et al. (2012) observed that the latent heat of vaporization ranged from 2,775.87 to 2,468.14 kJ kg^{-1} in corn kernels with moisture contents ranging from 12.76 to 23.26 (% db). The most significant variation in the latent heat of vaporization of crambe fruit is related to the chemical composition and moisture content used. Brooker et al. (1992) emphasized that the latent heat of vaporization of a product is primarily affected by moisture content and temperature.

Figure 2 shows the values of differential desorption entropy (ΔS) and enthalpy (Δhst) as a function of the equilibrium moisture content (% db).

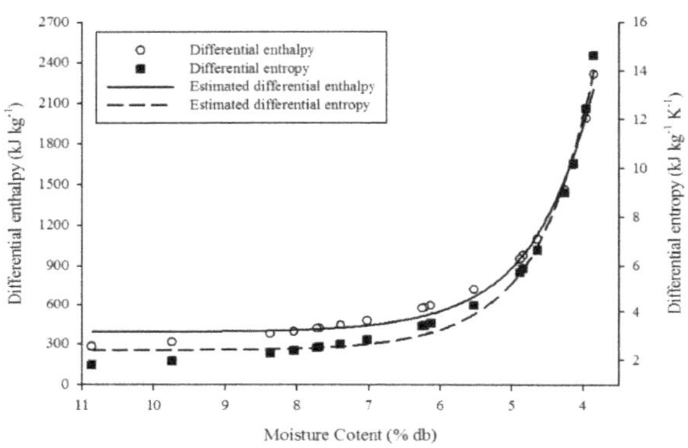

Figure 2. Experimental and estimated values of the differential sorption entropy and enthalpy of crambe (*Crambe abyssinica*) fruits.

The values of differential enthalpy and entropy noticeably increase as the moisture content of crambe fruits decreases (Figure 2), corroborating the results assessed using sesame seeds (Kaya & Kahyaoglu, 2006), okra seeds (Goneli et al., 2010b), Brazilian pine seeds (Cladera-Olivera et al., 2008; Thys et al., 2010), cocoa beans (Oliveira, Corrêa, Santos, Treto, & Diniz, 2011) and physic nut seeds (Oliveira et al., 2014b). Additionally, the differential enthalpy and entropy varied from 279.81 to 2,313.19 kJ kg^{-1} K^{-1} and from 1.7827 to 14.612 kJ kg^{-1} K^{-1}, respectively, as moisture content varied from 3.9 to 10.9 (% db). The high variations of differential enthalpy and entropy are related to the analysis of the equilibrium moisture content range and are also functions of the chemical composition of the crambe fruits.

Additionally, Figure 2 shows that the differential enthalpy and entropy showed similar trends in relation to the moisture content, and the magnitudes also showed a tendency to stabilize at high moisture contents. Tunç and Duman (2007), Thys et al. (2010) and Oliveira et al. (2014b) reported the same behavior in cottonseeds (*Gossypium hirsutum* L.), Brazilian pine (*Araucaria angustifólia*) and physic nut seeds (*Jatopha curvas* L.), respectively.

Table 4 shows the equations used to assess the differential enthalpy and entropy and the coefficients of determination of crambe fruits. All equation parameters were significant at the 1% significance level based on Student's t-test and showed high coefficients of determination ($R^2 > 99\%$), highlighting the suitability of the equations to represent the experimental data.

Table 4. Equations and coefficients of determination of differential desorption entropy (ΔS) and enthalpy (Δhst) of crambe (*Crambe abyssinica*) fruit.

Thermodynamic Properties	Equations	R^2 (%)
Differential enthalpy	Δhst=386.6590** + 144724.4935**. exp (-1.1345** . Xe)	99.09
Differential entropy	ΔS=2.4014** + 1300.8145** . exp (-1.2243** . Xe)	99.16

**Significant at 1% based on a t test.

Figure 3 shows that the curve representing the relationship between enthalpy and entropy regarding the desorption process, which were assessed using Equation 11 for each moisture content, was linear, considering that the differential enthalpy and entropy are invariable with temperature at a specific moisture content.

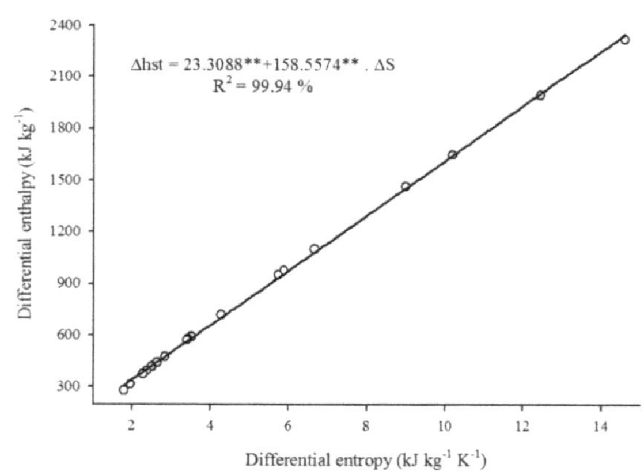

**Significant at 1% based on a t test.

Figure 3. Relationship between enthalpy and entropy regarding the water desorption process of crambe (*Crambe abyssinica*) fruits.

This relationship may be represented by a linear equation, which showed a high coefficient of determination (99.94%). Thus, the theory of enthalpy-entropy compensation or isokinetic theory is valid to describe the water sorption phenomena of crambe fruits because a high degree of linearity exists between the values of enthalpy and entropy. Goneli et al. (2010a) and Oliveira et al. (2013) observed the same behavior when studying the thermodynamic properties of millet and corn kernels, respectively.

The isokinetic temperature was compared with the harmonic mean (T_{hm}) of the studied temperature range used to determine the desorption process of crambe fruits towards testing the validity of the theory of enthalpy-entropy compensation. As reported by Telis-Romero, Kohayakawa, Silveira Júnior, Pedro, and Gabas (2005), linear enthalpy-entropy compensation is typically observed when $T_B \neq T_{hm}$. This process is controlled by enthalpy if $T_B > T_{hm}$; otherwise, the process is controlled by entropy (Telis-Romero et al., 2005).

The isokinetic temperature, which is the regression slope of the relationship between the enthalpy and entropy regarding the desorption process of crambe fruits, was 343.27 ± 5.11 K, and the isokinetic temperature ranged from 336.54 to 350.00 K. Conversely, the calculated harmonic mean temperature was 306.68 K, which is significantly different and higher than values of the reported isokinetic temperature, confirming the enthalpy-entropy compensation phenomenon regarding the desorption process of crambe fruits and that the process is controlled by enthalpy. Similar behaviors were observed in other agricultural products (Telis-Romero et al., 2005; Tunç & Duman, 2007; Goneli et al., 2010a; Oliveira et al., 2014a).

Figure 4 shows experimental and estimated values of Gibbs free energy as a function of the moisture content at each temperature during the desorption process of crambe fruits.

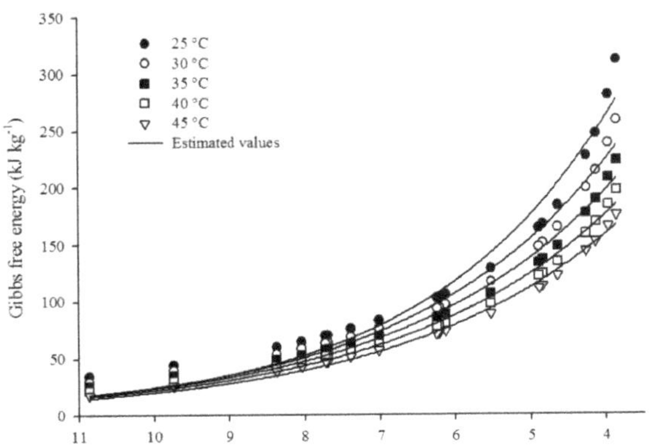

Figure 4. Gibbs free energy as a function of the moisture content (Xe) and temperature (T) of crambe fruits (*Crambe abyssinica*).

Changes in Gibbs free energy are related to the amounts of energy required to transfer water molecules from a solid state to a vapor state (Corrêa et al., 2010). Thus, the Gibbs free energy significantly increases with moisture content decreases and tends to stabilize at high levels of equilibrium moisture content and as temperature increases. Additionally, Gibbs free energy was noticeably positive at all tested temperatures, thus indicating that water desorption is a non-spontaneous process (Oliveira et al., 2013). These results corroborate the findings of Goneli et al. (2010b), Oliveira et al. (2011) and Goneli et al. (2013) in their studies of okra seeds, coconuts and coffee cherries, respectively.

Figure 5 shows that the parameters α and β of Gibbs free energy decrease as temperature increases, and both equations showed high coefficients of determination and the coefficients were significance.

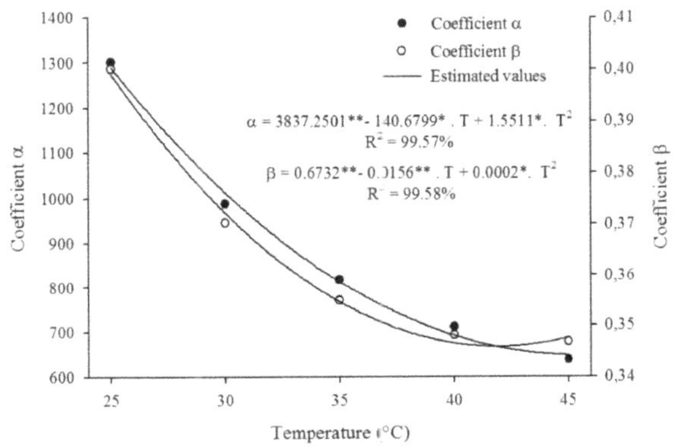

Figure 5. Coefficients α and β of the regression equation of Gibbs free energy for crambe (*Cramle abyssinica*) fruits.

Thus, Equation 17, which may be used to calculate the Gibbs free energy of crambe fruits in the studied temperature range, is created by substituting the equations of parameters α and β (Figure 5) into Equation 9:

$$\Delta G = \left(3837.2501 - 140.6799 \cdot T + 1.5511 \cdot T^2\right) \text{Exp}\left[\left(0.6732 - 0.0156 \cdot T + 0.0002 \cdot T^2\right) \cdot Xe \star\right] \tag{17}$$

Conclusion

The thermodynamic properties of crambe fruits are affected by equilibrium moisture content and temperature.

The isokinetic theory is valid for the desorption process, which is controlled by enthalpy.

Gibbs free energy is positive at all studied temperatures, indicating that water desorption is not a spontaneous process.

References

Beristain, C. I., Garcia, H. S., & Azuara, E. (1996). Enthalpy-entropy compensation in food vapor adsorption. *Journal of Food Engineering*, *30*(3-4), 405-415.

Brasil, Ministério da Agricultura e Reforma Agrária. Secretaria Nacional de defesa Agropecuária. (2009). *Regras para Análise de Sementes*. Brasília, DF: Mapa/ACS.

Brooker, D. B.; Bakker-Arkema, F. W.; Hall, C. W. (1992). *Drying and storage of grains and oilseeds*. Westport, CT: The Avi Publishing Company.

Cladera-Olivera, F., Pettermann, A. C., Noreña, C. P. Z., Wada, K., & Marczak, L. D. F. (2008). Thermodynamic properties of moisture desorption of raw pinhão (*Araucaria angustifolia* seeds). *International Journal of Food Science e Technology*, *43*(5), 900-907.

Cladera-Olivera, F., Marczak, L. D. F., Noreña, C. P. Z., & Pettermann, A. C. (2011). Modeling water adsorption of pinhão (*Araucaria angustifolia* seed) flour and thermodynamic analysis of the adsorption process. *Journal of Food Process Engineering*, *34*(3), 826-843.

Corrêa, P. C., Christ, D., Martins, J. H., & Mantovani, B. H. M. (1998). Curvas de dessorção e calor latente de vaporização para as sementes de milho pipoca (*Zea mays*) [Desorption curves and latent heat of vaporization for popcorn seeds (*Zea mays*)]. *Revista Brasileira de Engenharia Agrícola e Ambiental*, *2*(1), 7-11.

Corrêa, P. C., Oliveira, G. H. H., Botelho, F. M., Goneli, A. L. D., & Carvalho, F. M. (2010). Modelagem matemática e determinação das propriedades termodinâmicas do café (*Coffea arabica* L.) durante o processo de secagem [Mathematical modeling and determination of thermodynamic properties of coffee (*Coffea arabica* L.) during the drying process]. *Revista Ceres*, *57*(5), 595-601.

Corrêa, P. C., Botelho, F. M., Botelho, S. C. C., & Goneli, A. L. D. (2014). Isotermas de sorção de água de frutos de *Coffea canephora* [Sorption isotherms of fruits of *Coffea canephora*]. *Revista Brasileira de Engenharia Agrícola e Ambiental*, *18*(10), 1047-1052.

Costa, L. M., Resende, O., & Oliveira, D. E. C. (2015). Determinação das isotermas de equilíbrio higroscópico de frutos de crambe pelo método dinâmico [Determination of hygroscopic equilibrium isotherms of crambe fruit by dynamic method]. *Bioscience Journal*, *31*(2), 382-391.

Goneli, A. L. D., Corrêa, P. C., Oliveira, G. H. H., Gomes, C. F., & Botelho, F. M. (2010a). Water sorption isotherms and thermodynamic properties of pearl millet grain. *International Journal of Food Science and Technology*, *45*(4), 282-383.

Goneli, A. L. D., Corrêa, P. C., Oliveira, G. H. H., & Botelho, F. M. (2010b). Water desorption and thermodynamic properties of okra seeds. *Transaction of the ASAE*, *53*(1), 191-197.

Kaya, S., & Kahyaoglu, T. (2006). Influence of dehulling and roasting process on the thermodynamics of moisture adsorption in sesame seed. *Journal of Food Engineering*, *76*(2), 139-147.

Koua, B. K., Koffi, P. M. E., Gbaha, P., & Toure, S. (2014). Thermodynamic analysis of sorption isotherms of cassava (*Manihot esculenta*). *Journal of Food Science Technology*, *51*(9), 1711-1723.

Krug, R. R., Hunter, W. G., & Grieger, R. A. (1976). Enthalpy-entropy compensation: 1 - some fundamental statistical problems associated with the analysis of Van't Hoff and Arrhenius data. *Journal of Physical Chemistry*, *80*(21), 2335-2341.

Ojediran, J., Raji, A. O., & Owamah, H. I. (2013). Isosteric heats of water vapor sorption in two castor varieties. *Chemical Engineering & Process Technology*, *4*(2), 1-6.

Oliveira, D. E. C., Resende, O., Smaniotto, T. A. S., Sousa, K. A., & Campos, R. C. (2013). Propriedades termodinâmicas de grãos de milho para diferentes teores de água de equilíbrio [Thermodynamic properties of maize grains for different equilibrium moisture contents]. *Pesquisa Agropecuária Tropical*, *43*(1), 50-56.

Oliveira, D. E. C., Resende, O., Campos, R. C., & Sousa, K. A. (2014a). Propriedades termodinâmicas de sementes de tucumã-de-Goiás (*Astrocaryum huaimi* Mart.) [Thermodynamic properties of seeds tucumã-de-goiás (*Astrocaryum huaimi* Mart.)]. *Revista Caatinga*, *27*(3), 53-62.

Oliveira, D. E. C., Resende, O., Chaves, T. H., Sousa, K. A., & Smaniotto, T. A. S. (2014b). Propriedades termodinâmicas das sementes de pinhão-manso [Thermodynamic properties of seeds jatropha]. *Bioscience Journal*, *30*(3), 147-157.

Oliveira, G. H. H., Corrêa, P. C., Santos, E. S., Treto, P. C., & Diniz, M. D. M. S. (2011). Evaluation of thermodynamic properties using GAB model to describe the desorption process of cocoa beans. *International Journal of Food Science & Technology*, *46*(10), 2077-2084.

Othmer, D. F. (1940). Correlation vapor pressure and latent heat data: A new plot. *Industrial and Engineering Chemistry*, *32*(6), 841-845.

Resende, O., Corrêa, P. C., Goneli, A. L. D., & Ribeiro, D. M. (2006). Isotermas e calor isostérico de sorção do feijão [Isotherms and isosteric heats of sorptions of the edible bean]. *Ciência e Tecnologia dos Alimentos*, *26*(3), 626-631.

Rizvi, S. S. H. (1995). Thermodynamic properties of foods in dehydration. In M. A. Rao & S. S. H. Rizvi (Eds.), *Engineering properties of foods* (p. 223-309). New York, US: Academic Press.

Rosa, G. S., Moraes, M. A., & Pinto, L. A. A. (2010). Moisture sorption properties of chitosan. *Food Science and Technology*, *43*(3), 415-420.

Smaniotto, T. A. S., Resende, O., Oliveira, D. E. C., Sousa, K. A., & Campos, R. C. (2012). Isotermas e calor latente de dessorção dos grãos de milho da cultivar AG 7088 [Isotherms and latent heat of desorption of corn]. *Revista Brasileira de Milho e Sorgo*, *11*(3), 312-322.

Sousa, K. A., Resende, O., Goneli, A. L. D., Smaniotto, T. A. S., & Oliveira, D. E. C. (2015). Thermodynamic properties of water desorption of forage turnip seeds. *Acta Scientiarum. Agronomy*, *37*(1), 11-19.

Souza, A. D. V., Fávaro, S. P., Ítavo, L. C. V., & Roscoe, R. (2009). Caracterização química de sementes e tortas de pinhão-manso, nabo-forrageiro e crambe [Chemical characterization of seeds and presscakes of physic nut, radish and crambe]. *Pesquisa Agropecuária Brasileira*, *44*(10), 1328-1335.

Telis-Romero, J., Kohayakawa, M. N., Silveira Júnior, V., Pedro, M. A. M., & Gabas, A. L. (2005). Enthalpy-entropy compensation based on isotherms of mango. *Ciência e Tecnologia de Alimentos*, *25*(2), 297-303.

Thys, R. C. S., Noreña, C. P. Z., Marczak, L. D. F., Aires, A. G., & Cladera-Olivera, F. (2010). Adsorption isotherms of pinhão (Araucaria angustifolia seeds) starch and thermodynamic analysis. *Journal of Food Engineering*, *100*(3), 468-473.

Tunç, S., & Duman, O. (2007). Thermodynamic properties and moisture adsorption isotherms of cottonseed protein isolate and diferente forms of cottonseed samples. *Journal of Food Engineering*, *81*(1), 133-1443.

Wang, N., & Brennan, J. G. (1991). Moisture sorption isotherm characteristics of potato at four temperatures. *Journal of Food Engineering*, *14*(4), 269-287.

15

Water balance model and eucalyptus growth simulation in the rio doce basin, Brazil

Welliam Chaves Monteiro da Silva[1], Aristides Ribeiro[1*], Júlio César Lima Neves[2], Nairam Félix de Barros[2] and Fernando Palha Leite[3]

*[1]Departamento de Engenharia Agrícola, Universidade Federal de Viçosa, Av. P. H. Rolfs, s/n, 36570-000, Campus Universitário, Viçosa, Minas Gerais, Brazil. [2]Departamento de Solos, Universidade Federal de Viçosa, Viçosa, Minas Gerais, Brazil. [3]Celulose Nipo Brasileira S.A., Ipatinga, Minas Gerais, Brazil. *Author for correspondence. E-mail: ribeiro@ufv.br*

ABSTRACT. Although the 3-PG model is widely used for forest productivity calculations, there are processes that do not present appropriate physical treatment. The aim of this study was to generate a tool to improve the water balance calculation in the model to enhance the energy balance and transpiration process. The calculation of transpiration was modified to account for variations in solar radiation with the inclination and azimuth of the terrain; the vapor pressure deficit was changed based on the relative humidity and air temperature; and the stomatal conductance varied according to solar radiation, vapor pressure deficit and air temperature. The water storage in the soil varied with the depth of the root system and the total water availability (TWA) in the soil. The assessment was also changed from a monthly to an hourly basis. The study was conducted in areas surrounding Cenibra, and the data were collected from the Rio Doce river basin, in the Brazilian state of Minas Gerais. Taken together, these modifications improved growth - modeling processes and enhanced the capacity of this analytical tool to differentiate intra - region productivity.

Keywords: tilt surface, solar radiation, stomatal resistance, ecophysiological model.

Introduction

Modeling has become a highly important tool for the design and prediction of the performance of plants under different conditions to improve the understanding of the ecophysiology of this organism.

Plant growth modeling is also an organized and structured method to examine the procedures and interactions that determine crop yields.

One of the advantages of using models is the ability to estimate the values of the systems they represent. However, because of the complexity of plant growth and development, models cannot address all of the processes and variables involved in crop growth. Thus, in spite of the excellent results obtained, models are only a simplification of reality.

Process - based models predict forest productivity based on the physiological processes that control plant growth (photosynthesis, biomass allocation, respiration, transpiration, nutrition and leaf and twig drop). Landsberg and Waring (1997) developed a forest growth model called "Physiological Principles in Predicting Growth" (3-

PG), which considers constant and variable physiological relationships.

The forest growth 3-PG model uses physiological principles to estimate forest productivity in even - aged and monospecific plantations. Forest growth can be realistically estimated and easily standardized for a particular forest type with only a few adjustments (LANDSBERG; WARING, 1997; PÉREZ-CRUZADO et al., 2011); therefore, 3-PG has become an appropriate and widely used model for the prediction of forest development.

One advantage of the 3-PG model is that it incorporates the following submodels:

a) carbohydrate assimilation (biomass production); b) biomass distribution among leaves, stem (wood) and roots; c) soil water balance; d) biomass allocation to variables of interest for forest growth; and e) dendrometric characteristics.

These submodels and their interactions largely represent procedures that influence plant development (DYE et al., 2004; ESPREY et al., 2004; NIGHTINGALE et al., 2008a).

The 3-PG model estimates the amount of radiant energy absorption in the canopy and the carbohydrate conversion and allocation to different tree parts (leaves, bark, wood and roots). It also describes plant growth, based on energy, temperature and nutrition level (soil fertility). The model incorporates empirical relationships derived from the experimental measurements and physiological processes used in models of carbon balance, thereby facilitating the calculation of complex physiological processes and reducing the amount of data.

The 3-PG model primarily requires the following input data in simulations. 1 - Climate variables: Monthly average of the maximum and minimum air temperatures (°C), solar radiation (MJ m^{-2} day^{-1}), rainfall (mm), vapor pressure deficit (mbar) and number of frost days per month. 2 - Soil variables: Fertility, soil texture and water availability. 3 - Plant variables: Initial leaf, wood and root biomass, tree populations and allometric relationships.

Although in recent years 3-PG has become one of the most widely used models to calculate forest productivity (ALMEIDA et al., 2004; RODRÍGUEZ-SUÁREZ et al., 2010; STAPE et al., 2004; NIGHTINGALE et al., 2008a and b), the physical treatment of some procedures, such as the water balance calculation, are not adequately represented in the model.

The standardization of various forest growth models to region - specific conditions, such as soil - climate, must also be improved to make this method even more useful for understanding forest growth. Thus, improved estimates of these processes might result in significant gains in the simulation of forest plantation growth.

The aim of this study was to generate a new tool to compute water balance using the 3-PG model to improve the transpiration process with respect to stomatal control, which is regulated through variations in air temperature, vapor pressure deficit and energy balance, and to enhance the treatment of energy balance with respect to variations in solar energy according to the face of exposure and slope of the terrain. The time scale for the water balance calculation was changed from a monthly to an hourly basis.

Material and methods

The 3-PG model was originally designed on a monthly basis. In our study, the time scale for the water balance calculation was converted to an hourly basis. Simulations were performed using the 3-PG model in its "original" format with modifications developed in this study.

The radiation and water balance calculations were modified to simulate the growth of eucalyptus plantations based on the concepts of 3-PG (radiation use efficiency, carbon partitioning and allometric equations).

Changes with regard to radiation balance were based on the variation of solar energy in relation to the exposed surface and land slope (IQBAL, 1983; KAMALI et al., 2006; MEFTI et al., 2003). For the water balance calculation, the treatment of transpiration was modified with regard to the stomatal control through variations in air temperature, vapor pressure deficit (VPD) and global solar radiation. The water storage capacity (WSC), which varies with the depth of the root system, and the total soil water availability (TWA) were also modified.

The variables in the original model were utilized on a monthly scale. In this study, the variables involved in the water and energy balance calculations were used on an hourly scale and later converted to a monthly scale to apply the simulations for 3-PG eucalyptus growth models.

Estimation of Transpiration

For the estimation of transpiration (Eq. 1), the Penman - Monteith equation (ALLEN et al., 1998) was used on an hourly basis, which includes aerodynamic components and energy balance. The solar radiation was corrected, i.e., the radiation was adjusted, according to the inclination and azimuth angle of the areas.

The treatment of surface stomatal conductance was corrected according to the vapor pressure deficit and variations in solar radiation and air temperature, based on equation (4) (CARNEIRO et al., 2008), which was developed for the same study region.

In the original 3-PG model, the VPD is treated as dependent on air temperature only; therefore, Allen et al. (1998) proposed the treatment of the VPD with relation to the relative humidity and air temperature.

The reflection coefficient (α) was also included in this simulation of transpiration and corrected according to the zenith angle, based on equation (27).

$$T = \frac{\Delta(R_n - G) + M'\rho_a C_p (e_s - e_a)/r_a}{[\Delta + \gamma(1 + r_d / r_a)]\lambda} \quad (1)$$

where:

T = canopy transpiration (mm h^{-1});

λ = latent evaporation heat flux (MJ m^{-2} s^{-1});

Δ = curve declivity of vapor saturation pressure (kPa °C^{-1});

Rn = net radiation (MJ m^{-2} s^{-1});

G = soil heat flux (MJ m^{-2} s^{-1});

M' = time conversion factor = 3600;

ρ_a = absolute air density (kg m^{-3});

γ = psychrometric coefficient (kPa °C^{-1});

C_P = specific heat at constant pressure, 1013 (10^{-3}) MJ Kg^{-1}°C^{-1};

(e_s - e_a) = vapor pressure deficit (kPa);

r_a = aerodynamic resistance (s m^{-1});

r_d = surface resistance (s m^{-1}).

In this model, stomatal conductance was used to determine the plant resistance against water loss through transpiration.

The canopy resistance was calculated according to the equation below.

$$r_d = \frac{r_e^*}{LAI} \quad (2)$$

r_d = surface resistance (s m^{-1}).

r_e^* = stomatal resistance, as determined through limits in soil water availability(s m^{-2}); and

LAI = leaf area index (m^2 m^{-2}).

$$r_e^* = \frac{r_e}{K_S} \quad (3)$$

Stomatal resistance (r_e) was calculated using the equation of Carneiro et al. (2008):

$$r_e = 1694.6\left(\frac{(VPD\,T)}{Rgi\;277.78}\right)^{1.0568} \quad (4)$$

$$Ks = \frac{\log(WS_i + 1.0)}{\log(CAD + 1.0)} \quad (5)$$

Ks = coefficient of soil moisture (adimensional);

WS_i = water storage on day in mm (equation 35);

WSC = water storage capacity in mm (equation 36);

Ln = Neperian logarithm;

Rgi = correction of the global incident solar radiation (diffuse and direct) in MJ m^{-2} h^{-1} (equation 23).

Incident solar radiation at the Top of the Atmosphere (R_0) in MJ m^{-2} h^{-1}:

$$R_0 = \frac{12(60)}{\pi} G_{SC} d_r [(\omega_2 - \omega_1) sen\phi\, sen\delta + + \cos\phi \cos\delta (sen\omega_2 - sen\omega_1)] \quad (6)$$

G_{SC} = solar constant = 0.0820 MJ m^{-2} min^{-1};

δ = solar declination (rad), equation 26; and

ϕ = latitude (rad).

$$dr = 1 + 0.033\cos\left(\frac{2\pi}{365} J\right) \quad (7)$$

dr = relative earth - sun distance (rad); and

J = number of day of the year.

$$\omega_1 = \omega - \frac{\pi\, t_1}{24} \quad (8)$$

$$\omega_2 = \omega + \frac{\pi\, t_1}{24} \quad (9)$$

t_1 = duration of calculation period (hour), i.e., 1 for a one - hour period or 0.5 for half - an - hour period;

ω_1 = solar time angle at the beginning of the period (rad); and

ω_2 = solar time angle at the end of the period (rad);

The time angle (ω) is the angle between the position of the sun and the local meridian. Due to 24 - hour Earth rotation, this angle varies 15° every 60 minutes and is negative in the morning, positive in the afternoon and numerically equal to zero at noon.

$$\omega = \frac{\pi}{12}[(T_1 + 0.06667(L_Z - L_m) + S_C) - 12] \quad (10)$$

T_1 = (hour + 0.5)

T_1 = standard clock time at the midpoint of the period [hour]. For example, for a period between 14 and 15 h, t = 14.5;

ω = solar time angle at the midpoint of one - hour or shorter period (rad); and

L_Z = longitude of the center of the local time zone (degrees west of Greenwich).

$$Lz = -\left(\frac{L_m}{15.0}\right)15 \quad (11)$$

L_m = longitude of the measurement site (degrees west of Greenwich);

$$S_C = 0.1645\sin(2b) - 0.1255\cos(b) - 0.025\sin(b) \quad (12)$$

Sc = seasonal correction for solar time (hour).

$$b = \frac{2\pi(J-81)}{364} \quad (13)$$

where:

J = number of day of the year; and

b = fit equation.

The correction for the variation in solar radiation during the day with regard to terrain inclination and exposed surface areas was based on the vertical angle of incidence of direct sun rays (θ) (IQBAL, 1983), where:

$$\cos\theta = A\cos\omega_1 + B\sin\omega_1 + C \quad (14)$$

$$A = \cos i\cos\phi\cos\delta + \sin i\cos A_Z\sin\phi\cos\delta \quad (15)$$

$$B = \sin i\cos\delta\sin A_Z \quad (16)$$

$$C = \cos i\sin\delta\sin\phi - \sin i\cos A_Z\sin\delta\cos\phi \quad (17)$$

where:

δ = solar declination (equation 26);

ϕ = latitude in radians;

θ = incidence angle of direct sun rays (degrees);

ω_1 = time angle (degrees);

A, B and C = parameters of simplification of the equation;

A_z = azimuth angle or surface orientation (rad); and

i = surface inclination angle (degrees).

$$A_z = A_{zd}\left(\frac{PI}{180}\right) \quad (18)$$

A_{zd} = azimuth angle or surface orientation (degrees).

The hourly diffuse solar radiation was estimated using a previously described mathematical model (FACCO et al., 2009). This model utilizes a combination of linear and nonlinear cubic equations, in which different intervals between global and atmospheric radiation are considered (R_o) (FACCO et al., 2009).

The direct and diffuse solar radiance was determined using the following equation:

$$I_d = M\,R_g, M_T = \frac{R_g}{R_0} \quad (19)$$

$M = 1 - 0.221M_T$, for $M_T \leq 0.20$ (20)

$M = (0.798 + 2.442M_T - 9.634M_T^2 + 6.9381M_T^3)$, (21)

for $0.20 > M_T \leq 0.80$

$M = 0.135$, para $M_T > 0.80$ (22)

where:

I_d = diffuse solar radiation on a plain surface (MJ $m^{-2}h^{-1}$);

M = ratio between i_d and R_g (adimensional);

M_T = ratio between R_g and R_0 (adimensional); and

R_g = global solar radiation (MJ $m^{-2}h^{-1}$).

$$Rgi = (Dn\,Cos\theta) + (I_d\,Cos\left(\frac{i}{2}\right)Cos\left(\frac{i}{2}\right)) + ((Dn\,CosZ + I_d)\,\alpha\sin\left(\frac{i}{2}\right)\sin\left(\frac{i}{2}\right)) \quad (23)$$

where:

Rgi = corrected incident global solar radiation (diffuse and direct) in MJ $m^{-2}h^{-1}$;

i = surface inclination angle;

θ = angle of incidence of the direct sun rays (degrees); and

α = coefficient of reflection (albedo), equation 27.

$$Dn = \frac{I_d}{Cos(Z)} \quad (24)$$

where:

Dn = normal direct solar radiation on the surface (corrected direct); and

Z = zenith angle.

$$Cos(Z) = Sin\phi\,Sin\delta + Cos\phi\,Cos\delta\,Cos\omega_1 \quad (25)$$

where:

ϕ = latitude (rad);

δ = solar declination (rad);

ϕ = (ϕ pi)/180 (conversion of the latitude to radians); and

ω_1 = solar time angle at the beginning of the period (rad).

The solar declination, for a particular day of the year, was calculated using the equation:

$$\delta = (23.45\sin(\frac{360}{365}(284 + J)))\frac{PI}{180} \quad (26)$$

J = number of day of the year;

$$\alpha = (0.1715\ \exp(0.0056Cos(Z)\frac{180}{PI})) \quad (27)$$

$$RS_W = 0.0036Rgi \quad (28)$$

RS_W = incident global solar radiation (W m^{-2}).

$$Rns = (1.0 - \alpha)Rgi \quad (29)$$

Rns = short - wave net radiation (MJ m^{-2} h^{-1}).

The long - wave net radiation was estimated using a regression equation, derived from the incident global solar radiation of the study region:

$$Rnl_W = -0.00002RS_W^2 - 0.0272RS_W - 1.3458 \quad (30)$$

Rnl_W = correction of the long - wave net radiation (W m^{-2})

$$Rnl = Rnl_W\ 0.0036 \quad (31)$$

Rnl = long - wave net radiation (MJ m^{-2} h^{-1})

$$Rn = Rns + Rnl \quad (32)$$

Rn = net radiation (MJ m^{-2} h^{-1})

If hours < 5 or hours >= 18, then

$$G = 0.1Rn \quad (33)$$

Or

$$G = (0.1131RS_W + 2.3949)0.0036\ (MJ\ m^{-2}\ h^{-1}) \quad (34)$$

G = soil heat flux (MJ m^{-2} h^{-1})

Experimental area

This study was conducted in the state of Minas Gerais in areas near Celulose Nipo - Brasileira SA (CENIBRA). Meteorological data from four regions of the Rio Doce basin in the state of Minas Gerais were used (Figure 1). The basin has a rather hilly relief, with altitudes ranging from 240 to 1273 m, which is representative of essential soil and climate differences at study sites (SOUZA et al., 2006) in the counties Belo Oriente (Rio Doce), Santa Barbara (Santa Barbara), Peçanha (Virginópolis) and Antonio Dias (Cocais) (Figure 1).

The region of Rio Doce belongs to the county of Belo Oriente, where the relief is plain and rather hilly, with a deep clayey soil medium of limited fertility. Annual average values of 1160 mm rainfall, 25.1°C air temperature and 66% relative humidity are observed. The meteorological station is located at a height of 240 m asl, and a annual water deficit of 402 mm is observed, using a climatic water balance (SOUZA et al., 2006).

The region of Santa Barbara, county of Santa Barbara, is a soft, wavy and mountainous relief, with a shallow to deep sandy - clayey soil medium of limited fertility. Annual average values of 1424 mm rainfall, 22.7°C air temperature and 59% relative humidity are observed. According to the climatic water balance, an annual water deficit of 119 mm was determined (SOUZA et al., 2006). The meteorological station is located at an altitude of 847 m asl.

The region of Virginópolis, county of Peçanha, is a soft and hilly relief, with deep, clayey soil medium of limited fertility and average annual values of 1,137 mm rainfall, 22.1°C of air temperature and 60% relative humidity. The annual water deficit, according to the climatic water balance, is 162 mm (SOUZA et al., 2006), and the meteorological station is located at an altitude of 1,015 m asl.

The relief in the region of Cocais, county Antonio Dias, is rather mountainous, with deep, sandy - clayey soil of limited fertility. Average annual values of 1258 mm rainfall, 21.2°C air temperature and 66% of air relative humidity are observed. An annual water deficit of 148 mm was determined using the climatic water balance (SOUZA et al., 2006). The altitude of the meteorological station is 1273 m.

The following meteorological data were used: mean temperature (°C), relative humidity (%), wind speed (m s^{-1}), solar radiation (MJ m^{-2}) and rainfall (mm). The meteorological data were collected on an hourly basis, using automatic weather stations, from April 2000 to April 2006.

The database of the forest inventory contained information on the month and year of planting, the trunk diameter at breast height (cm) and the total tree height (m).

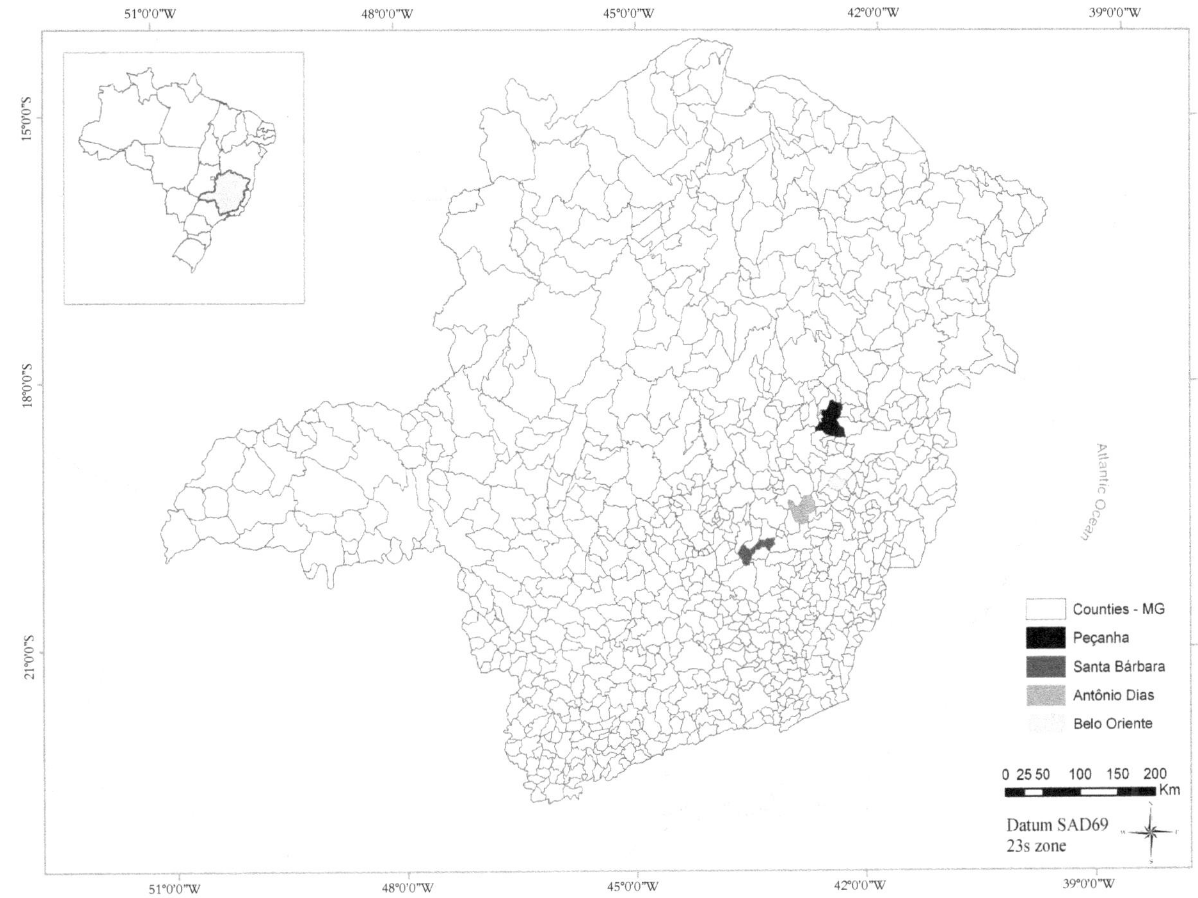

Region	County	Latitude	Longitude	Altitude (m)*	Mean temp. (°C)	Rainfall (mm)	Water stress (mm)**
Virginópolis	Peçanha	18°42'24"S	42°29'25"O	1015 m	22.1	1137	203
Cocais	Antônio Dias	19°29'19"S	42°51'54"O	1273 m	21.2	1258	148
Santa Bárbara	Santa Bárbara	19°59'00"S	43°18'19"O	847 m	22.7	1424	195
Rio Doce	Belo Oriente	19°18'50"S	42°23'38"O	240 m	25.1	1160	429

*Values correspond to the location of the meteorological stations. **Thornthwaite and Matter 1948.

Figure 1. Location of the studied counties and meteorological stations in the State of Minas Gerais.

Based on biometric models, which were calibrated for the reality of the stand areas and with input variables such as the trunk diameter at breast height (DBH) and tree height, the mean annual increment (MAI) was estimated in 2004 and 2005 for each plot.

Simulations

In this study, simulations were performed for three exposed terrain surfaces (azimuth): north (N), south (S) and plain (East - West). We considered two inclination classes of the field (0-15 and > 15°). Significant differences in the radiation incidence were observed in studies conducted in the same region (FACCO et al., 2009), with slopes between 0 and 15 and above 15°.

In this study, the simulations were performed using the original 3-PG (LANDSBERG; WARING, 1997) and the parametric model, which incorporated the water balance, coefficient of soil moisture, solar radiation and transpiration.

The proposed growth model was initially parameterized, according to the region - specific ecophysiological characteristics for the development of plantations. The parameterization of the original and proposed models was similar to the previously described values (GUIMARÃES et al., 2007), except for the transpiration value, which represents the relationship between the net (NPP) and gross (GPP) primary production and was obtained in a sensitivity analysis based on values of Landsberg and Waring (1997).

Water balance

A submodel of water balance (described below) was coupled with the 3-PG model to simulate the growth of eucalyptus plantations.

The water balance was calculated for the soil layer, which contains roots and varies depending on the plant growth stage. Once the depth of the root system is determined, the control volume and maximum quantity of plant - available water can be inferred. In this control volume, the water entry and exit is calculated to determine the variation in the water amount stored in the soil using the following formula:

$$WS_i = WS_{i-1} + PTH - IPC - PERC - ET \quad (35)$$

where:

WS_i = water storage on day i (mm);

WS_{i-1} = water storage on the day before (mm);

PTH = total rainfall on day i (mm);

IPC = total intercepted rainwater by the canopy and by the litter on day i (mm);

PERC= percolation into the useful soil layer where roots were found on day i (mm);

ET = canopy transpiration on day i (mm).

The calculation of the water balance began in the month when the accumulated rainfall was sufficient to reach the maximal capacity of available water in the soil.

The model was fed with hourly data of an automatic meteorological station.

The total water storage capacity of the soil must be calculated to the soil depth that corresponds to the real depth of the root system of the crop as follows:

$$WSC = TWA\ EDRS \quad (36)$$

where:

WSC = available water capacity in soil in mm m^{-1};

TWA = total water available, in mm m^{-1} of soil depth; and

EDRS = effective depth of the root system, in cm.

$$TWA = \left(\frac{CCamp - PmPer}{10} da \right) \quad (37)$$

where:

CCamp = field capacity, % in weight;

PmPer = wilting point, % in weight; and

da = apparent soil density, g cm^{-3}.

The field capacity values, wilting point and soil density were determined in the soil samples collected from the study areas.

For the effective depth of the root system (EDRS), a soil section containing at least 80% of the root system of the crop must be considered. This depth depends on the crop and the soil depth of the area. The effective depth of the root system was estimated using a previously proposed equation (FACCO et al., 2009):

$$EDRS = 287.31(1.0 - \exp(-0.19483\ age)^{1.0219} \quad (38)$$

Using the interception of precipitation, Moura et al. (2009) showed that the forest cover causes a reduction in the total rainwater that reaches the ground, which consequently influences the dynamics of the surface runoff that reaches the hydrologic network and the process of infiltration that feeds subsurface water.

The interception of the plant cover was estimated according to Costa et al. (2006):

$$IPC = PTH\ 0.12 \quad (39)$$

The water percolation was estimated using the following equation:

$$PERC = WS_{i-1} - WSC \quad (40)$$

where:

IPC = total intercepted rainwater from the canopy and the litter on day i (mm);

PTH = total rainfall on day i (mm);

PERC = percolation into the useful soil layer where roots were found on day i (mm);

WS_{i-1} = the initial storage was equal to TWA, that is, WS_{i-1} = TWA.

Determination of the mean annual increment

For calibration of the modified model, a forestry inventory database was used, which contained information on the month and year of planting, the trunk diameter at chest height (cm) and the total height of the trees (m). Based on biometric models, which were calibrated to the planted area, trunk diameter at breast height (DBH) and total height of the trees, the mean annual increment (MAI) was estimated for each plot. Thus, 28 trees were randomly selected, and the DBH and total height were measured. The total height and DBH of the other trees were estimated using adjusted biometric models.

To estimate the height of the trees, the following model was used:

$$LnHT = b_0 + \frac{b_1}{DBH} + b_2 * LnHd + \xi \qquad (41)$$

where,

HT = height of the tree which was not measured (m);

$b_0, b_1, b_2 =$ coefficients determined using regression analysis and the variables for DBH, Hd and observed height;

DBH = Diameter at Breast Height (cm);

Hd = mean height of the dominant trees of the plot; and

ξ = random error.

To estimate the mean annual increment (MAI), the following biometric model was used:

$$LnVarv = \beta_0 + \beta_1 LnDBH + \beta_2 * LnHT + \xi \qquad (42)$$

where:

$Varv$ = tree volume (m^3);

$\beta_0, \beta_1, \beta_2$ = coefficients for each location studied, obtained from the local company (Table 1);

Table 1. Values of the coefficients β0, β1, β2 for each location studied.

Location	β_0	β_1	β_2
Cocais	-10.142538	1.836628	1.106458
Virginópolis	-10.373936	1.809317	1.199962
Rio Doce	-10.143457	1.856938	1.095924
Santa Bárbara	-10.156326	1.870474	1.079379

DBH = Diameter at Breast Height (cm);

HT = height of the tree (m) – estimated by the LnHT equation (Eq.41); and

ξ = random error.

The trunk volume per hectare was calculated using the following formula:

$$\frac{V}{ha} = \frac{\Sigma\, Varv \;\; * 10000}{Plot\_Area} \;(m^3) \qquad (43)$$

and the mean volume by:

$$IMA = \frac{\frac{V}{ha}}{Age\,(years)} \; m^3\, ha^{-1} year^{-1} \qquad (44)$$

Results and discussion

Figure 2 shows the correlations between the MAI modeled and observed data in all four regions studied: Cocais, Virginópolis, Rio Doce and Santa Barbara.

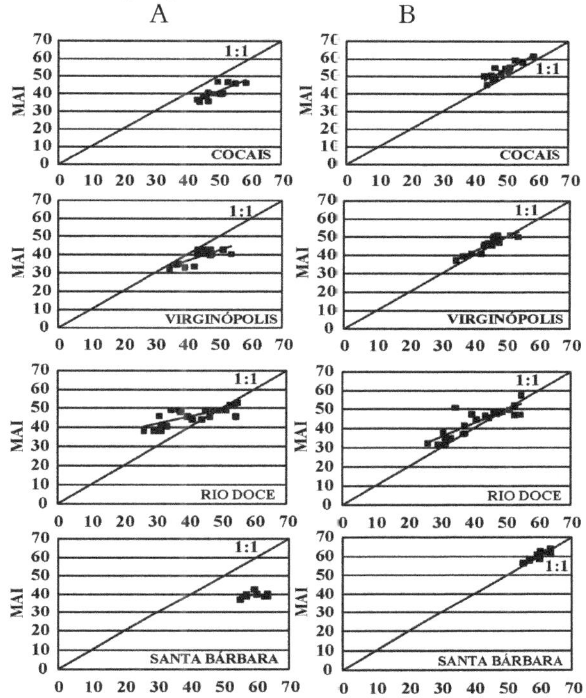

Figure 2. Correlation between the data from the mean annual increment (MAI – $m^3\ ha^{-1}\ yr^{-1}$) observed (axis x) and simulated (axis y) using the original model (a) and between the observed (axis x) and simulated (axis y) MAI using the model proposed (b) for the regions of Cocais, Virginópolis, Rio Doce and Santa Bárbara.

These diagrams illustrate the correlations between the observed data using the original model (a) and the proposed model (b). The determination coefficients for the original model varied from 0.06 to 0.66, and the angular coefficient (Table 2) varied from 0.13 to 0.75.

Using the original model in simulations for the same study area, Guimarães et al. (2007) was able to differentiate the productivity between regions, based on variations in the soil - climatic characteristics of the Rio Doce basin.

Table 2. Statistical indices used in the performance analysis of the models: MBE (mean bias error), MAE (mean absolute error), RMSE (root mean square error), d (agreement index of Willmott), R^2 (coefficient of determination) and "b" (angular coefficient for linear fit through zero).

	MBE	MAE	RMSE	d	R^2	b
Cocais Original	-8.49	8.49	8.87	0.73	0.66	0.75x
Cocais Proposed	3.82	3.82	4.37	0.93	0.78	0.91x
Virginópolis Original	-5.37	5.37	6.19	-0.34	0.59	0.58x
Virginópolis Proposed	0.86	1.69	2.06	0.84	0.84	0.79x
Rio Doce Original	4.49	5.78	7.52	0.29	0.56	0.38x
Rio Doce Proposed	1.94	3.19	4.86	0.71	0.73	0.70x
Santa Bárbara Original	-20.15	20.15	20.34	-14.10	0.06	0.13x
Santa Bárbara Proposed	0.45	1.31	1.44	0.78	0.74	0.71x

The coefficient of determination (Table 2) in the proposed model ranged from 0.74 to 0.84, respectively, for Santa Barbara and Virginópolis. The minor adjustment observed for Santa Barbara could reflect the smaller number of observational data available for analysis. A good fit was observed in Cocais and Rio Doce, with coefficients of determination of 0.78 and 0.73, respectively. The angular coefficient between the regions using the proposed model varied from 0.70 to 0.90.

Modifications to the transpiration calculations and improvements to the water balance, which, in contrast with the original model, considered variations in root growth with age, changes to the TWA in the soil and the "time step" permitted for obtaining better results. Thus, the proposed model obtained a significant gain in explanatory power, as demonstrated by the higher correlation coefficients and angular coefficients closer to 1. These improved results showed greater sensitivity to variations and the exponential gain of growth with age.

To improve the propose model, the transpiration calculation was adjusted, i.e., the methodology for calculating the stomatal resistance and vapor pressure deficit was modified. With these modifications, the model demonstrated that the total annual transpiration for Cocais ranged from 370 to 1,060 mm and that for Rio Doce ranged from 803 to 1,298 mm. In Santa Barbara, this value ranged from 451 to 1,478 mm, and in the region of Virginópolis, the total annual transpiration ranged from 707 to 1,024 mm. The original model showed annual transpiration totals ranging from 400 to 818 mm for Cocais, 465 to 683 mm for Santa Barbara, 559 to 802 mm for Rio Doce and 528 to 796 mm for Virginópolis.

The results obtained with the propose model are consistent with the values obtained in studies performed in the same region, and higher transpiration values than those obtained by the original model were observed (SOUZA et al., 2006). Because these regions are located in the tropics, i.e., regions with greater rain and atmospheric demands, high transpiration values and consequently productivity were expected.

With changes in transpiration, improvements in water balance with consideration for age - related root growth variations, and changes in the TWA of the soil and 'time step', a greater sensitivity to variations and an exponential gain in growth with age were observed.

Table 2 presents the statistical analyses (FEIKEMA et al., 2010; PEREIRA et al., 2009) of the original and proposed 3-PG models compared with the observed data at the four locations studied. The mean bias error (MBE) for the regions of Cocais, Virginópolis and Santa Barbara was underestimated with the original model, with values of -8.49, -5.37 and -20.15, respectively, and overestimated with the proposed model, with values of 3.82, 0.86 and 0.45, respectively. For the Rio Doce region, both models overestimated the values, with values of 4.49 and 1.94 for the original and proposed models, respectively.

For the mean absolute error (MAE), the values of the two models in the four regions ranged from 5.37 to 20.15 for the original model and from 1.31 to 3.82 for the proposed model. For the root mean square error (RMSE), however, the values ranged from 6.19 to 20.34 for the original and 1.44 to 4.86 for the proposed model.

The values of the Willmott index of agreement (D) for the original model varied from 14.10 to 0.73, indicating significant deviations in the regions of Rio Doce, Virginópolis and Santa Barbara. The results were significantly improved for the proposed model, with a level of agreement ranging from 0.71 to 0.93.

For the original model, the coefficient of determination (R^2) ranged from 0.06 to 0.66 and from 0.73 to 0.84 for the proposed model, indicating a generally good fit to the proposed model.

Although the fit of the proposed model to the line of equal values (1:1) was not perfect, the values of the evaluated indices were generally much better than those obtained using the original model.

Conclusion

Improving the calculation of environmental factors, such as vapor pressure deficit and solar irradiance, and modifying the water balance calculation to an hourly scale, which considers the TWA of the soil and the variation of root system depth with forest age, were useful in improving the estimation of the biomass gain in eucalyptus trees. These alterations had an important effect on wood volume gain.

Moreover, the changes to the 3-PG model improved the treatment of the procedures, and the computer simulation tool showed improvements in the differentiation of regional and intra - regional productivity. The proposed model is therefore no longer a tool for strategic planning only, but seems increasingly promising for use in forest management.

Acknowledgements

The authors would like to thank the Celulose Nipo - Brasileira (CENIBRA SA), the *Conselho Nacional de Desenvolvimento Científico e Tecnológico* (CNPq – no.

481195/2004-1 and no. 311922/2006-6), the Coordenação de Aperfeiçoamento de Pessoal de Nível Superior (CAPES) and the *Fundação de Amparo à Pesquisa do Estado de Minas Gerais* (CAG 61879/05) for support and financial assistance in developing and conducting this research.

References

ALLEN, R. G.; PEREIRA, L. S.; RAES, D.; SMITH, M. Crop evapotranspiration: Guidelines for computing crop water requirements. **Irrigation and Drainage**. Paper 56. Italy: UN-FAO, 1998.

ALMEIDA, A. C.; LANDSBERG, J. J.; SANDS, P. J. Parameterisation of 3-PG model for fast-growing *Eucalyptus grandis* plantations. **Forest Ecology and Management**, v. 193, n. 1-2, p. 179-195, 2004.

CARNEIRO, R. L. C.; RIBEIRO, A.; HUAMAN, C. A. M. Y.; LEITE, F. P.; SEDIYAMA, G. C.; BARROS, N. F. Consumo de água em plantios de eucalipto - Parte 1: Determinação da condutância estomática em tratamentos irrigados e não irrigados. **Revista Árvore**, v. 32, n. 1, p. 1-10, 2008.

COSTA, A. C. L.; BRAGA, A. P.; GONÇALVES, P. H. L.; COSTA, R. F.; SILVA JÚNIOR, J. A.; MALHI, Y. S.; ARAGÃO, L. E. O. E. C.; MEIR, P. Estudos hidrometeorológicos em uma Floresta Tropical chuvosa na Amazônia - projeto Esecaflor. **Revista Brasileira de Meteorologia**, v. 21, n. 3b, p. 283-290, 2006.

DYE, P. J.; JACOBS S.; DREW, D. Verification of 3-PG growth and water-use predictions in twelve Eucalyptus plantation stands in Zululand, South Africa. **Forest Ecology and Management**, v. 193, n. 1-2, p. 197-218, 2004.

ESPREY, L. J.; SANDS, P. J.; SMITH, C. W. Understanding 3-PG using a sensitivity analysis. **Forest Ecology and Management**, v. 193, n. 1-2, p. 235-250, 2004.

FEIKEMA, P. M.; MORRIS, J. D.; BEVERLY, C. R.; COLLOPY, J. J.; BAKER, T. G.; LANE, P. J. Validation of plantation transpiration in south-eastern Australia estimated using the 3PG+ forest growth model. **Forest Ecology and Management**, v. 260, n. 5, p 663-678, 2010.

FACCO, A. G.; RIBEIRO, A.; SEDIYAMA, G. C.; LEITE, F. P.; BARROS, N. F. Saldo de radiação em plantios de eucalipto em áreas de relevo ondulado. **Revista Árvore**, v. 33, n. 3, p. 471-480, 2009.

GUIMARÃES, D. P. G.; SILVA, G. G. C.; SANS, L. M. A.; LEITE, F. P. Uso do modelo de crescimento 3-PG para o zoneamento do potencial produtivo do eucalipto no estado de Minas Gerais. **Revista Brasileira de Agrometeorologia**, v. 15, n. 2, p. 192-197, 2007.

IQBAL, M. **An introduction to solar radiation**. Toronto: Academic Press, 1983.

KAMALI, G. H. A.; MORADI, I.; KHALILI, A. Estimating solar radiation on tilted surfaces with various orientations: a study case in Karaj (Iran). **Theoretical Applied Climatology**, v. 84, n. 4, p. 235-241, 2006.

LANDSBERG, J. J.; WARING, R H. A generalized model of forest productivity using simplified concepts of radiation-use efficiency, carbon balance and partitioning. **Forest Ecology and Management**, v. 95, n. 3, p. 209-228, 1997.

MEFTI, A.; BOUROUBI, M. Y.; ADANE, A. Generation of hourly solar radiation for inclined surfaces using monthly mean sunshine duration in Algeria. **Energy conversion and Management**, v. 44, n. 19, p. 3125-3141, 2003.

MOURA, A. E. S. S.; CORREA, M. M.; SILVA, E. R.; FERREIRA, R. L. C.; FIGUEIREDO, A. C.; POSSAS, J. M. C. Interceptação das Chuvas em um Fragmento de Floresta da Mata Atlântica na Bacia do Prata, Recife, PE. **Revista Árvore**, v. 33, n. 3, p. 461-469, 2009.

NIGHTINGALE, J. M.; HILL, M. J.; PHINN, S. R.; DAVIES, I. D.; HELD, A. A.; ERSKINE, P. D. Use of 3-PG and 3-PGS to simulate forest growth dynamics of Australian tropical rainforests: I. Parameterisation and calibration for old-growth, regenerating and plantation forests. **Forest Ecology and Management**, v. 254, n. 2, p. 107-121, 2008a.

NIGHTINGALE, J. M.; HILL, M. J.; PHINN, S. R.; DAVIES, I. D.; HELD, A. A. Use of 3-PG and 3-PGS to simulate forest growth dynamics of Australian tropical rainforests: II. An integrated system for modelling forest growth and scenario assessment within the wet tropics bioregion. **Forest Ecology and Management**, v. 254, n. 2, p. 122-133, 2008b.

PEREIRA, D. R.; YANAGI, S. N. M.; MELLO, C. R.; SILVA, A. M.; SILVA, L. A. Desempenho de métodos de estimativa da evapotranspiração de referência para a região da Serra da Mantiqueira, MG. **Ciência Rural**, v. 39, n. 9, p. 2488-2493, 2009.

PÉREZ-CRUZADO, C.; MUÑOZ-SÁEZ, F.; BASURCO, F.; RIESCO, G.; RODRÍGUEZ-SOALLEIRO, R. Combining empirical models and the process-based model 3-PG to predict Eucalyptus nitens plantations growth in Spain. **Forest Ecology and Management**, v. 262, n. 6, p. 1067-1077, 2011.

RODRÍGUEZ-SUÁREZ, J. A.; SOTO, B.; IGLESIAS, M. L.; DIAZ-FIERROS, F. Application of the 3PG forest growth model to a Eucalyptus globulus plantation in Northwest Spain. **European Journal of Forest Research**, v. 129, n. 4, p. 573-583, 2010.

SOUZA, M. J. H.; RIBEIRO, A.; LEITE, H. G.; LEITE, F. P.; MINUZZI, R. B. Disponibilidade hídrica do solo e produtividade do eucalipto em três regiões da bacia do rio Doce. **Revista Árvore**, v. 30, n. 3, p. 399-410, 2006.

STAPE, J. L.; RYAN, M. G.; BINKLEY, D. Testing the utility of the 3-PG model for growth of *Eucalyptus grandis* X *urophylla* with natural and manipulated supplies of water and nutrients. **Forest Ecology and Management**, v. 193, n. 1-2, p. 219-234, 2004.

16

Model to support the evaluation of the impact of the use and management of the soil on the water balance

Eloy Lemos de Mello[1*], Fernando Falco Pruski[2], Ildegardis Bertol[3], Lucas Mucida Costa[2], Gleiph Ghiotto Menezes[2] and Nívia Carla Rodrigues[2]

*[1]Universidade Estadual do Oeste do Paraná, Rua Universitária 2069, 85819-110, Jardim Universitário, Cascavel, Paraná, Brazil. [2]Universidade Federal de Viçosa, Viçosa, Minas Gerais, Brazil. [3]Universidade do Estado de Santa Catarina, Lages, Santa Catarina, Brazil. *Author for correspondence. E-mail: eloy.mello@unioeste.br*

ABSTRACT. The purpose of this work was to develop a hydrological model for the simulation of the water balance in agricultural areas. The model uses a double exponential equation to represent the rainfall profile and uses the Green-Ampt-Mein-Larson equation to estimate the soil water infiltration rate. To evaluate the model, monthly and yearly rainfall and runoff data were obtained from four experimental plots located at the State University of Santa Catarina, UDESC, in the period from 2003 to 2008. The four treatments were bare soil (BS), conventional tillage (CT), minimum tillage (MT) and no-tillage (NT). The simulation was carried out using two synthetic rainfall series, one adjusted on a monthly basis (SAMB) and the other adjusted on a daily basis (SADB). The parameters for the Green-Amp equation modified by Mein-Larson (GAML) were adjusted, and the best combination was found to be the replacement of the saturated hydraulic conductivity (K) by the constant infiltration rate (fc) and the estimation of the wetting front head pressure as a function of the hydraulic conductivity in the saturation zone ($\psi(K_W)$). There was not a significant difference between the observed annual mean runoff and the annual mean runoff estimated by the model.

Keywords: runoff, infiltration, water erosion, hydrological modeling.

Introduction

The knowledge of the various processes that compose the hydrological cycle is important for many reasons, for example, to estimate the water balance in different soil usage scenarios (LEGESSE et al., 2003), to support the planning and management of water resources (BAIGORRIA; ROMERO, 2007), to study the soil water erosion (MOHAMMAD; ADAM, 2010) and the soil and water management in agricultural areas (JI et al., 2007), and to promote the efficient use of and minimize the conflicts for the water resources (JIE et al., 2010).

Hydrological models are important analysis tools because they promote a better understanding of the hydrological processes; by analyzing the performance of soil management practices, the models provide support to decision-making in the implementation of agricultural systems (BAIGORRIA; ROMERO, 2007; MELLO et al., 2008). The major difficulties in the simulation of the water balance are the complex relationships among the hydrological processes, the climate and the vegetation dynamics (BOULAIN et al., 2009), and the various spatial and time scales that may need to be considered (MAAYAR et al., 2009).

There are various models that have been developed to perform hydrological simulations (NOTTER et al., 2007), but the main limitations include obtaining reliable data, especially climatic data, and determining some of the soil properties in specific situations (MELLO et al., 2008; BORMANN et al., 2007). The application and the adjustment of these models for the Brazilian edaphoclimatic conditions have been a great challenge to the professionals and researchers in the area.

This challenge is especially true when using the Green-Ampt equation modified by Mein-Larson, GAML (MEIN; LARSON, 1973), to estimate the soil water infiltration. The application of the GAML equation in its original form is not recommended because the input parameters do not properly represent the real field conditions, and the methods for obtaining the input parameters are not yet sufficiently reliable (CECÍLIO et al., 2007).

In the international literature, many works are dedicated to methods to obtain or adjust the GAML parameters (BARRY et al., 2005; CHU; MARIÑO, 2005; FUJIMAKI et al., 2004; GÓMEZ et al., 2009; MA et al., 2010; MOODY et al., 2009). These adjustments are necessary to take into consideration the natural changes in the soil and changes induced by human actions, such as the operations of soil management and tillage (RISSE et al., 1995).

The aim of this work was to develop and parameterize a hydrological model for the simulation of the water balance in agricultural areas. The GAML equation was parameterized to evaluate the capacity of the hydrological model to predict the runoff in agricultural areas.

Material and methods

Pruski et al. (2001) proposed a hydrological model that enables the estimation of the components in the water balance in agricultural areas. The model was based on the following assumptions: (I) the rainfall reaches the soil surface only after the canopy interception capacity has been completed, and (II) the depressional storage capacity does not vary with time. In this work, carried out from 2008 to 2010 at the Federal University of Viçosa, we made a few but significant changes to this model to improve its simulations capabilities.

Hydrological model

Rainfall

The first change implemented to the original model was the replacement of the rainfall profile based on the IDF (intensity, duration, frequency) curve by a double exponential function (Figure 1). The double exponential profile consists of an increasing exponential from the start of the event up to the time when the maximum rainfall intensity occurs, and a decreasing exponential after this point that describes the behavior of the profile until the end of the event (NICKS et al., 1995). The function is expressed by the equation:

$$i_i = \begin{cases} ae^{bt} & 0 < t < t_p \\ ce^{-dt} & t_p < t \le t_d \end{cases} \quad (1)$$

where:

i_i = the instantaneous rainfall intensity (mm h^{-1}); t = the time from the start of the rainfall (h); t_p = the time to the peak intensity (h); t_d = the rainfall duration (h); a, b, c, d = the adjustable parameters (dimensionless).

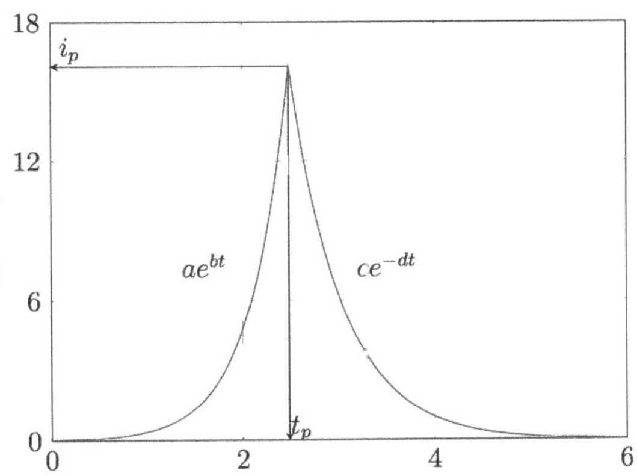

Figure 1. Example of a rainfall profile described by a double exponential function, where i_i is the instantaneous rainfall intensity, t is the time from the start of the rainfall, i_p is the peak intensity, and t_p is the time to the peak intensity.

The total rainfall (P, in mm) that occurs during the considered event, with the duration t_d, is obtained by the equation:

$$P = \int_0^{t_p} ae^{bt}dt + \int_{t_p}^{t_d} ce^{-dt}dt \quad (2)$$

Soil water infiltration

When the rainfall begins, it is immediately intercepted by the vegetation canopy. After the amount of rainfall reaches the amount that can be potentially intercepted by the vegetation cover, the rainfall begins to infiltrate the soil. From this moment, the infiltration rate (*f*) is equal to the instantaneous rainfall intensity (i_i). This condition is maintained until the instantaneous rainfall intensity exceeds the soil infiltration capacity and the depressional storage at the soil surface starts. The cumulative infiltration (*F*, in mm), which occurs between the end of the canopy interception and the beginning of the depressional storage, can be estimated by the equation:

$$F = \int_{t_{SL}}^{t_{iDS}} i_i dt \tag{3}$$

where:

tSL = the duration of the canopy interception (min.); $tiDS$ = the time when the depressional storage begins (min.).

From the time the depressional storage begins ($tiDS$), the infiltration rate is calculated by the Green-Amp equation modified by Mein-Larson (GAML):

$$f = K\left[1 + \frac{\psi(\theta s - \theta i)}{I}\right] \tag{4}$$

where:

f = the infiltration rate (mm h^{-1}); K = the saturated hydraulic conductivity (mm h^{-1}); θs = the saturated water content (cm^3 cm^{-3}); θi = the initial water content (cm^3 cm^{-3}); ψf = the wetting front suction head (mm); $F(t)$ = the cumulative infiltration (mm).

This equation for the infiltration rate holds until the rainfall intensity (ii) becomes equal to the infiltration rate (*f*), i.e., until the runoff ends (tfR). $tiDS$ and tfR are determined using the equation:

$$K\left[1 + \frac{\psi(\theta s - \theta i)}{I}\right] = \begin{cases} ae^{bt} & 0 < t < t_p \\ ce^{-dt} & t_p < t \le t_d \end{cases} \tag{5}$$

The cumulative infiltration (*F*, in mm) is obtained by the sum of the infiltration that occurs during the different phases associated with the water balance and is expressed by the equation:

$$F = \int_{t_{SL}}^{t_{iDS}} i_i dt + \int_{t_{iDS}}^{t_{fDS}} f dt + \int_{t_{fDS}}^{t_d} i_i dt \tag{6}$$

where:

t_{fDS} = the time when depressional storage ends (h).

If the rainfall ends before the total depressional storage capacity is filled, infiltration will continue to be expressed by equation 4 until the depressional storage capacity in the soil is completely filled (t_{fDS}). In this case, it is necessary to compute the infiltration that occurred after the end of the rainfall in equation 6.

Runoff

The runoff begins when the depressional storage capacity is filled. The runoff rate (qR, in mm h^{-1}) is expressed by the equation:

$$q_R = i_i - f \tag{7}$$

The total runoff (*R*, in mm) is obtained by equation 8. The depressional storage is not considered in this equation because it is added to the infiltration.

$$R = P - S_L - F \tag{8}$$

where:

SL = the canopy interception (*mm*).

Equations 7 and 8, used in the runoff calculation, are the same as those presented by Pruski et al. (2001). However, there are differences in the relationships among the water balance components due the change in the rainfall profile (Figure 2).

Figure 2a presents the components of the water balance in the soil surface using the IDF-based rainfall profile (PRUSKI et al., 2001). The rainfall intensity is at the maximum in the beginning and continually decreases until the end of the event.

Figure 2b presents the components of the water balance at the soil surface using a rainfall profile based on the double exponential equation proposed in this work. The rainfall intensity increases until it reaches its peak and subsequently decreases until the end of event.

In both cases, the water infiltration starts after the canopy interception capacity is completely filled. However, in Figure 2a, the infiltration rate is constantly decreasing, whereas in Figure 2b it increases until the rainfall intensity exceeds the soil infiltration capacity.

The equations for estimating the canopy interception, depressional storage, percolation and evapotranspiration were those used in the original model presented by Pruski et al. (2001).

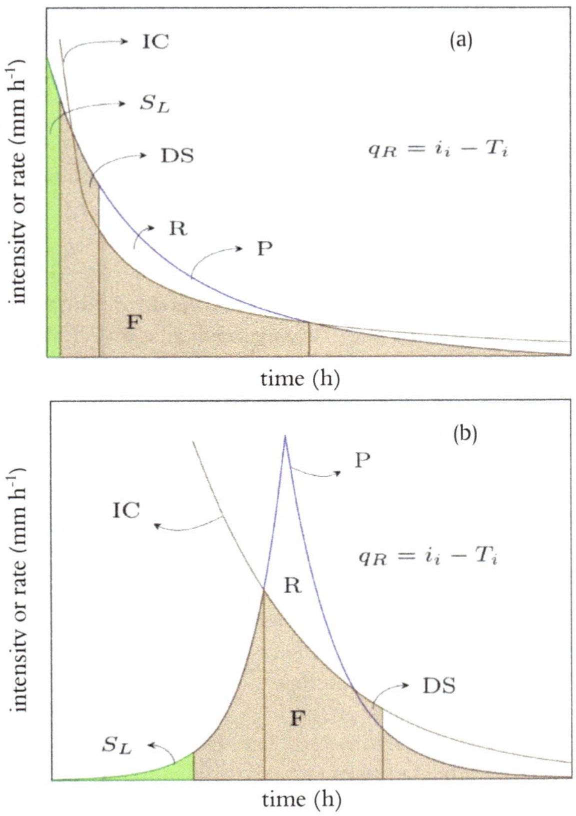

Figure 2. Components associated with the water balance in the soil surface proposed by Pruski et al. (2001), with a rainfall profile based on the IDF equation (a), and the water balance proposed in this work, with the rainfall profile based on a double exponential equation (b). IC = the infiltration capacity, S_L = the canopy interception, DS = the depressional storage, R = the runoff, F = the infiltration, P = the precipitation.

Model parameterization and evaluation

Experimental data

In the model parameterization, the data were obtained from four experimental plots located in the experimental area of the Agriculture and Veterinary Center – CAV of the State University of Santa Catarina (Brazil). The soil in the experiment site is an Inceptisol consisting of 420 g kg^{-1} of clay, 170 g kg^{-1} of sand and 410 g kg^{-1} of silt.

The plots were 22.1 m in length (in the direction of the slope) by 3.5 m in width, and the mean slope is 0.1 m m^{-1}. Each plot was delimited laterally and at the upper end by galvanized plaques of 2 x 0.2 m that were stuck in the soil up to 10 cm deep. At the lower end, a collecting gutter guided the runoff to a sedimentation tank with capacity of 750 *L*. This tank, is connected to a second tank using a Geib-type divider. With the aid of the divider, only 1/9 of the exceeding volume of the first tank is transferred to the second tank.

The following soil tillage treatments were applied: bare soil with a plowing + disking operation (BS), conventional tillage with a plowing + two disking operation (CT), minimum tillage with chiseling + disking operation (MT), and no-tillage (NT) where the soil did not receive any preparation. The three last treatments were cultivated under the rotation of soya, wheat, vetch, corn, oats and forage turnip. The historical series of rainfall and runoff used in this study covered the period from January 2003 to November 2008.

Rainfall synthetic series

The mean annual rainfall registered in the experimental area, in the period from 2003 to 2008, was 1182 mm, with a standard deviation of 140 mm. The lowest annual rainfall observed was 1027 mm in 2006, and the highest was 1352 mm in 2007.

To represent the rainfall registered in the historical series, the weather generator ClimaBR (ZANETTI et al., 2006) was used. Two synthetic series of rainfall were generated, each with a duration of 6 years.

Initially, a series adjusted on a monthly basis (SAMB) was generated, where the monthly rainfall was adjusted to be equal to the monthly rainfall observed in the historical series.

After composing the SAMB, a daily adjustment was carried out such that both the synthetic and historical series had the same calendar days with rain and the same total rainfall on each day, generating a series adjusted on a daily basis (SADB). In the SADB, it was not possible to adjust the total duration of each rainfall event.

Therefore, the basic difference between the two series is that in the SAMB, the distribution of the rainy days during the month is not required to be equal to the historical series, while in the SADB, both the monthly and the daily rainfall values are identical to those of the historical series.

Management and vegetation cover

The leaf area index (*LAI*), the random roughness (*RR*) and the crop coefficient (K_c) were selected according to the recommendations found in the literature for cultivations using the rotation system for different soil tillages (ALBERTS et al., 1995; ALLEN et al., 2006).

Soil

Four combinations of methods to determine the saturated hydraulic conductivity (*K*) and the wetting front head pressure (ψf) were investigated. Combinations 1 and 2 were obtained from Cecílio et al. (2007).

Combination 1

- The hydraulic conductivity in the saturated zone (*Kw*) is equal to the saturated hydraulic conductivity (*K*, in mm h^{-1}):

$$K_w = K \tag{9}$$

- The wetting front head pressure (*ψf*, in mm) is obtained from the soil porosity and texture, as suggested by Risse et al. (1995):

$$\psi_f = 0.01e^x \tag{10}$$

here,

$$x = 6.531 - 7.326\phi + 15.8C^2 + 3.809\phi^2 + 3.44SC - 4.989S\phi + 16.1S^2\phi^2 + 16C^2\phi^2 - 13.6S^2 - 34.8C^2\phi - 7.99S^2\phi \tag{11}$$

where:

ϕ = the soil porosity (cm^3 cm^{-3}), C = the clay fraction (kg kg^{-1}), and S = the sand fraction (kg kg^{-1}).

Combination 2

- The saturated hydraulic conductivity (*K*) is replaced with the hydraulic conductivity in the saturation zone (*Kw*), which is equal to the constant infiltration rate (*fc*, in mm h^{-1}), as proposed by Silva and Kato (1998):

$$K_w = f_c \tag{12}$$

The wetting front head pressure is obtained using equation 10.

Combination 3

- The saturated hydraulic conductivity (*K*) is replaced by the hydraulic conductivity in the saturation zone (*Kw*, in mm h^{-1}), which is obtained as a function of the soil clay fraction (*C*, in kg kg^{-1}), as suggested by Alberts et al. (1995):

$$K_w = 0.0066e^{2.44C} \tag{13}$$

- The wetting front head pressure (*ψf*, in mm) is obtained from the hydraulic conductivity in the saturated zone (*Kw*, in mm h^{-1}), as proposed by Rawls et al. (1996):

$$\psi_f = 49.03(K_w + 0.02)^{-0.4932} \tag{14}$$

Combination 4

- The saturated hydraulic conductivity (*K*) is replaced by the hydraulic conductivity in the saturation zone (*Kw*), which is equal to the constant infiltration rate (*fc*), as in equation 12.

- The wetting front head pressure (*ψf*) is obtained from equation 14.

The saturated hydraulic conductivity (*K*) was defined to be 18 mm h^{-1}, using the data obtained from Costa et al. (2006). According to the recommendation from Bertol et al. (2001), the constant infiltration rate (*fc*) for the no-tillage treatment in the experimental area is approximately 16.7 mm h^{-1}. For the conventional tillage (CT) and minimum tillage (MT) treatments, the infiltration rate was reduced to half of the *fc*, and for the bare soil (BS) treatment, the infiltration rate was reduced to one-third of the *fc*.

Analysis of the results

The results of the model parameterization were evaluated by comparing the observed and estimated annual averages using Student's *t*-test. To verify if the model presents overestimation or underestimation tendencies, we analyzed the coefficients of the regression equations between the observed and estimated values (monthly and annual). The adjustment of the equations was analyzed using the coefficient of determination (r^2).

Results and discussion

Model parameterization

Combination 4 (equations 12 and 14) resulted in the best GAML parameterization. The values of *fc* and *ψf* obtained with this combination are presented in Table 1.

Table 1. Values of the constant infiltration rate (f_c), the wetting front head pressure (ψ_f) and the saturated hydraulic conductivity (K) used for the model evaluation.

Soil tillage	f_c (mm h^{-1})	ψ_f (mm)	K (mm h^{-1})
Bare soil (BS)	5.5	65.0	18.0
Conventional tillage (CT)	8.4	53.0	18.0
Minimu tillage (MT)	8.4	53.0	18.0
No-tillage (NT)	16.7	38.0	18.0

Monthly runoff values

Figure 3 presents the scatter plots for the observed and estimated values of the monthly runoff using the series adjusted on a monthly basis (SAMB). In this figure, the monthly values from the period from 2003 to 2008 were grouped in a single diagram for each type of soil tillage.

The intercepts of the regression equations in Figure 3 show that the model had a tendency of overestimating the monthly runoff values in the months during which low magnitude events occurred. The highest intercept was observed in the bare soil treatment, (BS, 14.4), and the lowest in the no-tillage treatment (NT, 5.68). The intercepts were similar in the conventional tillage (CT, 7.24) and in the minimum tillage (MT, 7.55) treatments.

The lowest angular coefficients were observed in conventional tillage (CT, 0.35) and minimum tillage (MT, 0.33) treatments, and these values indicate the tendency to underestimate the runoff in the months during which the largest events occur. Although the no-tillage (NT) treatment was where the lowest r^2 occurred (0.27), it was also where the angular coefficient was closest to 1 and the intercept was closest to zero, among all the treatments.

Figure 4 presents the regression equations for the observed and estimated monthly runoff values using the series adjusted on a daily basis (SADB), in the period from 2003 to 2008. In this figure, the monthly values for all the years were grouped in a single diagram for each type of soil tillage.

In comparison with the SAMB results (Figure 3), it can be observed that the utilization of the SADB (Figure 4) resulted in modifications to the regression coefficients, and increased the coefficients of determination in all types of soil tillage. The intercepts became smaller, and the angular coefficients became closer to 1, indicating an improvement in the estimation of the monthly runoff.

For the bare soil (BS) treatment, the r^2 value increased from 0.55 using the SAMB to 0.66 using the SADB. For the no-tillage (NT) treatment, the increase was from 0.27 to 0.55. The highest increase occurred in the no-tillage (NT) treatment, due to the prevalence of observed values of lower magnitude. With the use of the SADB, there appeared to be an increase in the overestimation of the runoff, causing the slope of the line to increase and the value of the intercept to decrease. This difference may also have favored the lower dispersion of the values around the regression line (Figure 4).

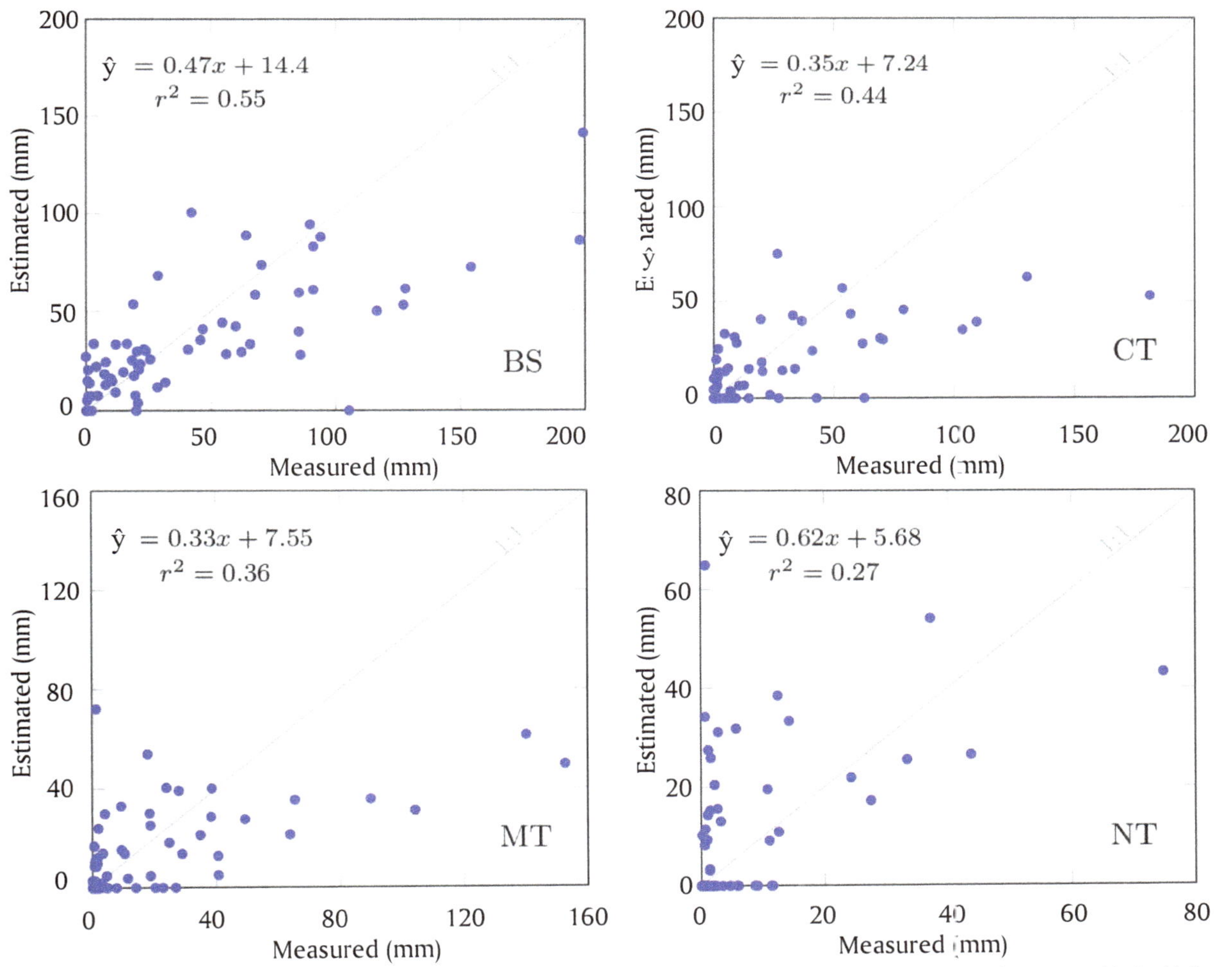

Figure 3. Regression between the observed and estimated monthly runoff values with the SAMB (series adjusted on a monthly basis) for the bare soil (BS), conventional tillage (CT), minimum tillage (MT) and no-tillage (NT) treatments. The period was from 2003 to 2008. Months with rainfall equal to zero were not considered.

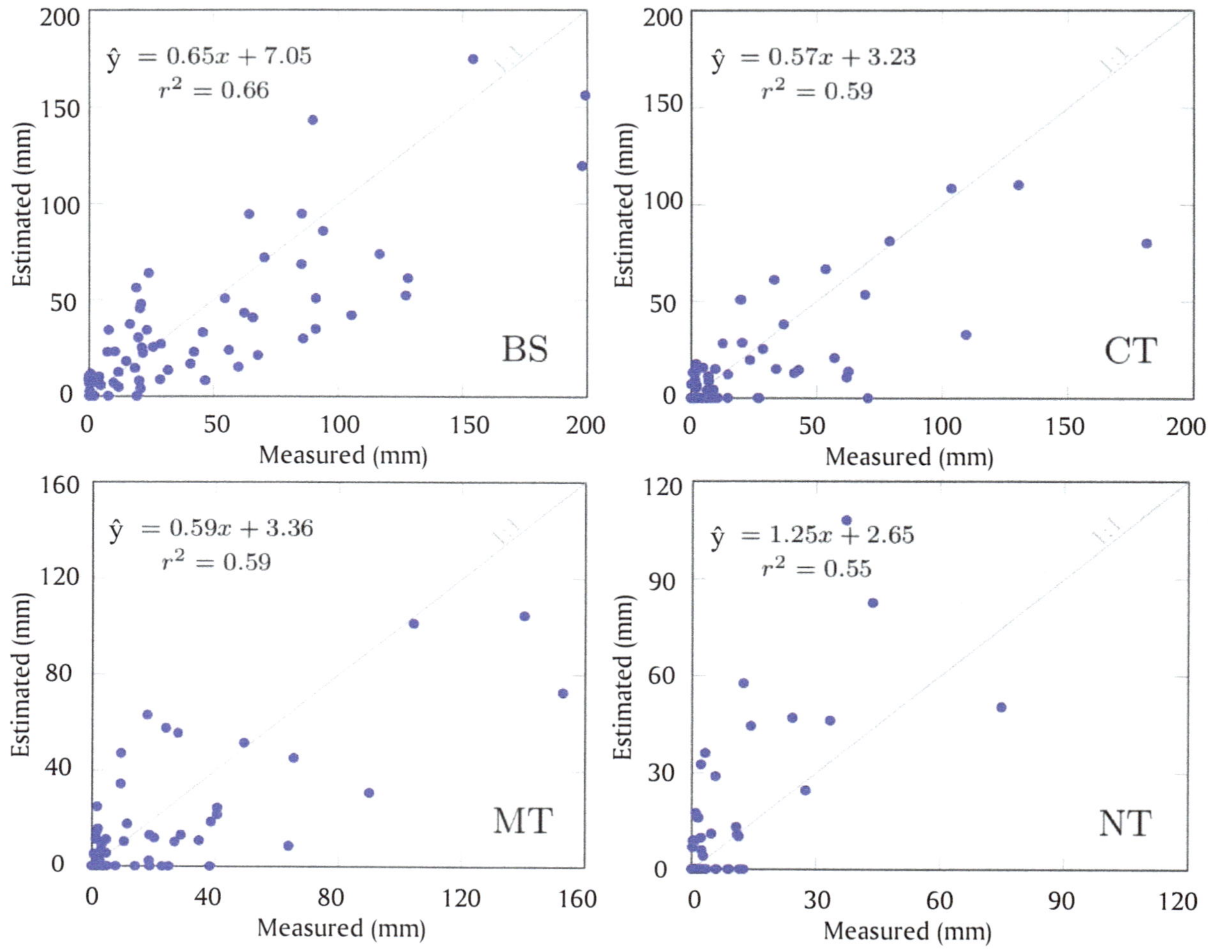

Figure 4. Regression between the observed and estimated monthly runoff values with the SADB (series adjusted on a daily basis) for the bare soil (BS), conventional tillage (CT), minimum tillage (MT) and no-tillage (NT) treatments. The period was from 2003 to 2008. Months with rainfall equal to zero were not considered.

This increase in the overestimation of the runoff also occurred in the other soil tillages. However, the overestimation of the smallest monthly runoff events was partially compensated for by the underestimation of the largest events.

The differences observed in the regression coefficients or in the r^2 values between the SADB (Figure 3) and the SAMB (Figure 4) can be explained by the differences between the two synthetic rainfall series.

The two synthetic series had monthly rainfall values equal to those observed in the historical series, but the SAMB (series adjusted on a monthly basis) did not produce the same distribution of rainy days or the same total daily rainfall that was observed in the historical series. However, the SADB (series adjusted on a daily basis) was identical to the historical series, with respect to both the number of rainy days and the total daily rainfall.

Figure 5 shows the differences between the two synthetic series, and represents the distribution of rainfall observed in the historical series and the distribution obtained with the synthetic rainfall series for September 2004. Only one month was selected at random to simplify the visualization in the graphic.

The rainfall recorded in the experimental area during this month was equal to 289.8 mm. Both the total rainfall obtained with the SAMB and the SADB were equal to the rainfall recorded in the historical series. However, Figure 5 shows that with the use of the SAMB, the rainfall was more evenly distributed along September, with a small predominance of larger events during the first 10 days of the month.

The SADB coincided exactly with the distribution observed in the historical series, presenting a lower number of rainy days, and consequently, higher concentrations of the total precipitation during those days. As shown in Figure 5, five events of large magnitude occurred in September, and they were responsible for 90% of the total monthly rainfall.

In the SADB, the rainy days and the total daily rainfall coincided with the historical series; thus, the monthly runoff estimation was better than the estimation performed with the SAMB.

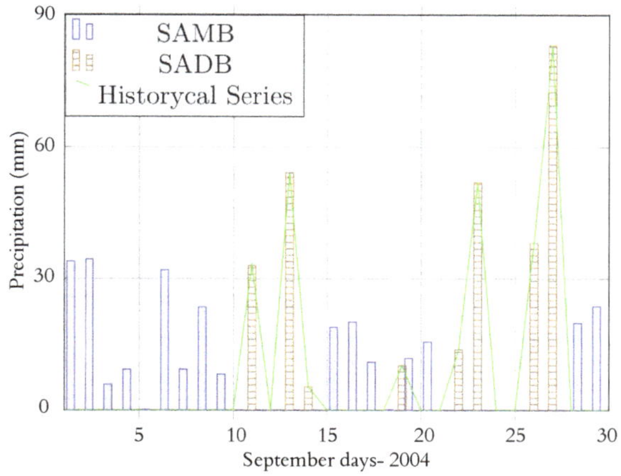

Figure 5. Distribution of the daily rainfall for September 2004 and the distribution obtained with the synthetic series SAMB (series adjusted on a monthly basis) and SADB (series adjusted on a daily basis).

Mean annual runoff

Figure 6 presents the mean annual runoff measured and estimated through the model with the utilization of the SAMB and the SADB for the different types of soil tillage.

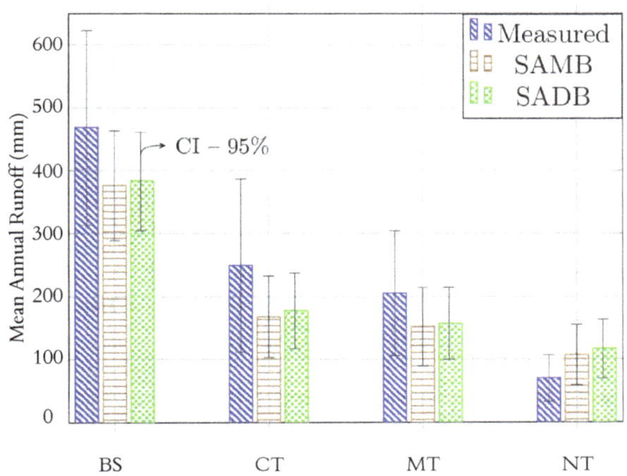

Figure 6. Observed and estimated annual runoff with the SAMB (series adjusted on a monthly basis) and the SADB (series adjusted on a daily basis), in bare soil (BS), conventional tillage (CT), minimum tillage (MT) and no-tillage (NT) treatments. Averages of 6 years and confidence intervals (CI) associated with 95% probability are shown.

As expected, the confidence intervals for the means are broad, due to the large standard deviation that typically occurs in works related to water and soil losses. With the exception of the no-tillage treatment, the confidence intervals were higher for the measured averages than the estimated averages.

The mean annual runoff measured was 24, 48 and 35% higher than the mean annual runoff estimated with the SAMB for the bare soil (BS), conventional tillage (CT) and minimum tillage (MT) treatments, respectively, and 22, 41 and 31% higher than the mean annual runoff estimated with the SADB for the same types of soil tillage. In the no-tillage (NT) treatment, the estimated average runoff with SAMB and with SADB was 53 and 67% higher, respectively, than the observed annual average runoff.

However, the deviations presented in Figure 6 were not sufficient to confirm that significant differences existed between the observed and estimated average annual runoff or between the estimated average runoff determined with the use of the SAMB and SADB.

This result confirms the pre-assumption that the use of a synthetic rainfall series for a long simulation period must adequately represent the behavior of the historical series and, consequently, that the average of the results from the synthetic series must be similar to the observed averages. In other words, for long simulation periods, the SAMB may be sufficient to provide an adequate estimation of the annual runoff average.

For short simulation periods, it may be interesting to adopt a synthetic rainfall series in which the distribution of the rainy days and the total daily rainfall are closest to the actual series.

Conclusion

The combination of parameters of the Green-Ampt equation modified by Mein-Larson (GAML) that resulted in the best runoff simulation was replacing the saturated hydraulic conductivity (K) by the constant infiltration rate (fc), as suggested by Silva and Kato (1998), and obtaining the wetting front head pressure (ψf) from the hydraulic conductivity of the soil, as suggested by Rawls et al. (1996).

The model overestimated the monthly mean runoff for the months during which the smallest events occurred, and underestimated the monthly mean runoff for the months during which the largest events occurred.

The model underestimated the annual mean runoff for the bare soil (BS), conventional tillage (CT), and minimum tillage (MT) treatments, and overestimated the annual mean runoff for the no-tillage (NT) treatment.

There was not a significant difference between the annual mean runoff observed and the annual mean runoff estimated by the model.

References

ALBERTS, E. E.; NEARING, M. A.; WELTZ, M. A.; RISSE, L. M.; PIERSON, F. B.; ZHANG, X. C.; LAFLEN, J. M.; SIMANTON, J. R. Soil component. In: NSERL. (Ed.). **Water erosion prediction project**. USDA-ARS. West Lafayette: National Soil Erosion Research Laboratory, 1995. p. 7.1-7.47.

ALLEN, R. G.; PEREIRA, L. S.; RAES, D.; SMITH, M. **Evapotranspiración del cultivo**: guías para la determinación de los requerimientos de agua de los cultivos. Roma: Organización de las Naciones Unidas para la Agricultura y la Alimentación, 2006. (Estudio FAO: Riego y Drenaje, 56).

BAIGORRIA, G. A.; ROMERO, C. C. Assesment of erosion hotspots in a watershed: Integrating the WEPP model and GIS in a case study in the Peruvian Andes. **Environmental Modelling and Software**, v. 22, n. 8, p. 1175-1183, 2007.

BARRY, D. A.; PARLANGE, J. Y.; LI, L.; JENG, J. S.; CRAPPER, M. Green-Ampt approximations. **Advances in Water Resources**, v. 28, n. 10, p. 1003-1009, 2005.

BERTOL, I.; BEUTLER, J. F.; LEITE, D.; BATISTELA, O. Propriedades físicas de um Cambissolo Húmico afetadas pelo tipo de manejo do solo. **Scientia Agricola**, v. 58, n. 3, p. 555-560, 2001.

BORMANN, H.; BREUER, L.; GRAFF, T.; HUISMAN, J. A. Analyzing the effects of soil properties changes associated with land use changes on the simulated water balance: a comparison of three hydrological catchment models for scenario analyses. **Ecological Modelling**, v. 209, n. 1, p. 29-40, 2007.

BOULAIN, N.; CAPPELARE, B.; SÉGUIS, L.; FAVREAU, G.; GIGNOUX, J. Water balance and vegetation change in the Sahel: A case study at the watershed scale with an eco-hydrological model. **Journal of Arid Environments**, v. 73, n. 12, p. 1125-1135, 2009.

CECÍLIO, R. A.; MARTINEZ, M. A.; PRUSKI, F. F.; SILVA, D. D.; ATAÍDE, W. F. Substituição dos parametros do modelo de Green-Ampt-Mein-Larson para estimativa da infiltração em alguns solos do Brasil. **Revista Brasileira de Ciencia do Solo**, v. 31, n. 5, p. 1141-1151, 2007.

CHU, X.; MARIÑO, M. Determination of ponding condition and infiltration in layered soils under unsteady rainfall. **Journal of Hydrology**, v. 313, n. 3-4, p. 195-207, 2005.

COSTA, A.; ALBUQUERQUE, J. A.; ERNANI, P. R.; BAYER, C.; MERTZ, L. M. Alterações físicas e químicas num cambissolo húmico de campo nativo após a correção da acidez. **Revista de Ciências Agroveterinárias**, v. 5, n. 2, p. 118-130, 2006.

FUJIMAKI, H.; INOUE, M.; KONISHI, K. A multi-step flow method for estimating hydraulic conductivity at low pressure under the wetting process. **Geoderma**, v. 120, n. 3-4, p. 177-185, 2004.

GÓMEZ, S.; SEVERINO, G.; RANDAZZO, L.; TORALDO, G.; OTERO, J. M. Identification of the hydraulic conductivity using a global optimization method. **Agricultural Water Management**, v. 96, n. 3, p. 504-510, 2009.

JI, X. B.; KANG, E. S.; CHEN, R. S.; ZHAO, W. Z.; ZHANG, Z. H.; JIN, B. W. A mathematical model for simulating water balances in cropped sandy soil with flood irrigation applied. **Agricultural Water Management**, v. 87, n. 3, p. 337-346, 2007.

JIE, Z.; GUANG-YONG, L.; ZHEN-ZHONG, H.; GUO-XIA, M. Hydrological cycle simulation of an irrigation district based on a SWAT model. **Mathematical and Computer Modelling**, v. 51, n. 11-12, p. 1312-1318, 2010.

LEGESSE, D.; VALLET-COULOMB, C.; GASSE, F. Hydrological response of a catchment to climate and land use changes in Tropical Africa: case study South Central Ethiopia. **Journal of Hydrology**, v. 275, n. 1-2, p. 67-85, 2003.

MA, Y.; FENG, S.; SU, D.; GAO, G.; HUO, Z. Modelling water infiltration in a large layered soil column with a modified Green-Ampt model and HYDRUS-1D. **Computers and Eletronics in Agriculture**, v. 71, n. 1, p. S40-S47, 2010.

MAAYAR, M. E.; PRICE, D. T.; CHEN, J. M. Simulating daily, monthly and annual water balances in a land surface model using alternative root water uptake schemes. **Advances in Water Resources**, v. 32, n. 9, p. 1444-1459, 2009.

MEIN, R. G.; LARSON, C. L. Modeling infiltration during a steady rain. **Water Resources Research**, v. 9, n. 4, p. 384-394, 1973.

MELLO, C. R.; VIOLA, M. R.; NORTON, L. D.; SILVA, A. M.; WEIMAR, F. A. Development and application of a simple hydrologic model simulation for a Brazilian headwater basin. **Catena**, v. 75, n. 3, p. 235-247, 2008.

MOHAMMAD, A. G.; ADAM, M. A. The impact of vegetative cover type on runoff and soil erosion under different land uses. **Catena**, v. 81, n. 2, p. 97-103, 2010.

MOODY, J. A.; KINNER, D. A.; ÚBEDA, X. Linking hydraulic properties of fire-affected soils to infiltration and water repellency. **Journal of Hydrology**, v. 379, n. 3-4, p. 291-303, 2009.

NICKS, A. D.; LANE, L. J.; GANDER, G. A. Weather generator. In: FLANAGAN, D. C.; NEARING, M. A. (Ed.). **Water erosion prediction project (WEPP)**. West Lafayette: USDA-ARS/National Soil Erosion Research Laboratory, 1995. p. 2.1-2.22.

NOTTER, B.; MCMILLAN, L.; VIRIROLI, D.; WEINGARTNER, R.; LINIGER, H. P. Impacts of environmental change on water resources in the Mt. Kenya region. **Journal of Hydrology**, v. 343, n. 3-4, p. 266-278, 2007.

PRUSKI, F. F.; RODRIGUES, L. N.; SILVA, D. D. Modelo hidrológico para estimativa do escoamento superficial em áreas agrícolas. **Revista Brasileira de Engenharia Agrícola e Ambiental**, v. 5, n. 2, p. 301-307, 2001.

RAWLS, W. J.; DAVID, G.; VAN MULLEN, J. A.; WARD, T. J. Infiltration. In: MANAGEMENT GROUP D (Ed.). **Hydrology Handbook**. New York: ASCE, 1996. p. 75-124. (ASCE Manuals and Report on Engineering Practice, v. 28).

RISSE, L. M.; NEARING, M. A.; ZHANG, X. C. Variability in Green-Ampt effective conductivity under fallow conditions. **Journal of Hydrology**, v. 169, n. 1-4.

SILVA, C. L.; KATO, E. Avaliação de modelos para previsão da infiltração de água em solos sob cerrado. **Pesquisa Agropecuária Brasileira**, v. 33, n 7, p. 1149-1158, 1998.

ZANETTI, S. S.; OLIVEIRA, V. P. S.; PRUSKI, F. F. Validação do modelo ClimaBR em relação ao número de dias chuvosos e à precipitação total diária. **Engenharia Agrícola**, v. 26, n. 1, p. 96-102, 2006.

17

Simulation of bean crop growth, evapotranspiration and yield in Paraná State by the CROPGRO-Drybean model

Rivanildo Dallacort, Paulo Sergio Lourenço de Freitas*, Rogério Teixeira Faria, Antonio Carlos Andrade Goncalves, Roberto Rezende and Rachel Muylaert Locks Guimarães

*Departamento de Agronomia, Centro de Ciências Agrárias, Universidade Estadual de Maringá, Av. Colombo, 5790, 87020-900, Maringá, Paraná, Brazil. *Author for correspondence. E-mail: pslfreitas@uem.br*

ABSTRACT. The performance of the CROPGRO-Drybean model for the prediction of soil water balance, as well as growth components and bean crop yield, was assessed using data from two field experiments conducted at the State University of Maringá Iirrigation Technical Center, Paraná, Brazil, (latitude 23°27'S, longitude 51°57' and altitude 542 m) during the 2005 and 2006 growing seasons. This assessment aimed to analyze growth components, evapotranspiration and bean crop yield for the edaphoclimatic conditions of the area as simulated by the CROPGRO-Drybean model. This is a model for legumes, and it is grouped in a Decision Support System for Agrotechnology Transference (DSSAT). The model simulates growth, development and yield components of beans as a function of daily climatologic elements for each soil's specific conditions. It is possible to affirm that water deficit during flowering caused the model to penalize yield severely. It showed discrepant values in leaf area index simulation, and also accentuated discrepancies between measured and simulated values of soil moisture on the day following rainfall. Correlations between measured and simulated values of soil moisture exceeded 70% in both experiments. Yield was underestimated by the model in both experiments.

Keywords: water stress, soil moisture, agrometeorological models.

Introduction

Growth simulation models are valuable tools for estimating crop yield, considering various combinations of crop input, environmental factors and management practices. To estimate crop yield, agrometeorological-based models simulate crop development, available soil moisture and the effects of water stress on crop yield. Some models are classified as dynamic and mechanistic because they describe daily changes in the plant state variables, considering the main morphophysiological processes that occur during the period of simulation.Using simulations, the most suitable strategies can be assessed for each specific condition, such as irrigation management strategies, which can be simulated to predict effects on crop yield, and hydrologic components, such as evapotranspiration and crop water requirements.

Evangelista et al. (2006) evaluated the performance of the CLIGEN model in generating the climatic elements to utilize as input parameters for hydrological and agronomical models, comparing the generated values of climatic elements with the observed values. They concluded that the model showed good performance in generating climatic elements.

Wang et al. (2003) studied the performance of the CROPGRO-Soybean model for simulating site-specific crop growth, soil water content, and grain. The results indicated that CROPGRO-simulated soil water contents in the 15-90 cm soil profile agreed well with measured values. Simulated leaf area index and grain yield also agreed well with measured values during years of average precipitation but were underestimated during extremely dry years.

Sau et al. (2004) evaluated various potential evapotranspiration equations and different ways of partitioning evaporation between soil evaporation and crop transpiration in DSSAT models and particularly in CROPGRO faba bean (*Vicia faba* L.). They concluded that the Priestley–Taylor ET model used in CROPGRO-soybean to estimate ET tends to overestimate compared values with field values, especially in early crop growing stages, and Pennan-FAO24 is the that time, differences in E0 between crops were not well least adequate to simulate ET and biomass.

Freitas et al. (2005) evaluated the penalization imposed by the CERES-Maize model on corn yield variables. Based on simulations and on experimental conditions, it was concluded that the penalization imposed by the model severely influenced the corn crop yield for treatments with soil water deficit. The corn crop evapotranspiration calculated by the CERES-Maize model overestimated the real water needs of the crop.

In this context, DSSAT (Decision Support System for Agrotechnology Transfer) is a computer system that includes several crop growth models that are used by many researchers for the most varied climatic and soil conditions and are an excellent option for obtaining information that helps in agricultural planning and management. The DSSAT system uses a common soil-water balance module in models for several different crops: maize (Ceres-Maize), soybean (Cropgro-soybean), the common bean (Cropgro-drybean), wheat (Ceres-Wheat), sorghum (Ceres-Sorghum), rice (Ceres-Rice), barley (Ceres-Barley), sunflower (Ceres-Sunflower) and others (JONES; RITCHIE, 1990).

Meirelles et al. (2003) and Faria and Bowen (2003) pointed out inaccuracies of the DSSAT soil-water balance module and suggested changes to improve model prediction. Freitas et al. (2005) observed that CERES-Maize severely penalized yield during periods of water shortage. They also highlighted the need for research to test the soil-water balance model used, requiring research in experimental fields, monitoring of edaphoclimatic conditions and adjustments to the model when necessary and possible. Crop models, once calibrated, allow researchers to simulate multiple years of experiments, utilizing historical weather data, in a matter of hours.

Bastos et al. (2002) adapted the CROPGRO model for simulating cowpea growth and development under the soil and climate conditions of Piauí State, Brazil. The model showed excellent performance in the estimation of LAI, indicated by high R^2 (0.97) and low MAE (0.16 to 0.28) values, with maximum values of LAI, observed and simulated, varying from 3 to 4.3 in 1997 and 1998, respectively.

The aims of this research were to evaluate leaf area index, evapotranspiration, grain yield and the soil-water balance module in the CROPGRO-Drybean model against field data from experiments conducted with a bean crop under edaphoclimatic conditions in Maringá, Paraná State, Brazil.

Material and methods

Experimental procedures

The present study was carried out at the Irrigation Technical Center at the State University of Maringá, located in Maringá (latitude 23° 25' S, longitude 51° 57' W and altitude 542 m). The predominant climate in the region, according to the Köppen climate classification, is Cfa type, characterized by intense rainfall during the summer and a dry winter, with 1,673 mm average annual precipitation. The average annual temperature in the region is 23.2°C and the relative humidity of the air is 68%. The predominant soil in the experimental area is classified as Dystric Nitosols, with moderate A horizon and clay texture. The relief presented an average slope of 8%.

Two experiments were carried out with bean crops (*Phaseolus vulgaris* L.), cultivar IAPAR 81, during the 2005 and 2006 growing seasons, from May 4 to August 15, 2005, and April 12 to July 26, 2006, respectively. The experimental area (720 m^2) used was subdivided into 80 experimental units (3.0 x 3.0 m). After sowing, a fixed conventional sprinkler irrigation system was installed using Fabrimar sprinklers, model A232 ECO, at 18 x 15 m spacing, covering the entire area, both the experimental plots and the border.

Soil moisture was determined by TDR, using 80 probes of 0.30 m length in the first experiment and 80 probes of 0.20 m length in the second experiment, all of which were installed vertically from the soil surface at the center of each experimental unit. Dielectrics of the environment (Ka) were measured daily and were then converted into soil moisture θ (m^3 m^{-3}) using the following equation: $\theta = (0.842538 - 0.725175\ ds) + (-0.049601 + 0.060353\ ds)Ka + (0.001044 - 0.001061\ ds)Ka^2$, where ds is soil bulk density. Crop yield (kg ha^{-1}) was determined by multiplying plant yield, obtained from ten plant samplings collected randomly in each experimental unit, by plant population in each experimental unit. Seeds were manually detached from pods to determine seed mass, after drying in a ventilated oven at 70°C until the weight was constant.

Leaf dry mass data of the crop were determined by samplings in eight and seven distinct periods for the experiments, respectively. Leaf dry mass was collected on May 27, 2005; June 2, 9, 16, 26 and 30, 2005; and July 7 and 15, 2005 for the first experiment, and May 18 and 26, 2006; June 2, 9, 16 and 23, 2006; and July 7, 2006 for the second experiment. These samplings were carried out in all 80 experimental units, in which three plants were collected randomly to obtain leaf area and, later, the leaf area index.

To determine leaf area, the disk-leaf method was used according to the recommendations of Magalhães (1979), described and used by Freitas et al. (2003). A leaf puncher of 4.75 cm diameter and 17.69 cm^2 area was used to relate the dry weight of the leaf disk area with dry weight of all the leaves of each plant. The dry weight was used to determine leaf area; 12 leaf disks were collected from several parts of leaves that were placed in paper bags and kept in an oven at 65°C for drying until the weight was constant. The material taken from the oven was weighed on a balance with 0.01-g precision. The leaf area was estimated from relationships between dry weights of disks, total disk areas (sum of areas of all disks of the same plant), and total dry weights of collected leaves (used leaves, non-used leaves and disks). To determine the LAI, plant leaf area was divided by soil area available to the plant.

Model characteristics

The CROPGRO-Drybean model was developed to simulate growth, development and yield of a common bean crop. It considers the plant's main physical and physiological processes, such as photosynthesis, respiration, biomass accumulation and partitioning, phenology, soil water extraction, evapotranspiration and common bean growth and leaf area, for both stems and roots, as a function of daily climatological elements (rainfall, solar radiation, maximum and minimum temperatures), for the specific conditions of each soil. Water storage in the soil and its capacity to supply plant roots are predicted based on the processes of superficial runoff, water percolation and redistribution in the profile. The model is sensitive to the characteristics of each cultivar, sowing dates, crop spacing and irrigation management options (MEIRELLES et al., 2002).

Cultivar characteristics are expressed in the model by genetic coefficients, to express phonological traits related to growth and development of each genotype. The genetic coefficients for cultivar IAPAR 81 (Table 1) were obtained by interactively adjusting the values of the coefficients until flowering and maturity dates matched the experimental values and simulated dry matter coincided with measured grain yield.

Table 1. Calibrated genetic coefficients* for cultivar IAPAR 81.

Cultivar	CSDL	PPSEN	EM-FL	FL-SH	FL-SD	SD-PM	FL-LF
IAPAR 57	12.00	0.000	28.7	3.0	9.5	28.50	16.84
LFMAX	SLAVRN	SIZLF	XFRT	WTPSD	SFDUR	SDPDV	PODUR
1,00	308	133.0	1.00	0.251	12.5	4.12	9.1

*CSDL - critical day length, above which the reproductive development process is not affected (h); PPSEN - response inclination regarding development for the photophase with time (1/hour); EM-FL – period between plant emergence and the appearance of the first flower (R1) (photothermal days); FL-SH – period between the appearance of the first flower and the first pod (R3) (photothermal days); FL-SD - period between the appearance of the first flower and the start of seed formation (R5) (photothermal days); SD-PM – period between the start of seed formation and physiological maturity (R7) (photothermal days); FL-LF – period between the appearance of the first flower (R1) and the end of leaf expansion; LFMAX - maximum leaf photosynthesis rate at an optimal temperature rate of 30°C; SLAVARN - specific leaf area under standard growth conditions (cm^2); SIZLF - maximum size of completely expanded leaf (cm^2); XFRT - maximum fraction of the daily growth that is partitioned between the seed plots the pod; WTPSD - maximum weight per seed (g); SFDUR - duration of the grain swelling period in the pods, under standard growth conditions (photothermal days); SDPDV - mean seeds per pod, under standard growth conditions (photothermal days); PODUR - time necessary for the cultivar to reach ideal pod conditions (photothermal days).

In addition to the genetic coefficients of cultivar IAPAR 81, other inputs to the model were climatological and soil data. The set of daily climatological variables consisted of maximum air temperature, minimum air temperature, rainfall and sunshine hours, which were obtained from the INMET meteorological station, located close to the experimental field.

Soil samples were collected from the experimental area to determine the soil inputs (Table 2). Soil water retention and bulk density were determined from undisturbed O-ring samples (5-cm diameter by 5-cm tall metal rings) and particle size distribution was determined by the Bouyoucos method in the Soil Laboratory at the Department of Agriculture at the State University of Maringá. Samples were used to determine soil water retention curves, using the tension table for potentials between 0 and -6 kPa and a Richards Chamber for the other processes. Soil moisture θ (m^3 m^{-3}) was obtained as a function of pre-established tensions (ψm) in kPa. The upper moisture limit (field capacity) was considered as the water content retained at -6 kPa (REICHARDT, 1990) and the lower limit (permanent wilting point) at -1500 kPa.

Simulations were performed for both growth cycles, coinciding with the periods when the experiments were carried out.

Table 2. Soil physical characteristics for 0-30 cm soil depth at the experimental area.

Layer (cm)	Soil retention (m^3 m^{-3})			Bulk density (g cm^{-3})	Particle size distribution (%)		
	Saturation	Upper Limit	Lower Limit		Sand	Silt	Clay
0-15	0.57	0.46	0.30	1.33	12	12	76
15-30	0.58	0.44	0.33	1.39	12	12	76

Evapotranspiration calculated by the model

The CROPGRO-Drybean simulation model uses the Ritchie method to calculate evapotranspiration, which is used to calculate water balance in the soil. The model is based on works by Ritchie (1972) and Jones and Ritchie (1990), and consists of estimating crop evapotranspiration in mm day^{-1}, calculating water evaporation in the soil (Es) and transpiration (Ep) independently.

Simulations carried out

Simulations were carried out for both experimental periods, during the 2005 and 2006 growing seasons, from May 4 to August 15, 2005 and from April 12 to July 26, 2006, respectively, which made it possible to compare responses measured in the field during the experiments with data simulated by models.

Rainfall and irrigation

During the experimental periods, total rainfall plus irrigation for the 2005 and 2006 growing seasons were 332 and 303 mm (Figure 1), respectively. These values met crop water requirements during growing cycle found by some authors conducting experiments in Brazil (DANTAS NETO et al., 2005; NÓBREGA et al., 2001; SILVEIRA; STONE, 2001). Additionally, Doorenbos and Kassam (2000) concluded that maximum bean yield is achieved when 300 to 400 mm is available during the cycle, depending on the local edaphoclimatic conditions.

The coefficient of water application uniformity of the sprinkler system was 80% in the first experiment and 81% in the second experiment, values considered acceptable in a sprinkler irrigation system (BERNARDO et al., 2005). For the 2005 experiment, between emergence and flowering (May 10 to June 15), a total of 147.3 mm were concentrated, which corresponded to 44.4% of the total rainfall of the crop cycle. From flowering to the start of pod formation (June 16 to July 15), 83.5 mm of well-distributed rainfall occurred to fulfill crop water requirements.

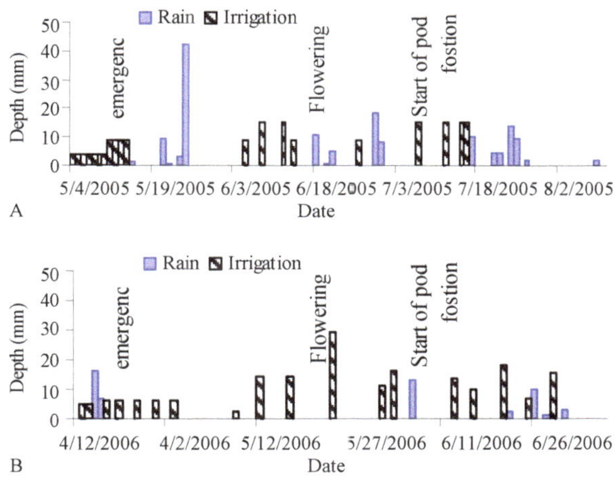

Figure 1. Rainfall and irrigation events during the experiments carried out in 2005 (A) and 2006 (B).

According to Nóbrega et al. (2001), water shortage at this phase would cause large reductions in yield. Finally, during grain filling, 77 mm of irrigation, together with well-distributed rainfall that occurred during the period, were sufficient to satisfy crop water requirements. For the experiment carried out in 2006, the crop received a total of 112.7 mm from emergence to flowering, corresponding to 37.2% of the total rainfall in the crop cycle. From flowering to the start of pod formation, the crop was supplied with 45.1 mm of irrigation, 34.6 mm less than the depth received for the 2005 experiment. Finally, from pod formation to maturity the crop received 105.4 mm.

Results and discussion

Simulated and measured soil moisture behavior

Moisture simulated by the CROPGRO-Drybean model showed great variations in amounts when compared to moisture measured through TDR (Figure 2). During the first experiment (from May 25 to July 20, 2005), soil moisture was above the permanent wilting point limit and values were higher than field capacity limits on only two days. In the experiment conducted in 2006, soil moisture was monitored using TDR from May 11 to July 11, 2006.

The correlation between values measured by TDR and simulated by the DSSAT model on the 47 days observed showed R = 74% and 0.018 $cm^3\ cm^{-3}$ standard deviation for the first experiment and R = 70.15% with 0.02 standard deviation on the 54 days on which moisture was monitored during the second experiment.

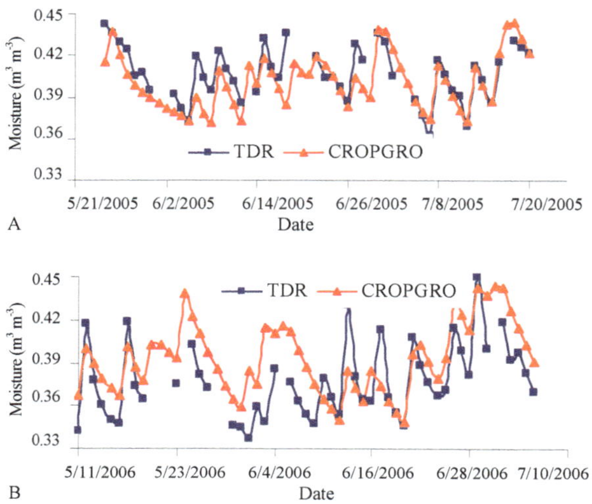

Figure 2. Variation of simulated and measured soil moisture at 20-cm depth during the experiment carried out in 2005 (A) and at 30 cm depth during the experiment conducted in 2006 (B).

Values presented in Figure 2 show the model's limitation in considering the available soil water. This was observed by Faria and Bowen (2003) when they evaluated the performance of the water balance module (WBM) of the DSSAT v 3.5 models. Soil moisture values measured and simulated by the model during the first experiment showed that 36.17% of simulated values were overestimated and 63.83% were underestimated, while results found by Wang et al. (2003) indicated that CROPGRO-simulated soil water contents in the 15–90 cm soil profile agreed well with measured values. In the second experiment, the same behavior was not observed; 74.07% of data were overestimated and 25.93% were underestimated. Sau et al. (2004) also found simulated evapotranspiration values to be higher than those observed in the field. Differences between the observed and simulated yield results were partly attributed to the semi-arid climatic conditions of the study area and to the CROPGRO-soybean ET model that overestimated simulated ET (DOGAN et al., 2007).

As suggested by Willmott (1981), the validation of simulated data are of great importance, and considering that numerical models are approximate rather than exact, data were tested due to errors by such proximity. Values of mean square error (MSE), which is associated with the tendency to underestimate or overestimate data in relation to the average, and the square root of mean square error (RMSE), which is associated with data dispersion, are presented in Table 3. Such data show that simulated data were overestimated in 2005 and underestimated in 2006. RMSE values observed in 2005 were lower than those observed in 2006, confirming lower data dispersion.

Table 3. Determined mean square error (MSE) and square root of mean square error (RMSE).

	MSE	RMSE
Growing season of 2005	0.0063	0.0142
Growing season of 2006	- 0.0152	0.0256

Evapotranspiration simulated by the model

Bean evapotranspiration simulated by the CROPGRO model, which uses the Ritchie method, allows the variation in observation of soil water evaporation and plant transpiration. In the initial phase of crop growth, soil water evaporation presented values close to those of evapotranspiration (Figure 3). For the 2005 experiment, bean transpiration values exceeded evaporation for June 15, 2005, showing values of 1.56 and 1.43 mm, respectively (Figure 3A). The simulated leaf area index (LAI) was 2.48 $m^2\ m^{-2}$. Maximum evaporation observed in the crop cycle was 4.8 mm day^{-1}, which happened during the initial phase of the crop cycle, and maximum transpiration was 2.47 mm day^{-1}, on July 2, 2005, during a crop phase with higher LAI.

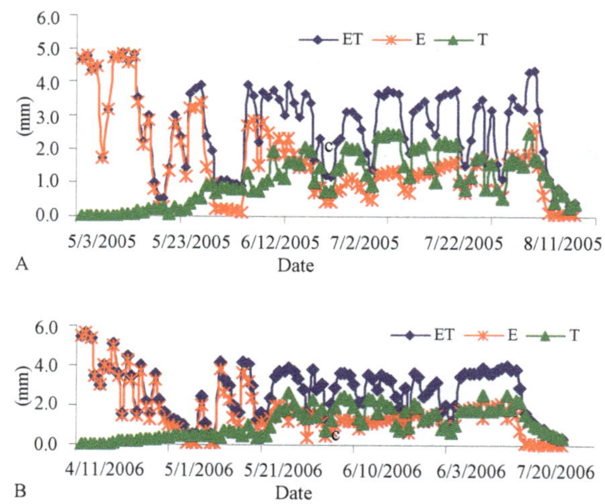

Figure 3. Evapotranspiration (ET), evaporation (E), and transpiration (T) of bean crop determined by the model during the experiments carried out in 2005 (A) and 2006 (B).

During the entire crop cycle, evapotranspiration was 260 mm: 156 mm from evaporation and 104 mm from transpiration. Evapotranspiration had maximal, mean and minimal values of 4.87, 2.71 and 0.31 mm day^{-1}, respectively. These values were similar to those found by Junqueira et al. (2004), who observed consumption of 2.1, 3.82, 2.52 and 1.94 mm day^{-1} in vegetative development stages, during flowering, during grain filling and at the beginning of maturation, respectively.

In the 2006 experiment, evaporation and evapotranspiration values exhibited the same behavior (Figure 3B). In this crop cycle, evapotranspiration was 283 mm, 168 mm from evaporation and 115 mm from transpiration. Evapotranspiration had maximal, mean and minimal values of 5.61, 2.70 and 0.34 mm day^{-1}, respectively.

The evaporation and transpiration totals found in both experiments were close to those found by Rodrigues et al. (1998), who observed 62.2% evapotranspiration due to evaporation and 37.8% due to transpiration, using the Ritchie model. Values found in this study were 59.29 and 40.71% in the first experiment, and 60.10 and 39.90% in the second experiment for evaporation and transpiration, respectively.

Measured and simulated soil moisture values (Figure 2A and B) showed great variation, interfering directly with soil water balance and its use in the evapotranspiration process of the bean crop. In Figures 4A and B, the behavior of evapotranspiration values simulated by the DSSAT model and evapotranspiration values calculated from daily variations in soil moisture in the layer are observed.

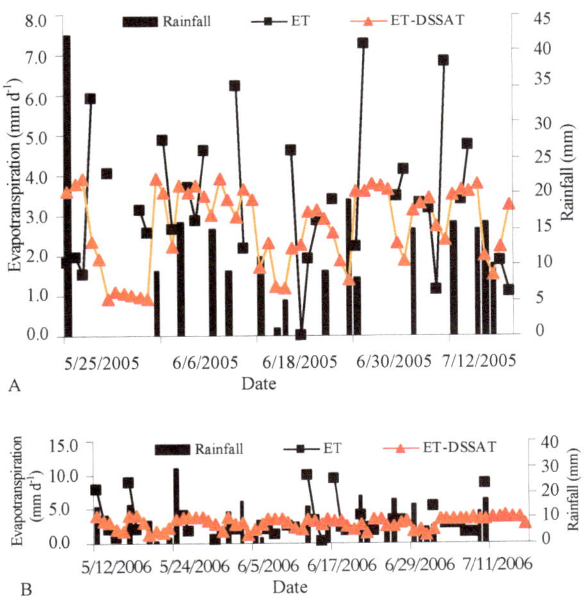

Figure 4. Evapotranspiration determined by the model and calculated from soil moisture values measured during the experiments carried out in 2005 (A) and 2006 (B).

It was not possible to measure soil moisture using TDR on rainy days. Therefore, it is not possible to determine evapotranspiration values by using differences in moisture during the whole cycle. On days following precipitation, evapotranspiration values determined by moisture measurements were higher. The values were closer to those determined by the CROPGRO model from the second day after the one with precipitation, which meant that evapotranspiration calculated by moisture data was overestimated on the day following precipitation, causing the high deviations that were found (Figure 4).

For example, 10 mm of precipitation fell on July 16, 2006. On the 17th, evapotranspiration calculated by the variation of moisture measurements determined by TDR was 9.48 mm day^{-1}, and that simulated by CROPGRO was 3.30 mm day^{-1}. The TDR method overestimated evapotranspiration approximately three times more than that simulated by the model. For the 18th day, two days following precipitation, evapotranspiration values were 2.32 and 3.27 mm day^{-1} for TDR and simulated by CROPGRO, respectively, which meant that TDR underestimated its measure, and a tendency towards measurement approximation was observed on subsequent dates.

Leaf area index

Figure 5A and B show leaf area index behavior determined in the field and simulated by the CROPGRO-Drybean model for both experiments. Simulated data were higher than measured values during the 2005 experiment. Simulated LAI_{Max} occurred on June 26, 2005, with a value of 2.56 m^2 m^{-2}. Measured LAI was 1.31 m^2 m^{-2} on this date, which was not the maximal value. Maximum leaf area index, according to the average of 80 experimental plots, was 2.14 m^2 m^{-2}, found on July 15, 19 days after the simulated LAI_{Max}. Similar values were found by Villa Nova et al. (2003), who observed that the LAI_{Max} measured in the field was 2.63 m^2 m^{-2} at 53 days after plant emergence, and by Bastos et al. (2002). These values were lower than those found by Meirelles et al. (2002).

Table 4 shows LAI behavior, which was determined in 80 experimental plots. Large variations in measured values can be observed; measured LAI_{Max} exceeded simulated values in some plots, though on average, simulated values were higher. High standard deviations of LAI values can be observed in the plots. Freitas et al. (2003) stated that the coefficient of water application uniformity influences LAI directly, and verified variation in the leaf area index as a function of the water application coefficient (CUC). For CUC > 86%, LAI_{Max} was close to 4.32 m^2 m^{-2}, and $LAI_{min.}$ was 3.10 m^2 m^{-2}. For CUC < 66%, LAI_{Max} was 2.84 m^2 m^{-2}, and $LAI_{min.}$ was 2.07 m^2 m^{-2}.

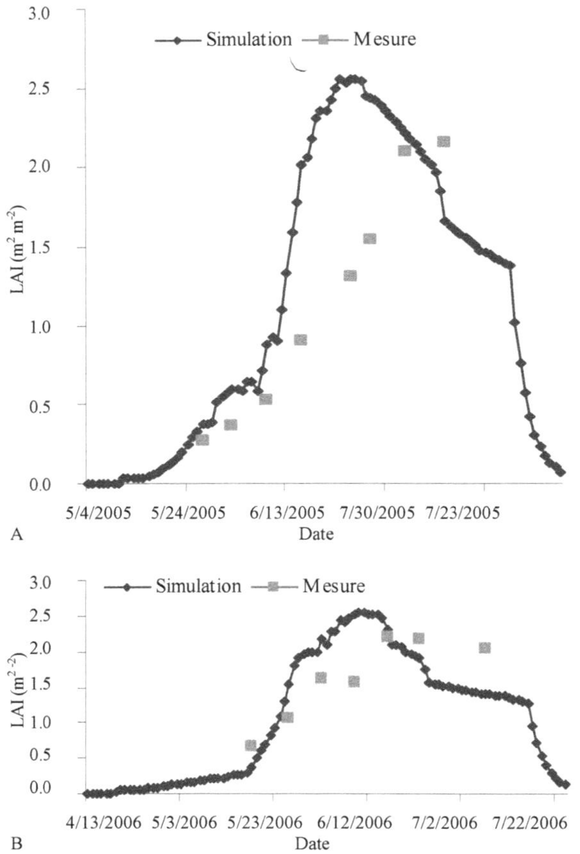

Figure 5. Leaf area indexes (LAI) determined in field and simulated by the model during the experiments carried out in 2005 (A) and 2006 (B).

Table 4. Leaf area indexes observed during the 2005 experiment.

LAI date	LAI_{mean}	LAI_{max}	LAI_{min}	Standard Deviation
5/27/2005	0.27	0.39	0.09	0.07
6/26/2005	0.36	0.63	0.12	0.11
6/9/2005	0.52	1.08	0.18	0.17
6/16/2005	0.89	1.79	0.29	0.29
6/26/2005	1.30	2.93	0.36	0.48
6/30/2005	1.54	4.37	0.45	0.61
7/7/2005	2.08	4.24	0.70	0.82
7/15/2005	2.14	6.73	0.67	1.01

Measured LAI values presented in Tables 3 and 4 showed that LAI_{max} values exhibited the same behavior, with 2.14 m^2 m^{-2} in 2005 and 2.21 m^2 m^{-2} in 2006. Several authors have developed studies for the determination of LAI average coefficients in various crop developmental stages. For the vegetative phase of development, Junqueira et al. (2004) observed LAI values of about 1.71 m^2 m^{-2} during the flowering-grain filling period. Doorenbos and Kassam (2000) found values between 1.05 and 1.20 m^2 m^{-2}. At the stage of physiological maturation, Rodrigues et al. (1998) observed values ranging from 2.3 to 3.1 m^2 m^{-2} in a study using the Ritchie model to determine bean crop transpiration and evaporation.

In the 2006 experiment, LAI_{Max} simulated by the model was 2.55 m^2 m^{-2}; this occurred on June 10 and 11, 2006. On June 16, 2006, an average LAI_{Max} of 2.21 m^2 m^{-2} was found in 80 experimental plots. For this date, the simulated LAI value was 2.30 m^2 m^{-2}.

In a study carried out using the same model with cultivar IAPAR 57, Dallacort et al. (2005) observed lower simulated yield during simulation with sowing on July 30, 1988, and leaf growth was severely reduced, showing a maximum LAI of 0.49 m^2 m^{-2} at 55 days after sowing. In the harvest with the greatest simulated yield, LAI was low in the initial phase due to water stress, but it was regulated during the crop cycle, reaching a maximum value of 3.34 m^2 m^{-2} at 68 days after sowing.

Yield

Measured yield standard deviation and coefficient of variation of bean yield from 80 experimental units in the 2005 experiment were higher than the values obtained in 2006 (Table 6). Although predicted values were within the range of mean ± standard deviation, the model underestimated crop yield by 19% in 2005 and by 29% in 2006 (Table 5). This can be attributed to the fact that the CROPGRO-Drybean model exaggerates the effect of water stress during the flowering phase, which leads to the underprediction of crop yield.

Table 5. Leaf area indexes observed during the 2006 experiment.

LAI date	LAI_{mean}	LAI_{max}	LAI_{min}	Standard Deviation
5/18/2006	0.67	1.37	0.25	0.24
5/26/2006	1.05	2.21	0.36	0.46
6/2/2006	1.62	3.54	0.40	0.69
6/9/2006	1.57	3.46	0.72	0.57
6/16/2006	2.21	4.63	0.66	0.81
6/23/2006	2.18	4.81	1.01	0.73
7/7/2006	2.04	3.83	1.11	0.53

Table 6. Simulated and measured mean, maximum, and minimum yields, standard deviations (SD) and coefficients of variation (CV) determined for both experiments.

Growing season	Yield					
	Simulated	Mean	Maximun	Minimun	SD	CV
	(kg ha^{-1})					(%)
2005	2,465	2,830	4,879	1,200	898.98	32
2006	1,748	2,462	4,197	1,478	549.75	22

Conclusion

Based on the results obtained with this work, it is possible to conclude that: water deficit during the flowering phase caused the model to penalize yield severely; the model showed discrepancies in simulated leaf area index values; there was a greater discrepancy between measured and simulated values of soil moisture on the day following rainfall; correlations found between measured and simulated values of soil moisture exceeded 70% in both experiments; yield was underestimated by the model in both growing seasons.

References

BASTOS, E. A.; FOLEGATTI, M. V.; FARIA, R. T.; ANDRADE JÚNIOR, A. S.; CARDOSO, M. J. Simulation of growth and development of irrigated cowpea in Piauí State by CROPGRO model. **Pesquisa Agropecuária Brasileira**, v. 37, n. 10, p. 1381-1387, 2002.

BERNARDO, S.; SOARES, A. A.; MANTOVANI, E. C. **Manual de irrigação**. 7. ed. Viçosa: UFV, 2005.

DALLACORT, R.; FREITAS, P. S. L.; RESENDE, R.; FARIA, R. T.; GONÇALVES, A. C. A. Utilização do modelo CROPGRO-drybean, na determinação das melhores épocas de semeadura da cultura do feijão, para a região de Maringá-PR. **Acta Scientiarum. Agronomy**, v. 27, n. 2, p. 353-359, 2005.

DANTAS NETO, F. S.; COSTA, J. M. N.; SOARES, A. A.; SEDIYAMA, G. C.; COSTA, L. C. Sistema computacionalaplicado ao cálculo da demanda hídrica georeferenciada para a cultura do feijão. **Revista Brasileira de Engenharia Agrícola e Ambiental**, v. 9, n. 2, p. 215-221, 2005.

DOGAN, E.; KIRNAK, H.; COPUR, O. Deficit irrigations during soybean reproductive stages and CROPGRO-soybean simulations under semi-arid climatic conditions. **Field Crops Research**, v. 103, n. 2, p. 154-159, 2007.

DOORENBOS, J.; KASSAM, A. H. **Efeito da água no rendimento das culturas**. 2. ed. Campina Grande: UFPB, 2000. (Estudos FAO. Irrigação e drenagem, 33).

EVANGELISTA, A. W. P.; FERREIRA, P. A.; PRUSKI, F. F.; SEDIYAMA, G. C. Avaliação do desempenho do modelo CLIGEN em gerar dados climáticos para modelos hidrológicos e agronômicos. **Revista Brasileira de Engenharia Agrícola e Ambiental**, v. 10, n. 1, p. 119-127, 2006.

FARIA, R. T.; BOWEN, W. T. Evaluation of DSSAT soil-water balance module under cropped and bare soil conditions. **Brazilian Archives of Biology and Technology**, v. 46, n. 4, p. 489-498, 2003.

FREITAS, P. S. L.; MANTOVANI, E. C.; SEDIYAMA, G. C.; COSTA, L. C. Penalização da produtividade da cultura do milho imposta pelo modelo CERES-Maize. **Acta Scientiarum. Agronomy**, v. 27, n. 1, p. 97-105, 2005.

FREITAS, P. S. L.; REZENDE, R.; MANTOVANI, E. C.; FRIZZONE, J. A. Viabilidade de inserção dos efeitos da uniformidade de irrigação em modelos de crescimento de culturas. **Revista Brasileira de Engenharia Agrícola e Ambiental**, v. 7, n. 3, p. 437-444, 2003.

JONES, J. W.; RITCHIE, J. T. Crop Growth Models. In: HOFFMAN, G. J.; HOWEL, T. A.; SOLOMON, K. H. (Ed.). **Management of farm irrigation systems**. St. Joseph: ASAE, 1990. p. 63-89.

JUNQUEIRA, A. M.; ANDRÉ, R. G. B.; PINHEIRO, F. M. A. Consumo de água pelo feijoeiro comum, cv. Carioca. **Revista Brasileira de Agrometeorologia**, v. 12, n. 1, p. 51-56, 2004.

MAGALHÃES, A. C. N. Análise quantitativa do crescimento. In: FERRI, M. G. (Ed.). **Fisiologia vegetal**. São Paulo: EPU/Edusp, 1979. v. 1, cap. 8, p. 331-350.

MEIRELLES, E. J. L.; PEREIRA, A. P.; SENTELHAS, P. C.; STONE, L. F.; ZIMMERMANN, F. J. P. Calibration and test of the cropgro-dry bean model for edaphoclimatic conditions in the savannas of central Brazil. **Scientia Agricola**, v. 59, n. 4, p. 723-729, 2002.

MEIRELLES, E. J. L.; PEREIRA, A. P.; SENTELHAS, P. C.; STONE, L. F.; ZIMMERMANN, F. J. P. Risco climático de quebra de produtividade da cultura do feijoeiro em Santo Antônio de Goiás, GO. **Bragantia**, v. 62, n. 1, p. 163-171, 2003.

NÓBREGA, J. Q.; RAO, T. V. R.; BELTRÃO, N. E. M.; FIDELES FILHO, J. Análise de crescimento do feijoeiro submetido a quatro níveis de umidade do solo. **Revista Brasileira de Engenharia Agrícola e Ambiental**, v. 5, n. 3, p. 437-443, 2001.

REICHARDT, K. A. **Água em sistemas agrícolas**. São Paulo: Manole, 1990.

RITCHIE, J. T. Model for predicting evaporation fron a row crop with a incomplete cover. **Water Resourse Research**, v. 8, n. 5, p. 1204-1213, 1972.

RODRIGUES, L. N.; SEDIYAMA, G. C.; SOCCOL, O. J.; MANTOVANI, E. C. Modelo de Ritchie para determinações da transpiração do feijoeiro (*Phaseolus vulgaris* L.) e da evaporação direta da água do solo. **Revista Brasileira de Agrometeorologia**, v. 6, n. 2, p. 179-185, 1998.

SAU, F.; BOOTE, K. J.; BOSTICK, W. M.; JONES, J. W.; MINGUEZ, M. I. Testing and improving evapotranspiration and soil water balance of the DSSAT crop models. **Agronomy Journal**, v. 96, n. 5, p. 1243-1257, 2004.

SILVEIRA, P. M. S.; STONE, L. F. **Irrigação do feijoeiro**. Santo Antonio de Goiás: Embrapa Arroz e Feijão, 2001.

VILLA NOVA, N. A.; SENTELHAS, P. C.; PEREIRA, A. B. Evapotranspiração máxima do feijoeiro, cv. Goiano precoce, e função do índice de área foliar e da evaporação do tanque classe A. **Publicatio UEPG. Ciências Exatas e da Terra, Ciências Agrárias e Engenharias**, v. 9, n. 2, p. 41-45, 2003.

WANG, F.; FRAISSEB, C. W.; KITCHENC, N. R.; SUDDUTH, K. A. Site-specific evaluation of the CROPGRO-soybean model on Missouri claypan soils. **Agricultural Systems**, v. 76, n. 3, p. 985-1005, 2003.

WILLMOTT, C. J. On the validation of models. **Physical Geography**, v. 2, n. 2, p. 184-94, 1981.

18

Chemical and instrumental characterization of pectin from dried pomace of eleven apple cultivars

Mariana de Fátima Sato[1], Dayana Carla Rigoni[1], Maria Helene Giovanetti Canteri[2], Carmen Lúcia de Oliveira Petkowicz[3], Alessandro Nogueira[1] and Gilvan Wosiacki[1*]

*[1]Departamento de Engenharia de Alimentos, Setor de Ciências Agrárias e de Tecnologia, Universidade Estadual de Ponta Grossa, Av. General Carlos Cavalcanti, 4748, 84030-900, Ponta Grossa, Paraná, Brazil. [2]Coordenação de Alimentos, Universidade Tecnológica Federal do Paraná, Ponta Grossa, Paraná, Brazil. [3]Centro Politécnico, Universidade Federal do Paraná, Curitiba, Paraná, Brazil. *Author for correspondence. E-mail: gilvan.wosiacki@pesquisador.cnpq.br*

ABSTRACT. Pectic substances from eleven samples of dried apple pomace were extracted using an acid procedure (100 mM HNO_3, 10 min., 80°C), dried, ground and stored in a P_2O_5 atmosphere. The degree of esterification was determined by the classical titrimetric method and confirmed by the instrumental FTIR method. The samples showed an average DE of 73% and the typical fingerprint in IR spectrogram. The steric exclusion chromatograms obtained with the multi-angle laser light scattering (MALLS) and refraction index (RI) detectors showed that all samples had a high molar fraction eluted at 36-38 min., another large polysaccharide family had an elution time around 43 min, and some samples showed yet another low molecular oligosaccharide that appeared at 58-60 min. Gas chromatography of alditol acetate allowed estimates of the proportion of the neutral sugar in the samples. The ratio of galacturonic acid and rhamnose may give the average size of the galacturonan fraction. These results showed that yields of pectin extracted by the same acid procedure were slightly different. However, the degree of esterification of pectin was similar, with both factors pointing to the best sources in this group of varieties.

Keywords: apple pomace, dietary fibers, pectin, degree of esterification.

Introduction

Pectin has been used in jelly manufacture for more than 200 years (WILLATS et al., 2006), but, *in praxis*, only two traditional sources have been economically important, citrus albedo and apple pomace, despite research into new raw materials (IGLESIAS; LOZANO, 2004; LEVIGNE et al., 2002). Apple pectin shows a similar rheological quality to that of citrus pectin, but due to the presence of phenol compounds, the latter is preferred for use in translucent or transparent foods.

The process of producing a colorless acidic polysaccharide from apple induces the separation of pectin from phenolic compounds by ion exchange chromatography (SCHIEBER et al., 2003). Closely related to cellulose and hemicelluloses, soluble pectin may be removed by water showing non-covalent bonds (THAKUR et al., 1997), and insoluble pectin, more strongly associated with cellular structure and known as protopectin, may become soluble after a controlled acid hydrolysis procedure (BERK, 1976).

Pectins are acid polysaccharides, and their composition depends on source, processing and environmental factors (IGLESIAS; LOZANO, 2004). The major backbone is a polymer of D-galacturonic acid with α(1 → 4) bonds between anhydrogalacturonic residues with varying amounts of methyl esterified carboxyl groups (PÉREZ et al., 2003). According to Renard et al. (1995), the insertion of deoxi-hexose into the main chain occurs as chained dimers ['GalA α (1 → 2)-Rha'] that may join together approximately 20 monomers. Usually, pectin contains 65% galacturonic acid and may show three kinds of forms: the linear homopolygalacturonan and the ramified, rhamnogalacturonan I and II (WILLATS et al., 2006). According to Novosel´Skaya et al. (2000), several authors reported a galacturonic acid:rhamnose ratio from 25 to 200. In the case of a citrus source, the ratio was 40, but only 15% of the rhamnose was in the linear form, and 85% was in the ramified fraction. In apple pectin (RENARD et al., 1995), 22% of the rhamnose is in the linear form, and 78% is in the ramified fraction. The hairy fraction has two different structural features. The first contains several GalA-Rham dimers in the main chain where the side neutral oligosaccharides composed of arabinose and galactose (RG I) are attached and the second, where the side neutral and rare monosaccharides are directly linked to the main chain in a locus containing approximately seven Galacturonic acid residues (RG II).

Pectin quality and acid extraction are related to many factors, including the type of fruit (THAKUR et al., 1997), and, concerning apple pomace, the variety and process (COSTENLA et al., 2002), including extraction, purification and drying. The type and concentration of acid (FERTONANI et al., 2006) and the time and temperature of reaction (FERTONANI et al., 2006; GARNA et al., 2007; SCABIO et al., 2007) play a fundamental role in the balance of pectin extraction and degradation.

In this work, dried pomace from eleven apple cultivars harvested during the 2006-2007 season in the agricultural region of Caçador (9 cultivars) and São Joaquim (2 cultivars) were used as raw material. The main objective of this study was to investigate the influence of variety on the yield and chemical characteristics of pectin.

Material and methods

Material

Samples of apple from eleven cultivars given by the Experimental Stations of Caçador and São Joaquim, which belong to the Empresa de Pesquisa Agrícola e de Extensão Rural de Santa Catarina – Epagri, were codified as (1) Catarina, (2) Joaquina, (3) M-11/00, (4) M-11/01, (5) M-1/01, (6) M-12/00, (7) M-13/00, (8) M-2/00, (9) M-8/01 and (11) MRC. Pectinolytic enzymes, produced by Novo, were donated by LNF Ltda., from Bento Gonçalves, Rio Grande do Sul State. Chemical products were all *pro analysis*.

Methods

Extraction of pectin. The pomace obtained from premium apple juice production was removed from the hydraulic press, washed with water (1:1) for five minutes at 22°C, centrifuged at 860 *g* until total liquid drainage and dried overnight in an air-circulated oven at 70°C. The dried pomace was ground through 60 MESH size and stored in polyethylene bags in a gel silica atmosphere at room temperature. Pectin from apple pomace was extracted according to the experimental conditions reported by Fertonani et al. (2006), with a nitric acid (HNO_3) solution (solid-liquid ratio 1:40) adjusted to 100 mM, at the boiling point for 10 min. Pectin was isolated after cooling to 4°C by ethanol 66% (v/v) precipitation, air-dried in a ventilated oven at 50°C and stored under a P_2O_5 atmosphere until use. The gravimetric yield was calculated as the ratio (%) of pectin obtained from the apple pomace raw material.

Analysis. All results were calculated on a dry matter basis. The moisture and mineral content were determined by gravimetric losses after 4h at 104 and 550°C, respectively (TANNER; BRUNNER, 1985). Total fat content was determined by Soxhlet extraction with n-hexane and the protein content (N x 6.25) by the Kjeldahl procedure (IAL, 2005). Reducing and total reducing sugar were determined by the classical method of Somogyi (1945) as modified by Nelson (1944). Glucose was determined by a glucose oxidase (GOD) kit, fructose by the difference between reducing sugar and glucose, and sucrose by the difference between total reducing sugar and reducing sugar. Total acidity was determined by titration with 0.1 N NaOH using the factor 0.64 to express the results as malic acid, in g 100 g^{-1} (TANNER; BRUNNER, 1985). Dotal dietary fiber was calculated using the enzymatic-gravimetric method (AOAC, 2000).

Pectic Substances. Pectic substances were characterized by titration of acidic functions of galacturonic acid before and after saponification. The total content of anhydrogalacturonic acid (AUA) and the number of methoxyl groups (MeO) were calculated by specific equations, shown as percentages (BOCHEK et al., 2001; FERTONANI et al., 2006). The sum of these fractions represents the acid polysaccharides – namely, pectinic acid – and the neutral are the remainder. Pectic substances were also characterized according to their fingerprint from the Fourier transform infrared (FTIR) method in the IR region 4000 cm^{-1} -400 cm^{-1}, with 4 cm^{-1} resolution. The FTIR spectra analyzed were obtained using the corrected peak areas at 1639.4 (COO-asymmetric stretching) and at 1747 cm^{-1} (C=O esterified), using the following expression $DE(\%) = 100\star (A_{1747}) / (A_{1747} + A_{1639})$ according to the literature in Gnanasambandam and Proctor (2000), Monsoor et al. (2001) and Faravash and Ashtiani (2008).

Individual neutral sugars were analyzed as their alditol acetates by gas-liquid chromatography (GLC) using an HP 5890 SII with FID system, equipped with a capillary column (025 mm Ø x 30 m, model B-210, with film 0.25 µm) (WOLFROM; THOMPSON, 1963a and b).

High performance size-exclusion chromatography (HPSEC) was carried out on the pectin solutions using multidetection equipment with a Waters 2410 differential refractometer (RI) and an on-line adapted Wyatt Technology Dawn F multi-angle laser light scattering (MALLS) detector. Four Waters Ultrahydrogel 2000/500/250/120 columns were connected in series and coupled to the multidetection equipment. A 0.1 M $NaNO_2$ solution containing NaN_3 (0.5 g L^{-1}) was used as eluent. The samples were previously filtered by a cellulose acetate membrane (0.22 µm; Millipore) and injected at 1.5 mg mL^{-1}. HPSEC data were collected and analyzed by a Wyatt Technology ASTRA program (HOKPUTSA et al., 2004).

Results and discussion

The raw material

The results of the simple statistical descriptive analysis of pectin are shown in Table 1. There was less variation in the ash and protein contents, with dispersion lower than 10%, and the sugar and malic acid contents were more dispersive. Total dietary fiber was more constant and less susceptible to the selected treatments, although the variation coefficient was 12.83%. Sudha et al. (2007) reported that the samples of commercial pomace showed contents of ash (0.50 g 100 g^{-1}), total fat (2.70 g 100 g^{-1}) and protein (2.06 g 100 g^{-1}), which represents 24.63, 156.06 and 66.88%, respectively, compared to our results. The average contents described by Paganini et al. (2005) were 1.52 g 100 g^{-1} of ash and 0.09 g 100 g^{-1} of acid to BelGolden, Gala and Fuji pomace apples, after first pressing to juice production. Cho and Hwang (2000) had the following contents from dried apple pomace produced in South Korea after juice manufacture: 11.4% moisture content, 3.7% protein, 4.5% fat, 1.8% ash and 70.9% carbohydrate.

Table 1. Approximate composition average of eleven dried pomace apple cultivars (dry basis)★.

Components	Average	Standard Deviation	Variation (%)	Confidence	
				-95%	+95%
Ash	2.03	0.13	6.40	1.95	2.12
Total Fat	1.73	0.40	23.12	1.46	1.99
Protein	3.08	0.21	6.82	2.94	3.23
Malic acid	1.16	0.20	17.24	1.02	1.29
Glucose	12.56	3.55	28.26	10.18	14.95
Fructose	17.93	2.90	16.17	15.98	19.88
Sucrose	7.10	4.81	67.78	3.87	10.33
Reducing sugar	30.34	4.03	13.28	27.45	33.22
Total sugar	39.13	8.05	20.57	33.72	44.53
Total dietary fibers	43.71	5.61	12.83	39.94	47.48
Soluble fibers	15.04	1.54	10.24	14.01	16.08
Insoluble fibers	28.67	5.69	9.85	24.85	32.49

Source: Sato et al. (2010).

Fiber fraction

In Table 2, the results of total dietary fiber are shown, both soluble and insoluble. Total dietary fiber (TDF) presented a total average of 43.71 g 100 g^{-1} with 15.02 g 100 g^{-1} (34%) from soluble fiber or pectic substances and 28.69 g 100 g^{-1} (66%) from insoluble fiber, mainly cellulose and hemicellulose. The results reported by Sudha et al. (2007) showed 14.60% soluble fiber and 36.50% insoluble fiber, similar if the standard deviations were considered. On the other hand, the values were lower than the contents of TDF reported by Figuerola et al. (2005) from apple fiber concentrate of Royal Gala, Granny Smith and Liberty varieties. These results were 78.2, 60.7 and 89.8%, respectively.

The yield of pectic substances had a variation coefficient around 10%, which indicated potential varieties of interest to be used for pectin extraction [coded (1), (3) and (5)].

Table 2. Fraction of alimentary fibers.

Sample	Fiber (g 100 g^{-1})		
	Total	Soluble	Insoluble
(1)	44.50	17.65	26.85
(2)	43.77	12.26	31.51
(3)	45.95	16.09	29.86
(4)	48.48	13.26	35.22
(5)	46.05	17.11	28.94
(6)	40.89	14.48	26.41
(7)	46.52	15.07	31.45
(8)	33.4	14.79	18.61
(9)	51.85	14.90	36.95
(10)	34.19	14.66	19.53
(11)	45.24	15.22	30.02
Average	43.71	15.04	28.67
Std Dev.	5.61	1.54	5.69
Variation	12.84	10.23	19.85

Therefore, phenolic compounds are retained in the extracted pectin of apple pomace and can contribute to the brown color after oxidation (SCHIEBER et al., 2003).

Titrimetric evaluation of pectic substances

In Table 3, titrimetric features of pectin are shown.

Table 3. Physicochemical characteristics of isolated pectin.

Cultivar	AUA (%)	MEO (%)	Acidic fraction (%)	Neutral fraction (%)	Degree of esterification (%)
(1)	46.90 ± 4.90	5.88 ± 0.59	52.79 ± 5.48	47.21 ± 5.48	71.24 ± 0.67
(2)	51.21 ± 2.97	6.49 ± 0.21	57.70 ± 3.15	42.30 ± 3.15	72.05 ± 2.63
(3)	50.86 ± 1.88	6.34 ± 0.27	57.19 ± 1.71	42.81 ± 1.71	70.86 ± 5.36
(4)	51.57 ± 4.53	6.67 ± 0.54	58.25 ± 4.95	41.75 ± 4.95	73.59 ± 4.46
(5)	51.57 ± 4.06	6.70 ± 0.24	58.27 ± 4.29	41.73 ± 4.29	74.01 ± 3.57
(6)	51.31 ± 4.34	6.66 ± 0.43	57.97 ± 4.73	42.03 ± 4.73	73.82 ± 3.03
(7)	53.38 ± 3.89	6.93 ± 0.24	60.32 ± 4.11	39.68 ± 4.11	73.90 ± 3.22
(8)	50.25 ± 2.41	6.35 ± 0.24	56.60 ± 2.59	43.40 ± 2.59	71.82 ± 2.47
(9)	50.41 ± 1.45	6.30 ± 0.49	56.72 ± 1.69	43.28 ± 1.69	70.99 ± 5.23
(10)	48.68 ± 1.54	6.06 ± 0.52	54.74 ± 2.04	45.26 ± 2.04	70.59 ± 3.88
(11)	52.85 ± 3.92	6.72 ± 0.28	59.57 ± 4.15	40.43 ± 4.15	72.33 ± 3.58
Average	50.82	6.46	57.28	42.72	72.29

The quantification of total uronic acid indicates the amount of free carboxyl and methoxylated total (AUA, %). The group of eleven genotypes had an average value of 50%, with a minimum of 48% and a maximum of 53% and the average value of methoxyl (MEO) is 6.46%, varying from 5.88 to 6.93%. Both results (AUA+MEO) represent the acidic fraction of pectin (pectinic acid), with an average value of 57.28%.

The neutral fraction, also named ballast (BERK, 1976), is composed of neutral sugars as arabinose and galactose attached to the rhamnose in the main chain as polymers (RG I) or directly to the GalA in the main chain (RG II) (WILLATS et al., 2006). The average value was 42.72%. The degree of esterification is also shown in Table 3, and the values found are similar, identifying the homogeneous group of high methoxyl pectin with an average DE of 72.29%. Constenla et al. (2002) extracted pectin from Granny Smith apple pomace, dried at 70°C. They found an average anhydrogalacturonic content of 58.8% and a degree of esterification of 73.9%.

FTIR analysis

In Figure 1, a typical spectrogram obtained from the analyzed pectin samples is shown. It shows the fingerprint of high methoxyl pectin, with a degree of esterification higher than 50%. However, the values found by this methodology were lower than those obtained by classical titrimetry. Such an approach is effective to confirm the pectin functional identity from the peaks concerned with esterified carboxyl groups (1747 cm^{-1}) and related to free carboxyl groups (1639.4 cm^{-1}), which is consistent with the literature (MARCON et al., 2005; FERTONANI et al., 2006).

In this study, the calibration procedure was not used to fit the results obtained from the titrimetric method to those from FTIR. The determination of the esterification degree by instrumental methods underestimated the values of DE (Figure 2). The calibration technique introduces the effect of stretching single C = O at 2900 cm^{-1}, and the results fit with a correlation $R^2 > 0.98$ (SCABIO et al., 2007). Another procedure to calculate DE by FTIR introduces the transmitance at 1820 cm^{-1} into the equation, giving still lower values (LEGENTIL et al., 1995).

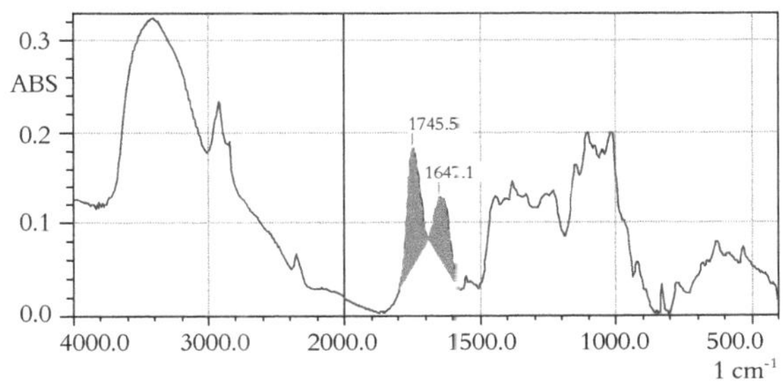

Figure 1. Spectrogram of a pectin sample showing the two wavelengths.

According to Gnanasambandam and Proctor (2000), a small difference in the structure and composition of a molecule can result in significant changes in the absorption peaks. Hence, samples from the same source/origin can be expected to have lower FTIR spectral variations as observed in this work.

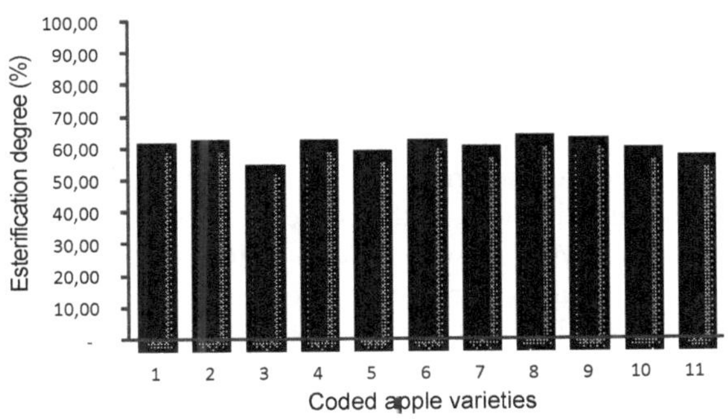

Figure 2. Esterification degree as determined by IR spectroscopy.

Neutral sugars composition

The sugar composition of the neutral fraction from all samples is shown as the results of simple descriptive statistical analysis. The results do not conflict with the literature, and even the high amounts of glucose observed are explained by Legentil et al. (1995) as being released from a glucan because all free sugars were removed during the alcohol extraction.

In Table 4, the proportional composition of the sugars from acid and neutral fractions is shown, including the methoxylated group. Such composition was calculated with respect to the amount of acidic and neutral fractions reported in Table 3, corresponding to AUA+MeO (pectinic acids) and the identified neutral sugar, respectively. Pectinic acids were quite homogeneous, with a variation lower than 4% and 5% for both components, also as seen in Table 3. However, all other components from the neutral fraction show a high dispersion, from 17% (glucose) to 72% (fucose). Rhamnose content was also very heterogeneous, but arabinose, xylose and galactose showed less dispersion among the samples, with a variation coefficient of approximately 24-27%. With these differences in neutral sugar composition, it is reasonable to accept differences in pectin structure as well.

Table 4. Proportional composition of the sugars from acid and neutral fractions.

Cultivar	Fraction (%)								
	AUA	MeO	Rham	Fuc	Ara	Xyl	Man	Gal	Glc
1	46.90	5.88	0.65	0.12	3.33	1.31	2.23	2.51	37.3
2	51.20	6.49	0.85	0.31	4.70	3.21	0.05	4.67	28.9
3	50.90	6.34	0.82	0.32	4.70	2.46	4.10	2.94	27.5
4	51.60	6.67	0.01	0.01	3.48	3.55	9.55	5.11	24.7
5	51.60	6.70	0.96	0.23	4.23	2.50	2.99	2.37	28.5
6	51.30	6.66	1.35	0.54	5.05	3.90	4.31	3.46	23.4
7	53.40	6.93	1.08	0.33	4.88	3.00	3.25	4.38	22.8
8	50.30	6.35	2.09	0.74	8.54	3.78	3.95	3.14	21.2
9	50.40	6.30	0.34	0.00	4.79	3.86	3.43	4.31	25.4
10	48.70	6.06	0.91	0.20	3.76	2.00	2.78	3.27	32.4
11	52.90	6.72	1.08	0.35	5.03	2.90	3.96	4.68	22.4
Average	50.80	6.46	0.92	0.29	4.77	2.95	3.69	3.71	26.8
Standard Deviation	1.72	0.30	0.51	0.21	1.33	0.79	2.17	0.91	4.60
Coefficient of Variation (%)	3.39	4.62	55.57	72.00	27.86	26.82	58.92	24.44	17.21

The composition of neutral sugars of apple pectin from Poland were 22 $\mu g\ mg^{-1}$ of Rha, 10 $\mu g\ mg^{-1}$ of Ara, 14 $\mu g\ mg^{-1}$ of Xyl, 26 $\mu g\ mg^{-1}$ of Gal and 36 $\mu g\ mg^{-1}$ of glucose in a total sugars of 10.8% (ZALESKA et al., 2000). Apple pomace from France presented similar composition of neutral sugar to some sugars but was significantly different in the contents of Ara (higher at 7.4%) and Man (lower at 1.5%) (RENARD et al., 1996). In Table 5, the ratios related to HG (GalA /Rham), to RGI ((Gal+Ara)/(Rham) and to RGII ((Gal+Ara)/(GalA)) are shown. The mean value found for the ratio of uronic acid:rhamnose was 67.81 ± 35.71. Even with a high variation coefficient of 52.66%, these values are in agreement with the literature, where different authors reported values between 25 and 200 (NOVOSEL´SKAYA et al., 2000). Legentil et al. (1995) found the same value, although they expressed it as rhamnose:uronic acid (1/67.81 = 0.014). Values represent the total amounts found in each sample of pectin and are supposed to be dispersed among three types of structures of the macromolecule. Indeed, as stated previously, in apple pectin, 22% of the rhamnose is in the linear form, and 78% is in the ramified fraction (RENARD et al., 1995). Therefore, the figures of galacturonan could increase to an average of 308 units of galacturonic residue to 1 unit of rhamnose. Also, the hairy structure would rise from 9.96 to 12.77 units of neutral sugar to 1 unit of rhamnose.

Table 5. Usual neutral sugars in pectin and their relationships.

Cultivar	GalA: Rham	Gal+Ara: Rham	Gal+Ara: GalA
1	81.20	8.93	0.12
2	67.87	11.06	0.18
3	69.80	9.32	0.15
4	58.27	Nd	0.17
5	60.73	6.88	0.13
6	42.93	6.31	0.17
7	55.86	8.57	0.17
8	27.11	5.59	0.23
9	166.76	26.47	0.18
10	60.18	7.77	0.14
11	55.20	8.96	0.18
Average	67.81 ± 35.71	9.96 ± 5.70	0.17 ± 0.03

Molecular mass dispersion

The steric exclusion chromatography coupled to the interferometric refractometry detector and to the multi-angle light laser scattering detector was effective for characterizing the homogeneity and molecular mass dispersion of this batch of pectin. The first sensor gives a response according to the concentration of macromolecules and the second to the size and concentration, providing information about the molecular weight and distribution. In Figure 3, a chromatogram of the pectin isolated from one apple pomace (coded 1) is shown, whose profile suggests the presence of four groups of polysaccharides. In the first set, minor components with the highest molecular weight are detected in a small volume at 38 min of elution by a MALLS sensor with high light scattering. The next set is the largest predominant and shows an elution peak volume at 43 min, followed by another set of low molecular weight in low concentration and ultimately, one final set of large oligosaccharides.

The results obtained by Scabio et al. (2007) were similar, showing the same retention times and polysaccharide profiles. Similar results were shown by Hokputsa et al. (2004) in their report regarding molecular dispersion of hydrosoluble acid polysaccharides from *Felinus gilvus*. Indeed, the mixture of these polysaccharides allows the gel formation.

Figure 4 shows all polysaccharide fractions of all apple pomace pectins identified by an RI detector. All samples have the high molecular weight fractions and the next low molecular weight fraction, but only some have the third polysaccharide fraction with still lower molecular weight.

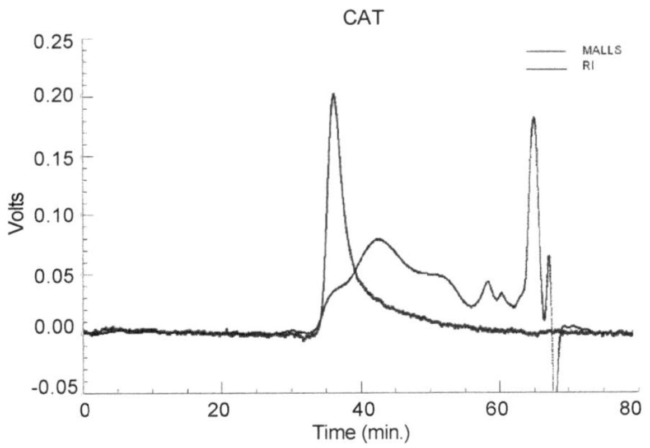

Figure 3. Molecular mass profile provided by multi-angle light laser scattering and by interferometric refractometry index detector in the high performance steric extrusion chromatography of a sample of pectin extracted from apple pomace of cv. Catarina.

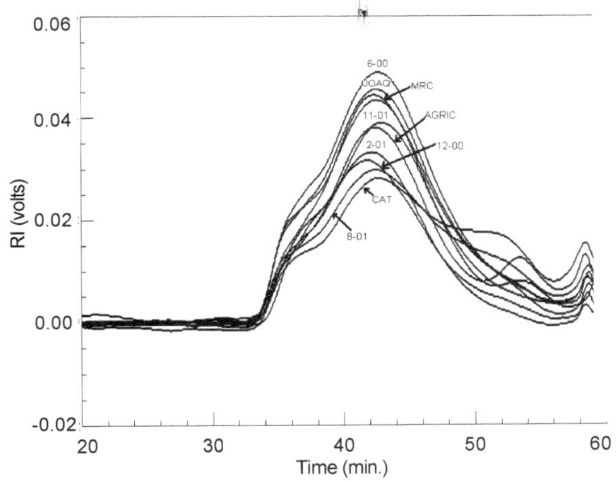

Figure 4. Molar mass distribution profiles of the eleven samples of pectin detected by refratometry index.

Conclusion

Acid-soluble polysaccharide extracted with an average yield of 15.04% ± 1.53 (variation = 10%) reflects the small difference of hydrocolloids in the raw material, and the average esterification degree of 72.29% points to a good processing. The mean value of the pectic fraction was 57.28%, and the neutral sugar from the hairy region (ballast) was 42.72%. The high glucose content suggests simultaneous extraction of cellulose or other glucan. The titrimetric, spectrometric and chromatographic characterizations indirectly provide structural information about the homogalacturonan and hairy regions. According to HPSEC, there is a group of very high molecular weight detected by the MALLS sensor at 38 min. of elution and a major set of lower molecular weight detected by the RI sensor at 43 min. The 3rd polysaccharide group was not observed in all samples, nor were the peaks corresponding to large oligosaccharides of around 58-60 min. of elution. The analysis by FTIR confirms the identity of pectic substances by the similarity with the pectin fingerprint, showing peaks at 1747 and 1639 cm^{-1} related to methoxylated and free carboxyl groups, respectively.

Acknowledgements

The authors are grateful to CNPq, FA, Capes, UEPG and UFPR for the support given to this project.

References

AOAC-Official Methods of Analysis. **AOAC official method 985.29**: total dietary fiber in foods. 30th ed. Washington, D.C., 2000.

BERK, Z. **Braverman´s introduction to the biochemistry of foods**. Amesterdam: Elsevier, 1976.

BOCHEK, A. M.; ZABIVALOVA, N. M.; PETROPAVLOVSKII, G. A. Determination of the esterification degree of polygalacturonic acid. **Russian Journal of Applied Chemistry**, v. 74, n. 5, p. 775-777, 2001.

CONSTENLA, D.; PONCE, A. G.; LOZANO, J. E. Effect of pomace drying on apple pectin. **LWT-Food Science and Technology**, v. 35, n. 3, p. 216-221, 2002.

CHO, Y. J.; HWANG, J. K. Modeling the yield and intrinsic viscosity of pectin in acidic solubilization of apple pomace. **Journal of Food Engineering**, v. 44, n. 2, p. 85-89, 2000.

FARAVASH, R. S.; ASHTIANI, F. Z. The influence of acid volume, ethanol-to-extract ratio and acid-washing time on the yield of pectic substances extraction from peach pomace. **Food Hydrocolloids**, v. 22, n. 1, p. 196-202, 2008.

FERTONANI, H. C. R.; SCABIO A.; CANTERI-SCHEMIN, M. H.; CARNEIRO, E. B. B.; NOGUEIRA, A.; WOSIACKI, G. Influence of acid concentration on extraction and quality of apple pomace pectin. **Semina: Ciências Agrárias**, v. 27, n. 4, p. 599-612, 2006.

FIGUEROLA, F.; HURTADO, M. L.; ESTÉVEZ, A. M.; CHIFFELLE, I.; ASENJO, F. Fibre concentrates from aplle pomace and citrus peel as potential fibre sources for food enrichment. **Food Chemistry**, v. 91, p. 395-401, 2005.

GARNA, H.; MABON, N.; ROBERT, C.; CORNET, C.; NOTT, K.; LEGROS, H.; WATHELET, B.; PAQUOT, M. Effect of extraction conditions on the yield and purity of apple pomace pectin precipitated but not washed by alcohol. **Journal of Food Science**, v. 72, n. 1, p. 1-4, 2007.

GNANASAMBANDAM, R.; PROCTOR, A. Determination of pectin degree of esterification by diffuse reflectance Fourier transform infrared spectroscopy. **Food Chemistry**, v. 68, n. 3, p. 327-332, 2000.

HOKPUTSA, S.; GERDDIT, W.; PONGSAMART, S.; INNGJERDINGEN, K.; HEINZE, T.; KOSCHELLA, A.; HARDING, S. E.; PAULSEN, B. S. Water-soluble polysaccharides with pharmaceutical importance from Durian rinds (*Durio Zibethinus Murr.*): isolation, fractionation, characterisation and bioactivity. **Carbohydrate Polymer**, v. 56, n. 4, p. 471-481, 2004.

IGLESIAS, M. T.; LOZANO, J. E. Extraction and characterization of sunflower pectin. **Journal of Food Engineering**, v. 62, n. 3, p. 215-223, 2004.

IAL-Instituto Adolfo Lutz. **Métodos físicos e químicos para análise de alimentos**: normas analíticas. 5. ed. São Paulo: Instituto Adolfo Lutz, 2005.

LEGENTIL, A.; GUICHARD, I.; PIFFAUT, B.; HALUK, J. P. Characterization of strawberry pectin extracted by chemical means. **LWT-Food Science Technology**, v. 28, n. 6, p. 569-576, 1995.

LEVIGNE, S.; RALET, M. C.; THIBAULT, J. F. Characterisation of pectins extracted from fresh sugar beet under different conditions using an experimental design. **Carbohydrate Polymers**, v. 49, n. 2, p. 145-153, 2002.

MARCON, M. V.; VRIESMANN, L. C.; WOSIACKI, G.; BELESKI-CARNEIRO, E.; PETKOWICZ, C. L. O. Pectins from apple pomace. **Polímeros: Ciência e Tecnologia**, v. 15, n. 2, p.127-129, 2005.

MONSOOR, M. A.; KALAPATHY, U.; PROCTOR, A. Improved method for determination of pectin degree of esterification by diffuse reflectance fourier transform infrared spectroscopy. **Journal Agricultural and Food Chemistry**, v. 49, n. 6, p. 2756-2760, 2001.

NELSON, N. A photometric adaptation of the somogyi method for determination of glucose. **The Journal of Biological Chemistry**, v.153, p. 375-380, 1944.

NOVOSEL´SKAYA, I. L.; VOROPAEVA, N. L.; SEMENOVA, L. N.; RASHIDOVA, S. Sh. Trends in the science and applications of pectins. **Chemistry of Natural Compounds**, v. 36, n. 1, p. 1-10, 2000.

PAGANINI, C.; NOGUEIRA, A.; SILVA, N. C.; WOSIACKI, G. Utilization of apple pomace for ethanol production and food fiber obtainment. **Ciência e Agrotecnologia**, v. 29, n. 6, p. 1231-1238, 2005.

PÉREZ, S.; RODRÍGUEZ-CARVAJAL, M. A.; DOCO, T. A complex plant cell wall polysaccharide: rhamnogalacturonan II. A structure in quest of a function. **Biochimie**, v. 85, n. 2, p. 109-121, 2003.

RENARD, C. M. G. C.; CRÉPEAU, M. J.; THIBAULT, J. F. Structure of the repeating units in the rhamnogalacturonic backbone of apple, beet and citrus pectins. **Carbohydrate Research**, v. 275, n. 1, p. 155-165, 1995.

RENARD, C. M. G. C.; ROHOU, Y.; DELLA VALLE, G.; THIBAULT, J. F.; HUBERT, C.; SAVINA, J. P. Bleaching of apple pomace by hydrogen peroxide in alkaline conditions: Optimisation and characterisation of the products. **Food Science and Technology**, v. 30, n. 4, p. 398-405, 1996.

SATO, M. F.; VIEIRA, R. G.; ZARDO, D. M.; FALCÃO, L. D.; NOGUEIRA, A.; WOSIACKI, G. Apple pomace from eleven cultivars: an approach to identify sources of bioactive compounds. **Acta Scientiarum**. **Agronomy**, v. 32, p. 29-35, 2010.

SCABIO, A.; FERTONANI, H. C. R.; SCHEMIN, M. H. C.; PETKOWICZ, C. L. O.; CARNEIRO, E. B. B.; NOGUEIRA, A.; WOSIACKI, G. A model for pectin extraction from apple pomace. **Brazilian Journal of Food Technology**, v. 10, n. 4, p. 259-265, 2007.

SCHIEBER, A.; HILT, P.; STREKER, P.; ENDREß, H. U.; RENTSCHLER, C.; CARLE, R. A new process for the combined recovery of pectin and phenolic compounds from apple pomace. **Innovative Food Science and Emerging Technologies,** v. 4, n. 1, p. 99-107, 2003.

SOMOGYI, M. A new reagent for the determination of sugars. **The Journal of Biological Chemistry**, v. 160, n. 1, p. 61-68, 1945.

SUDHA, M. L.; BASKARAN, A. V.; LEELAVATHI, K. Apple pomace as a source of dietary fiber and polyphenols and its effect on the rheological characteristics and cake makin. **Food Chemistry**, v. 104, n. 2, p. 686-692, 2007.

TANNER, H.; BRUNNER, H. R. **Getränke analytik**: untersuchungsmethode für die labor: und betriebspraxis. Wädenswill: Helles, 1985.

THAKUR, B. R.; SINGH, R. K.; HANDA, A. V. Chemistry and uses of pectin – A Review. **Critical Reviews in Food Science and Nutrition**, v. 37, n. 1, p. 47-73, 1997.

WILLATS, W. G. T.; KNOX, J. P.; MIKKELSEN, J. D. Pectin: new insights into an old polymer are starting to gel. **Trends in Food Science and Technology**, v. 17, n. 3, p. 97-104, 2006.

WOLFROM, M. L.; THOMPSON, A. Reduction with sodium borohydride. **Methods in Carbohydrate Chemistry**, v. 2, n. 1, p. 65-67, 1963a.

WOLFROM, M. L.; THOMPSON, A. Acetylation. **Methods in Carbohydrate Chemistry**, v. 2, n. 2, p. 211-215, 1963b.

ZALESKA, H.; RING, S. G.; TOMASIK, P. Apple pectin complexes with whey protein isolate. **Food Hydrocolloids**, v. 14, n. 4, p. 377-382, 2000.

19

Comparison of methodologies for degree-day estimation using numerical methods

Adilson Pacheco de Souza[1*], Clóvis Manoel Carvalho Ramos[2], Adriano Dawison de Lima[3], Helenice de Oliveira Florentino[3] and João Francisco Escobedo[4]

*[1]Instituto de Ciências Agrárias e Ambientais, Universidade Federal de Mato Grosso, Campus de Sinop, Av. Alexandre Ferronatto 1200, Distrito Industrial, 78557-267, Sinop, Mato Grosso, Brazil. [2]Departamento de Engenharia Rural, Faculdade de Ciências Agronômicas, Universidade Estadual Paulista "Júlio de Mesquita Filho", Botucatu, São Paulo, Brazil. [3]Departamento de Bioestatística, Instituto de Biociências, Universidade Estadual Paulista "Júlio de Mesquita Filho", Botucatu, São Paulo, Brazil. [4]Departamento de Recursos Naturais e Ambientais, Faculdade de Ciências Agronômicas, Universidade Estadual Paulista "Júlio de Mesquita Filho", Botucatu, São Paulo, Brazil. *Author for correspondence. E-mail: adilsonpacheco@ufmt.br*

ABSTRACT. The development of projects related to the yield of various crops has been greatly enhanced with the incorporation of mathematical models as well as essential and more consistent equations which enable a prediction and greater approximation to their actual behavior, thus reducing error in estimate. Among the operations requiring further investigation are those related to crop growth, characterized by the ideal temperature for addition of dry matter. Due to the wide use of mathematical methods for representing, analyzing and attaining degree-day estimation as well as the great importance of sugarcane in the Brazilian economy, we carried out an evaluation of the mathematical models and numerical integration methods commonly used for estimating the availability of degrees-day for this crop in the region of Botucatu, in São Paulo State, Brazil. Integration models with discretization every 6 hours have shown satisfactory results in degree-day estimation. Conventional methodologies have shown satisfactory results when the estimation of degrees-day was based on the time-temperature curve for each day and for groups of 3, 7, 15 and 30 days. Through numerical integration method, the region of Botucatu showed a annual thermal availability average from 1,070.6 degrees-day for the sugarcane.

Keywords: vegetative development, heat units, numerical integration methods, sugar cane.

Introduction

The development of projects related to the yield of various crops has been greatly enhanced with the incorporation of mathematical models, as well as more consistent equations, which enable a prediction and greater approximation to their actual behavior, thus reducing the error in model. Among the operations requiring further investigation are those related to crop growth, characterized by the ideal temperature for addition of organic matter. In this scenario, due to the great variation of climatic elements, many problems have arisen and become more complex, particularly those associated with the maturation process. Hence, a study on the variation of temperature in relation to crop yield which enables farmers to program their agricultural activities and no be surprised by meteorological extreme effects.

Managers of farming companies, together with research agencies, must resort to all tools available that can help in decision making. Thus, mathematical modeling can be useful in this area, since it is a technique that integrates knowledge from different disciplines in a common work structure, being able to define clear goals and compare objectives that are often conflictive and enabling to predict a system's behavior, which is particularly important for situations that have not yet been experienced (HANKS; RITCHIE, 1991).

Various studies have been developed with the purpose to estimate the crop productivity, aiming high yield and low cost as well as rationalizing the relationships between the different crop phases, thus searching the maximum performance by means of degree-day models (GALAN et al., 2001; PRELA; RIBEIRO, 2002; SCARPARI; BEAUCLAIR, 2008; SOUZA et al., 2009; VILLA NOVA et al., 1972). However, crop's phenological stages can be predicted based on previous knowledge of the temperature historic series, the location and the caloric sum required by the plant to achieve its maturity point (LITSCHMANN et al., 2008; DAY et al., 2008). During the development of the degree-day approach, estimation processes have incorporated various modeling concepts aiming at enhancing phenomenological representation. Among the models that have been developed, applied and presented, those by Arnold (1959), Villa Nova et al. (1972), Ometto (1981), Snyder (1985) and Scarpari and Beauclair (2004) can be pointed out.

Currently, sugarcane (*Saccharum* spp.) is one of the best options among the springs of renewable energy, presenting big importance in the Brazilian agricultural setting and a promising future in the world setting. Brazil is the greatest and more efficient producer of sugar, alcohol and byproducts of the world. Among the Brazilian States, São Paulo with approximately 66% of area planted in Center-South Brazilian is the main producer followed by the states of Alagoas, Paraná, Minas Gerais, Pernambuco, Mato Grosso, Goiás and Mato Grosso do Sul (VIANA et al., 2008).

The sugarcane Brazilian is cultivated in various soil types under the influence of different climates, resulting in various types of environments for the production, which associated with the management of culture and cultivar chosen interfere in the production and planning of activity sugarcane (FRIZZONE et al., 2001; MAULE et al., 2001; SNYDER et al., 2001; THOMPSON; CKARK, 2005). Decisions based on pre-established criteria, allows the possibility to define risk or economic success of this activity, depending on the decision from the information available, such as the availability of thermal in a given region.

Due to the wide use of mathematical methods for representing, analyzing and calculate degrees-day as well as the great importance of sugarcane crops in the Brazilian economy, we accomplished an evaluation of mathematical models and numerical methods of integration commonly used for estimating the availability of degrees-day for this crop in the region of Botucatu, in São Paulo State, Brazil.

Material and methods

Temperature data from the Meteorological Station at the School of Agronomic Sciences at Botucatu, São Paulo State, Brazil, were used. The Meteorological Station is located at latitude 22° 51' S, longitude at 48° 26' W and at 786 m of altitude. The predominant climatic type based on the Köppen's classification is the Cfa type, hot temperate climate (mesothermal) wet. The average temperature in the coldest month (July) is 17.1°C, and in the hottest month (February), 23.1°C, with an total annual of rainfall and potential evapotranspiration, respectively, of 1,428.4 and 945.2 mm (CUNHA; MARTINS, 2009; ROLIM et al., 2007).

The data were collected from the station at intervals of 1-hour, from 2002 to 2005, between January 1 and December 31 of each year. Thermal availability in day-degrees was estimated by five traditional methodologies used for sugarcane (ARNOLD, 1959; VILLA NOVA et al., 1972; OMETTO, 1981; SNYDER, 1985; SCARPARI and BEAUCLAIR, 2004). Conjointly were studied three numerical integration rules of the quadrature method by Newton-Cotes, which were discretized at different intervals during each observation daily (ARENALES; DAREZZO, 2008), thus enabling twelve different forms of degree-day of estimate by numerical integration. The expressions are listed below.

Conventional methods

- Arnold (1959), apud Souza et al. (2009):

$$GD = \sum_{i=1}^{n}(Ti - Tb) \quad (1)$$

where:

Ti = mean temperature; Tb = basal temperature; n = number of days in the period.

- Villa Nova et al. (1972), apud Lemos Filho et al. (1997):

Situation 1: $Tb \geq Tm$

$$GD = \frac{(TM - Tb)^2}{2.(TM - Tm)} \tag{2}$$

Situation 2: $Tb < Tm$

$$GD = \frac{(Tm - Tb) + (TM - Tm)}{2} \tag{3}$$

Situation 3: $Tb \geq TM$

$$GD = 0$$

where:

TM = daily maximum temperature, (°C); Tm = daily minimum temperature, (°C); Tb = basal temperature, (°C).

- Scarpare and Beauclair (2004), with adjustament of previous approaches for correction by photoperiod:

Situation 1: $TM > Tb > Tm$

$$GD = \frac{(TM - Tb)^2 + (TM - 25)^2}{2.(TM - Tm)}.f.n \tag{4}$$

Situation 2: $Tb < Tm$

$$GD = Tm - Tb + \left(\left(\frac{TM - Tm}{2}\right) + \frac{(TM - 25)^2}{2.(TM - Tm)}\right).f.n \tag{5}$$

where:

$$f = \left(\frac{N}{24 - N}\right)^2 \tag{6}$$

where:

n = length of day, (hours); TM = daily maximum temperature, (°C); Tm = daily minimum temperature, (°C); Tb = basal temperature, (°C); f = correction in function of the relation to the number of sunshine hours (N) over 24 hours; n = number of days of the month.

- Ometto (1981), apud Souza et al. (2009):

Situation 1: $TB > TM > Tm > Tb$ (corrected)

$$GD = \frac{(TM - Tm)}{2} + (Tm - Tb) \tag{7}$$

Situation 2: $TB > TM > Tb \geq Tm$

$$GD = \frac{(TM - Tb)^2}{2\,(TM - Tm)} \tag{8}$$

Situation 3: $TB > Tb > TM > Tm$

$$GD = 0$$

Situation 4: $TM > TB > Tm > Tb$

$$GD = \frac{2[(TM - Tm)(Tm - Tb)] + (TM - Tm)^2 - (TM - TB)^2}{2(TM - Tm)} \tag{9}$$

Situation 5: $TM > TB > Tb > Tm$

$$GD = \frac{1}{2}\left(\frac{(TM - Tb)^2 - (TM - TB)^2}{(TM - Tm)}\right) \tag{10}$$

where:

TM = daily maximum temperature, (°C); Tm = daily minimum temperature, (°C); Tb = minimum basal temperature, (°C); TB = maximum basal temperature, (°C).

- Snyder (1985), apud Galán et al. (2001):

Situation 1: $Tb < Tm$

$$GD = \frac{(TM + Tm)}{2} - Tb \tag{11}$$

Situation 2: $Tb > Tm$

$$GD = \frac{\left[(M - Tb)\left(\frac{\pi}{2} - \theta\right) + (W\cos\theta)\right]}{\pi} \tag{12}$$

Situation 3: $Tb < Tm$; $TM > TB$

$$GDb = \frac{(TM + Tm)}{2} - Tb \tag{13}$$

$$GDB = \frac{\left[(M - TB).\left(\frac{\pi}{2} - \varphi\right) + (W.\cos\varphi)\right]}{\pi} \tag{14}$$

$$GD = GDb - GDB \tag{15}$$

Situation 4: $Tb < Tm$; $TM > TB$

$$GDb = \frac{\left[(M - Tb).\left(\frac{\pi}{2} - \theta\right) + (W.\cos\theta) \right]}{\pi} \quad (16)$$

$$GDB = \frac{\left[(M - TB).\left(\frac{\pi}{2} - \varphi\right) + (W.\cos\varphi) \right]}{\pi} \quad (17)$$

$$GD = GDb - GDB \quad (18)$$

where:

$$M = \frac{TM + Tm}{2} \quad (19)$$

$$W = \frac{TM - Tm}{2} \quad (20)$$

$$\theta = arcsen\left(\frac{Tb - M}{W}\right) \quad (21)$$

$$\varphi = arcsen\left(\frac{TB - M}{W}\right) \quad (22)$$

where:

TM = daily maximum temperature, (°C); Tm = daily minimum temperature, (°C); Tb = minimum basal temperature, (°C); TB = maximum basal temperature, (°C).

Numerical Integration Methods

- *Trapeze rule and trapeze rule with steps*

Situation 1: $h = 1/2$

$$GD = \frac{h}{2}.\left[T_0 + 2.T_{12} + T_{24}\right] - Tb \quad (23)$$

Situation 2: $h = 1/4$

$$GD = \frac{h}{2}.\left[T_0 + 2.\left(\sum_{i=6,12,18} T_i\right) + T_{24}\right] - Tb \quad (24)$$

Situation 3: $h = 1/8$

$$GD = \frac{h}{2}.\left[T_0 + 2.\left(\sum_{i=1}^{7} T_{(3\times i)}\right) + T_{24}\right] - Tb \quad (25)$$

Situation 4: $h = 1/24$

$$GD = \frac{h}{2}.\left[T_0 + 2.\left(\sum_{i=1}^{23} T_i\right) + T_{24}\right] - Tb \quad (26)$$

- Simpson's 1/3 numerical integration

Situation 1: $h = 1/2$

$$GD = \frac{h}{3}.\left[T_0 + 4.T_{12} + T_{24}\right] - Tb \quad (27)$$

Situation 2: $h = 1/4$

$$GD = \frac{h}{3}.\left[T_0 + 4.\left(\sum_{i=6,18} T_i\right) + 2.T_{12} + T_{24}\right] - Tb \quad (28)$$

Situation 3: $h = 1/8$

$$GD = \frac{h}{3}.\left[\begin{array}{l} T_0 + 4.\left(\sum_{i=3,9,15,21} T_i\right) \\ + 2.\left(\sum_{i=6,12,18} T_i\right) + T_{24} \end{array}\right] - Tb \quad (29)$$

Situation 4: $h = 1/24$

$$GD = \frac{h}{3}.\left[\begin{array}{l} T_0 + 4.\left(\sum_{i=1}^{12} T_{(2\times i)-1}\right) \\ + 2.\left(\sum_{i=1}^{12} T_{(2\times i)}\right) + T_{24} \end{array}\right] - Tb \quad (30)$$

- Simpson's 3/8 numerical integration

Situation 1: $h = 1/3$

$$GD = \frac{3}{8}.h.\left[T_0 + 3.\sum_{i=8,16} T_i + T_{24}\right] - Tb \quad (31)$$

Situation 2: $h = 1/12$

$$GD = \frac{3}{8}.h.\left[\begin{array}{l} T_0 + 3.\left(\sum_{1}^{4} T_{(6\times i)-4} + \sum_{1}^{4} T_{(6\times i)-2}\right) \\ + 2.\sum_{1}^{3} T_{(6\times i)} \end{array}\right] - Tb \quad (32)$$

Situation 3: $h = 1/24$

$$GD = \frac{3}{8}.h.\left[\begin{array}{l} T_0 + 3.\left(\sum_1^8 T_{(3\times i)-2} + \sum_1^8 T_{(3\times i)-1}\right) \\ + 2.\sum_1^7 T_{(3\times i)} \end{array}\right] - Tb \quad (33)$$

where:

h = discretization interval time; Ti = temperature in determined time, with variation between 0 a 24 hour; T_o = temperature in 0 hour; T_{12} = temperature in 12 hours; T_{24} = temperature in 24 hours?; Tb = minimum basal temperature, (°C);

The thermal sum in degrees-day was given by the value of the area between the variation of temperature hourly for each day and the basal temperatures. The minimum and maximum basal temperatures used were of 18 and 38°C, respectively (TERUEL et al., 1997).

Hourly discretizations (h = 1/24) were considered as those presenting the smallest errors, since the greater the discretization interval, the greater the expected error (ARENALES; DAREZZO, 2008). Nevertheless, Simpson's 1/3 numerical integration with hourly discretization was taken as reference based on the analysis of monthly degrees-day by the Scott-Knott test with 1% of probability. The different discretizations in the interval hourly time for possibilities at twelve numerical integration were evaluated by the same statistical test by grouping months from October to April (considered to be hot months) and from May to September (cold months) during the four years of observation. Simple linear regression equations were adjusted between the traditional methodologies and Simpson's 1/3 integration with hourly discretization (h = 1/24). These equations were submitted to an evaluation of the standard error of the estimate (SEE), agreement index (d) and performance index (c), according to the methodology by Barros et al. (2009), in the estimation of degrees-day used by numerical integration adopted as reference.

Results and discussion

When comparing the accumulated values of degrees-day monthly (Table 1) for each of the numerical integration rules used, two groups statistically different were found, because of the difference in temperatures over the seasons of the year. The smallest availability of monthly sums thermal was observed in the period from May to September, with a daily variation of 0.903 to 2.492 DD (degree-day). However, in the spring and summer months we found the greatest daily value for the sugarcane crop (3.17 DD in November to 4.24 DD in March). The rules of numerical integration with hourly discretization did not differ from each other for the hot months either in the monthly estimation of the accumulated degrees-day.

Table 1. Average availability of monthly thermal sums (degrees-day accumulated) in Botucatu, São Paulo State for sugarcane, using methodologies of numerical integration rules with hourly.

Months	Trapeze rule of integration	Simpson's 1/3 rule of integration	Simpson's 3/8 rule of integration
January	117.75 Aa	117.75 Aa	117.75 Aa
February	127.75 Aa	128.00 Aa	128.00 Aa
March	131.25 Aa	131.25 Aa	131.25 Aa
April	117.75 Aa	117.50 Aa	117.50 Aa
Mai	43.25 Ba	43.25 Ba	43.25 Ba
June	41.50 Ba	41.50 Ba	41.50 Ba
July	28.00 Ba	28.00 Ba	28.00 Ba
August	62.50 Ba	62.50 Ba	62.50 Ba
September	74.75 Ba	74.75 Ba	74.75 Ba
October	109.75 Aa	109.25 Aa	109.75 Aa
November	95.00 Aa	94.50 Aa	95.00 Aa
December	123.25 Aa	122.75 Aa	123.00 Aa
Total	1,072.50	1,071.00	1,072.25

*Mean followed by the same uppercase and lowercase letters, column and row respectively, do not differ by Scott Knott test at 1% probability.

The greatest variation was observed for the month of December, but it was not higher than 0.5%. In general, any of these methodologies of hourly integration could be used to establish the comparison with those conventionally used for estimating the accumulated degrees-day. In this study, Simpson's 1/3 integration was used as a basis, it estimates the sum of degrees-day for a time, with at least three points and polynomials, which allows minor errors in the integration (ARENALES; DAREZZO, 2008).

By analyzing only the clustered hot and cold months, we observed that, among the variation possibilities of numerical integration for each method, given by the amplitude of the time discretization interval and its requirements, just in the cold months, the twelve integration methodologies studied did not differ from each other due to the low temperature values observed in these periods (Table 2). On the other hand, in the hot months, the methods trapezoid, Simpson's 1/3 and Simpson's 3/8 with time discretization in 0-24, 0-12-24, 0-8-16-24h differed from the other methodologies, since the greater the discretization interval larger the error found. Consequently, the distribution of the time in interval six hours showed similar results to those from hourly for both the hot and cold months, with variations of 2.87 and 0.06% observed for numerical integration by successive trapezoids and Simpson's 1/3, respectively, in hot months. Hence, it was demonstrated that this time interval can be an interesting possibility for estimating degrees-day when applied in equation 28 (Simpson's 1/3 with discretization every 6 hours), since, in Brazil, records of hourly temperature are not available in most meteorological stations. Also, most of time, long time series of collected data does not exist, which is largely a result from the limitations of instrumentation in rural areas and memory limitations of time storage. It is noteworthy that for moments when the temperature at the time of collection was higher than the maximum basal temperature (38°C), the temperature value to be considered must be TB itself, since no vegetative development occurs above of the limiting temperature.

Higley et al. (1986) discussed various classes of factors that can influence the capacity of different models to predict accumulated degrees-day and identified that errors associated with the adoption and approximation of the methods used to estimate development rates and basal temperatures, conjointly with the limitation in the evaluation of climatic data, have great impacts on the results found (PRELA et al., 2006). To Camargo et al. (1987), the degree-day method admits a linear relationship between temperature increase and plant development, as each species and/or variety is characterized by a distinct base temperature that can vary according to the plant's age or phenological phase (LITSCHMANN et al., 2008; SOUZA et al., 2009). It is common however, to adopt a single base temperature for the whole plant's cycle, which may lead to errors in predicting of the harvest and yield time of a given variety.

Table 2. Average of the hot months and cold by methodologies of numerical integrations with different time of discretization. Data from 2002 to 2005.

Methodology	Mean of hot months	Mean of cold months
Trapeze		
Hourly	97.92 b	29.02 a
0-24	50.82 c	10.68 a
0-12-24	112.56 a	34.7 a
0-6-12-18-24	100.73 b	28.71 a
0-3-6-9-12-15-18-21-24	101.39 b	30.11 a
1/3 Simpson		
Hourly	96.75 b	28.83 a
0-12-24	135.41 a	45.83 a
0-6-12-18-24	96.81 b	27.09 a
0-3-6-9-12-15-18-21-24	101.64 b	30.61 a
3/8 Simpson		
Hourly	97.49 b	28.9 a
0-8-16-24	110.99 a	32.88 a
0-2-4-6-8-10-12-14-16-18-20-22-24	101.57 b	29.7 a

*Mean followed by the same uppercase letters, column, do not differ by Scott Knott test at 1% probability.

As to this approach, Roltsch et al. (1999) cite that greater reliability on values of degree-day found can be attained with the use of data of hourly temperature, and, consequently, the actual daily and hourly temperature curves are not symmetrical, and their shape may vary considerably due to specific of climatic conditions (atmosphere) and time of the year. When temperature data of daily minimum and maximum (min/max) are used, suppositions of a profile daily temperature are obtained. Such profile is represented by a specific geometric shape (rectangle, triangle, double triangle, single senoidal and double senoidal, among others). However, an alternative is to perform approximations with simple empirical models based on local climatic variations, which can minimize the errors occurring in estimations with geometric shapes pre-defined (RAWORTH, 1994).

Nevertheless, various efforts have been made in the attempt to use correction measures for some methodologies that use such pre-defined geometric shapes based on daily temperature curves utilizing hourly databases, have presented little improvement (ROLTSCH et al., 1999; CESARACCIO et al., 2001; RULM et al., 2010). As example, the method developed by Parton and Logan (1981) can be cited. Their method has shown to be versatile; however, its use requires considerable investment in obtaining parameters for each season of the year and for the behavior of the hourly temperature in at least one year. Additionally, limitations to its use have been identified, since not all day the temperature curves are well represented by the method (ROLTSCH et al., 1999).

It was observed that the methodologies commonly used in Brazil for estimating degrees-day show a tendency to overestimate the monthly accumulated values over the year (Figure 1), except for the method proposed by correction photoperiod, which underestimates in hot months (SD = 9.79) and overestimates in cold months (SD = 8.83), however that variation to the long one of the year favored the obtaining of the most determination coefficient found ($R^2 = 0.9542$). This is explained by the fact that such methodology uses a correction in function of the relation to the number of sunshine hours (N) over 24 hours. The methodologies proposed by Ometto (1981) and Snyder (1985) presented identical results, showing similarity between the geometric figures considered by each estimation method.

When comparing the values degree-day obtained with the numerical integration of data of hourly temperature, in the average the methodologies of Arnold (1959), Villa Nova et al. (1972), Ometto (1981) and Snyder (1985) overestimated the annual availability of degrees-day for the sugarcane crop in 19.90, 17.44, 27.57 and 27.87%, respectively (Table 3), whereas the methodology with correction by photoperiod (Scarpari and Beauclair, 2004) underestimates the monthly availability in to 9.59%. The high standard-deviation values observed in six methodologies resulted from the low temperatures occurring in 2004, which presented an annual average of 19.89°C whereas, in the other years, the annual temperatures averages were of 21.27, 20.57 and 20.51°C, respectively for the years 2002, 2003 and 2005. In general, the region of Botucatu, is appropriate for sugarcane, since its annual average temperature is of 20.3°C, and its annual average hydric deficiency is of 5 mm (CUNHA; MARTINS, 2009).

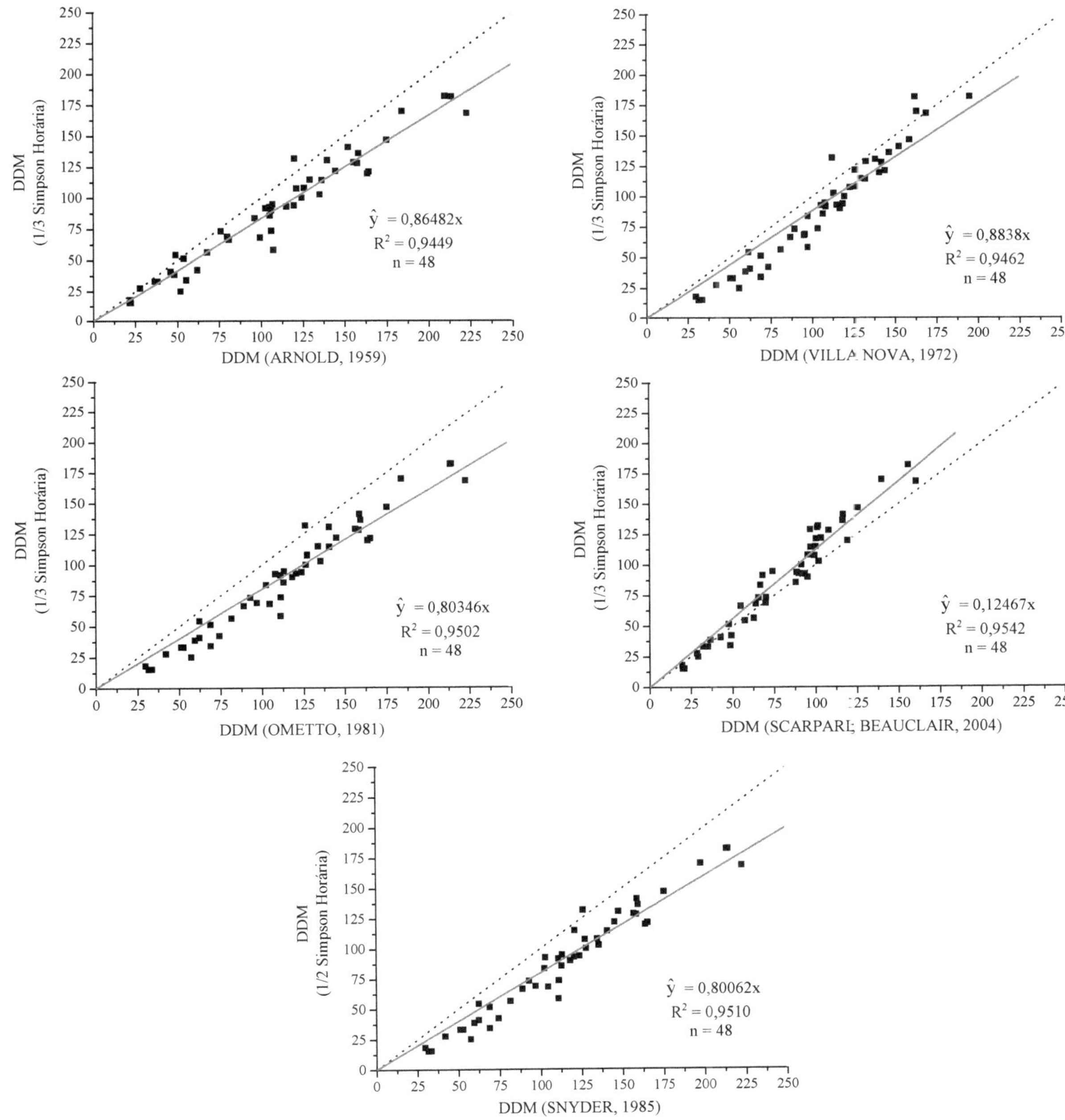

Figure 1. Correlation between monthly degree-day (DDM) by methodologies geometric figure and 1/3 Simpson rule for the numerical integration in hours. Data from 2002 to 2005.

Table 3. Annual availability of degrees-day calculated by six methodologies for the sugarcane crop in Botucatu, São Paulo State. Data from 2002 to 2005.

Year	Methodology					
	Arnold (1959)	Villa Nova et al. (1972)	Ometto (1981)	Snyder (1985)	Scarpari and Beauclair (2004)	1/3 Simpson Hourly
2002	1,656.29	1,477.96	1,703.57	1,704.44	1,178.44	1,324.23
2003	1,252.57	1,263.23	1,354.13	1,368.06	972.16	1,051.70
2004	1,039.43	1,088.67	1,128.48	1,124.25	799.60	873.22
2005	1,185.93	1,199.04	1,276.53	1,278.80	921.40	1,033.10
Mean	1,283.55	1,257.23	1,365.68	1,368.89	967.94	1,070.56
Standard deviation (SD)	263.95	163.86	243.93	245.33	157.90	187.13

The cultivation sugarcane in the Mid-Southern region occurs at two different times, known as "cane cultivation 1-year", performed in October, and "cane cultivation 1-year-and-half", performed in January/February (SCARPARI; BEAUCLAIR, 2004), the thermal sum accumulated during the whole crop cycle and in the different phenophases will depend on the planting type. Cane of 1-year reaches its maximum development from November to April, which then decreases due to adverse climatic conditions, whereas that of cane a-year-and-half is restricted from May to September in function of the climate. The greatest development occurs from October to April, and particularly after December, under favorable of rainfall conditions, which coincides with the variation of degree-day found for this crop in Botucatu, São Paulo State (Table 1).

Scarpari and Beauclair (2008) observed for the cultivars SP 80-3280 and RB 85-5156, in the region of Piracicaba, São Paulo State, using the methodology of the Villa Nova et al. (1972), the thermal sum accumulated by the crop was lower than 1600 DD, with the leaf index showing maximum values between 750 e 1000 DD.

Since the adoption of determination coefficient as the only criterion for defining the quality of a given method is not appropriate due to the fact that it does not establish the type or magnitude of the differences between a standard value and a value predicted for estimation of models or other different measurement mechanisms standard, a study on the performance of the methodologies (Table 4) was conducted according to Barros et al. (2009). It was found that only the methodology proposed by correction by photoperiod showed limitations as regards clusters. For all the methodologies, the greater the data clusters, the better results found in the estimation of degree-day. However, since this concept is widely used, both in crops and in pest management, it is important to evaluate these methods in shorter periods of time.

Table 4. Performance of the methodologies for calculated degrees-day compared with 1/3 Simpson rules for the numerical integration with hourly, in different groupings. Data from 2002 to 2005.

Methodology	a_1	SEE (GD)	R^2	d	C	Classes
		15 Days				
Arnold (1959)	0.8533	7.72	0.9112	0.9812	0.94	Excellent
Villa Nova et al. (1972)	0.9233	8.56	0.8780	0.9747	0.91	Excellent
Ometto (1981)	0.8246	8.06	0.9031	0.9776	0.93	Excellent
Scarpari and Beauclair (2004)	0.7579	8.11	0.8988	0.9807	0.93	Excellent
Snyder (1985)	0.8181	7.82	0.9095	0.9795	0.93	Excellent
		7 Days				
Arnold (1959)	0.8183	5.00	0.8638	0.9686	0.90	Very Good
Villa Nova et al. (1972)	0.8772	5.67	0.8211	0.9535	0.86	Very Good
Ometto (1981)	0.7942	5.17	0.8524	0.9635	0.89	Very Good
Scarpari and Beauclair (2004)	0.7801	5.40	0.8330	0.9729	0.89	Very Good
Snyder (1985)	0.7914	5.12	0.8555	0.9644	0.89	Very Good
		3 Days				
Arnold (1959)	0.9033	2.12	0.8924	0.9370	0.88	Very Good
Villa Nova et al. (1972)	0.9366	2.59	0.8429	0.9504	0.87	Very Good
Ometto (1981)	0.8712	2.26	0.8787	0.9246	0.87	Very Good
Scarpari and Beauclair (2004)	0.8859	1.93	0.8691	0.9336	0.88	Very Good
Snyder (1985)	0.8713	2.24	0.8802	0.9253	0.87	Very Good
		1 Day				
Arnold (1959)	0.8826	1.91	0.8133	0.9119	0.80	Good
Villa Nova et al. (1972)	0.9945	1.80	0.7764	0.9437	0.83	Good
Ometto (1981)	0.8713	1.03	0.8168	0.9446	0.85	Good
Scarpari and Beauclair (2004)	1.0803	1.69	0.8443	0,9239	0.83	Good
Snyder (1985)	0.8719	2.02	0.7745	0.9343	0.80	Good

a_1: regression coefficient for the linear regression equation: $DDM_{Integrated} = a_1 (DDM_{Methodologies})$; SEE: standard error of estimative (degrees-day); (R^2): coefficient of determination; (d) index of agreement; (c) perfomance index.

It was possible to verify that, in general, the methods from Ometto (1981) and Snyder (1985) satisfactorily estimated the degrees-day obtained by the hourly numerical integration, considering the performance rates over 70% for all the clusters analyzed (Table 4). Based on the results from the regression analysis with the data of degree-day clustered in periods of 3, 7 and 15-day, we found an increase in the standard error of the estimation (SEE) in relation to the daily analysis. As the time period examined increases, the accumulated values are higher, therefore, the values of SEE tend to be higher. However, were obtained greater correlation coefficients and, consequently, better performance rates of the analyzed methods.

Conclusion

The numerical integration with discretization every six hours showed satisfactory results in the estimation of degrees-day. Conventional methodologies showed satisfactory performances concerning the estimation of degrees-day based in the hourly temperature curve for each day for the clusters 3, 7, 15 and 30-day. Through numerical integration method, the region of Botucatu, in São Paulo State, Brazil, shows a thermal availability from 1,070.6 degrees-day annual for the sugarcane crop.

References

ARENALES, S.; DAREZZO, A. **Cálculo numérico**: aprendizagem com apoio de software. São Paulo: Thomson Learning, 2008.

ARNOLD, C. Y. The determination and significance of the base temperature in a linear heat unit system. **Journal of the American Society for Horticultural Science**, n. 74, p. 430-445, 1959.

BARROS, V. R.; SOUZA, A. P.; FONSECA, D. F.; SILVA, L. B. D. Avaliação da evapotranspiração de referência na região de Seropédica, Rio de Janeiro, utilizando lisímetros de pesagem e modelos matemáticos. **Brazilian Journal of Agricultural Sciences**, v. 4, n. 2, p. 198-203, 2009.

CAMARGO, M. B. P.; BRUNINI, O.; MIRANDA, M. A. C. Temperatura-base para o cálculo de graus-dia para cultivares de soja em São Paulo. **Pesquisa Agropecuária Brasileira**, v. 22, n. 2, p. 115-121, 1987.

CESARACCIO, C.; SPANO, D.; DUCE, P.; SNYDER, R. L. A improved model for determining degree-day values from daily temperature data. **International Journal of Biometeorology**, v. 45, n. 4, p. 161-169, 2001.

CUNHA, A. R.; MARTINS, D. Classificação climática para os municípios de Botucatu e São Manuel, SP. **Irriga**, v. 14, n. 1, p. 1-11, 2009.

DAY, K.; LOPEZ, G.; DEJONG, T. Using growing degree hours accumulated thirty days after bloom to predict peach and nectarine harvest date. **Acta Horticulturae**, v. 803, n. 1, p. 163-166, 2008.

FRIZZONE, J. A.; MATIOLI, C. S.; REZENDE, R.; GONÇALVES, A. C. A. Viabilidade econômica da irrigação suplementar da cana-de-açúcar, *Saccharum* spp., para a região Norte do Estado de São Paulo. **Acta Scientiarum. Agronomy**, v. 23, n. 5, p. 1131-1137, 2001.

GALÁN, C.; GARCÍA-MOZO, H.; CARIÑANOS, P.; ALCÁZAR, P.; DOMÍNGUEZ-VILCHES, E. The role of temperature in the onset of the *Olea eureopaea* L. pollen season in Southwestern Spain. **International Journal of Biometeorology**, v. 45, n. 1, p. 8-12, 2001.

HANKS, R. J.; RITCHIE, J. T. **Modeling plants and soil systems**. Madison: American Society of Agronomy; Crop Science Society of America; Inc. Soil Science Society of America, 1991. (Agronomy, 31).

HIGLEY, L. G.; PEDIGO, L. P.; OSTLIE, K. R. DEGDAY: a program for calculating degree-days, and assumptions behind the degree-day approach. **Environmental Entolomology**, v. 15, n. 6, p. 999-1016, 1986.

LEMOS FILHO, J. P.; VILLA NOVA, N. A.; PINTO, H. S. A model including photoperiod in degree days for estimating *Hevea* bud growth. **International Journal of Biometeorology**, v. 41, n. 1, p. 1-4, 1997.

LITSCHMANN, T.; OUKROPEC, I.; KIRZAN, B. Predicting individual phenological phases in peaches using meteorological data. **Horticultural Science**, v. 35, n. 2, p. 65-71, 2008.

MAULE, R. F.; MAZZA, J. A.; MARTHA JÚNIOR, G. B. Produtividade agrícola de cultivares de cana-de-açúcar em diferentes solos e épocas de colheita. **Scientia Agricola**, v. 58, n. 2, p. 295-301, 2001.

OMETTO, J. C. **Bioclimatologia vegetal**. São Paulo: Agronômica Ceres, 1981.

PARTON, W. J.; LOGAN, J. A. A model for diurnal variation in soil and air temperature. **Agricultural Meteorology**, v. 23, n. 1, p. 205-216, 1981.

PRELA, A. P.; RIBEIRO, A. M. A. Determinação de graus-dia acumulados e sua aplicação no planejamento do cultivo de feijão-vagem (*Phaseolus vulgaris* L.) para Londrina - PR. **Revista Brasileira de Agrometeorologia**, v. 10, n. 1, p. 83-86, 2002.

PRELA, A. P.; GARCIA, B. I. L.; PEREIRA, A. R. Variação de graus-dias em Piracicaba (SP), em anos de El Niño e La Niña. **Bragantia**, v. 65, n. 3, p. 527-532, 2006.

RAWORTH, D. A. Estimation of degree-days using temperature data recorded at regular intervals. **Environmental Entomology**, v. 23, n. 4, p. 893-899, 1994.

ROLIM, G. S.; CAMARGO, M. B. P.; LANIA, D. G.; MORAES, J. F. L. Classificação climática de Köppen e de Thornthwaite e sua aplicabilidade na determinação de zonas agroclimáticas para o Estado de São Paulo. **Bragantia**, v. 66, n. 4, p. 711-720, 2007.

ROLTSCH, W. J.; ZALOM, F. G.; STRAWN, A. J.; STRAND, J. F.; PITCAIRN, M. J. Evaluation of several degree-day estimation methods in California climates.

International Journal of Biometeorology, v. 42, n. 4, p. 169-176, 1999.

RULM, M.; VUKOVIC, A.; MILATOVIC, D. Evaluation of different methods for determining growing degree-day thresholds in apricot cultivars. **International Journal of Biometeorology**, v. 54, n. 4, p. 411-422, 2010.

SCARPARI, M. S.; BEAUCLAIR, E. G. F. Sugarcane maturity estimation through edaphic-climatic parameters. **Scientia Agricola**, v. 61, n. 5, p. 486-491, 2004.

SCARPARI, M. S.; BEAUCLAIR, E. G. F. Variação espaço-temporal do índice de área foliar e do brix em cana-de-açúcar. **Bragantia**, v. 67, n. 1, p. 35-41, 2008.

SNYDER, R. L. Hand calculating degree days. **Agriculture and Forest Meteorology**, v. 35, n. 1-4, p. 353-358, 1985.

SNYDER, R. L.; SPANO, D.; DUCE, P.; CESARACCIO, C. Temperature data for phonological models. **International Journal of Biometeorology**, v. 45, n. 4, p. 178-183, 2001.

SOUZA, A. P.; SILVA, A. C.; LEONEL, S.; ESCOBEDO, J. F. Temperaturas basais e soma térmica para a figueira podada em diferentes épocas. **Revista Brasileira de Fruticultura**, v. 31, n. 2, p. 314-322, 2009.

TERUEL, D. A.; BARBIERI, V.; FERRARO JÚNIOR, L. A. Sugarcane leaf área índex modeling under different soil water conditions. **Scientia Agricola**, v. 54, n. esp., p. 39-44, 1997.

THOMPSON, R.; CLARK, R. M. Spatio-temporal modeling and assessment of within-species phonological variability using thermal time methods. **International Journal of Biometeorology**, v. 50, n. 5, p. 312-322, 2005.

VIANA, R. S.; SILA, P. H.; MUTTON, M. A.; MUTTON, M. J. R.; GUIMARÃES, E. R.; BENTO, M. Efeito da aplicação de maturadores químicos na cultura da cana de açúcar (*Saccharum* spp.), variedade SP81-3250. **Acta Scientiarum. Agronomy**, v. 30, n. 1, p. 65-71, 2008.

VILLA NOVA, N. A.; PEDRO JÚNIOR, M. J.; PEREIRA, A. R.; OMETTO, J. C. **Estimativas de grus-dias acumulados acima de qualquer temperatura - base, em função das temperaturas máxima e mínima**. São Paulo: Universidade de São Paulo - Instituto de Geografia, 1972. (Caderno de Ciências da Terra, 30).

20

Modeling of the shrinkage kinetics of coffee berries during drying

Gabriel Henrique Horta Oliveira[1*], Paulo Cesar Corrêa[2], Fernando Mendes Botelho[3], André Luis Duarte Goneli[4], Paulo Cesar Afonso Júnior[5] and Sílvia Carvalho Campos[6]

*[1]Instituto Federal de Brasília, DF-480, Lote 1, Setor de Múltiplas Atividades, 72405-980, Brasília, Distrito Federal, Brazil. [2]Departamento de Engenharia Agrícola, Centro de Ciências Agrárias, Universidade Federal de Viçosa, Viçosa, Minas Gerais, Brazil. [3]Universidade Federal de Mato Grosso, Sinop, Mato Grosso, Brazil. [4]Faculdade de Ciências Agrárias, Universidade Federal da Grande Dourados, Dourados, Mato Grosso do Sul, Brazil. [5]Empresa Brasileira de Pesquisa Agropecuária, Centro Nacional de Pesquisa do Café, Brasília, Distrito Federal, Brazil. [6]Empresa Brasileira de Pesquisa Agropecuária, Embrapa Agrossilvipastoril, Sinop, Mato Grosso, Brazil. *Author for correspondence. E-mail: gabriel.oliveira@ifb.edu.br*

ABSTRACT. The effects of drying conditions on the shrinkage of *Coffea arabica* L., cultivar Catuaí Vermelho, were studied. An exponential model derived from the kinetic model was used to describe the shrinkage kinetics. In addition, an equation was proposed to describe the relationship between the superficial areas and weights of the coffee berries. Coffee berries were dried at temperatures (T) of 35, 45, 55 and 65°C and at relative humidities (φ) of 25, 35, 45 and 55%, forming 16 different sets of conditions, with an initial moisture content for all conditions of 2.27 dry basis (d.b.), dried until 0.11 d.b. Changes in the product dimensions were measured continuously during drying. Surface area-to-volume ratio values increased with moisture content decrease. The model was suitable for representing the shrinkage kinetics, as well the empirical equation for predicting surface area. Temperature had a greater influence over the constant rate of change of the characteristic dimension (k values) than φ. Activation energies were also determined to have magnitudes of 39.31, 37.32, 36.28 and 36.22 kJ mol^{-1} for relative humidities of 25, 35, 45 and 55%, respectively.

Keywords: surface area-to-volume ratio, post-harvest, activation energy, *Coffea arabica* L, shrinkage.

Introduction

The drying or dehydration of foods is one of the most common methods for assuring the quality and stability of agricultural products, decreasing the water activity of the material, reducing the biological activity, and reducing the chemical and physical changes of the material during storage. Moisture loss leads to modifications in cellular structure, consequently altering product shape and dimension. This information is particularly important to coffee because it presents a high level of moisture content during the initial stages of drying.This moisture loss, according to Yan et al. (2007), influence the products' physical properties and modify the final texture and transport properties of the dry foods.

The volumetric behavior of a grain or fruit, during the drying process, is important to estimate tension levels that the product is submitted, as well when developing equipment, processes and simulation models. This behavior is dependent of temperature and humidity gradients. Additionally, the superficial area, shape and dimensions of an agricultural product are crucial data for mass and heat transfer studies as well as for airflow within a granular mass of the product. These parameters and additional information, such as moisture content, are used for predicting storage and drying conditions of grain and cereals and quality losses during the post-harvest period.

Agricultural products with high initial moisture contents such as coffee (approx. 60% w.b.) (RESENDE et al., 2009) easily lose moisture via diffusion to the surrounding environment, due to the great water vapor pressure difference between the product and the air. Depending upon the evaporative demand of drying air, the rate of moisture loss contributes to a marked reduction of the moisture content, material dimensions and, consequently, volume.

Shrinkage is the ratio between the initial volume and the volume at a certain time after the moisture loss processes. According to Ramos et al. (2005), this trend is due to the reduction of the product size (macroscopic phenomenon), which is a reflection of the reduction of its cellular dimensions (microscopic phenomena) because of the loss of moisture. Aguilera (2004) reported that shrinkage during the dehydration of agricultural products occurs when the viscoelastic matrix contracts into the space previously occupied by the water removed from the cells.

Bennamoun and Belhamri (2006), Ramos et al. (2005) and Mayor and Sereno (2004) reported that shrinkage is an important phenomenon that is frequently neglected during the development of drying models. Neglect of this phenomenon leads to inefficient machinery and handling. The correct quantification of this trend is vital because it allows for the analysis of drying kinetics and, consequently, of the whole drying process (HASHEMI et al., 2009).

Shrinkage has been correlated linearly to moisture content (DISSA et al., 2008; BATISTA et al., 2007). Hatamipour and Mowla (2002) reported a linear correlation for the volume change in carrots during the drying process. Ramos et al. (1994) studied microstructural changes in grape cells during the first stage of convective air drying and concluded that even though cell dimensions suffered modifications during drying, the cell shapes remained unchanged.

Several authors have studied the influence of different process conditions on the volume change of materials during dehydration. McMinn and Magee (1997) analyzed the effect of temperature on the drying of potatoes; Ratti (1994) investigated the influence of air velocity and relative humidity on the drying of potatoes, apples and carrots; and Afonso Júnior et al. (2003) examined the effect of varying moisture content on the reduction of superficial area, volume and equivalent sphere diameter of coffee berries. However, these studies have not satisfactorily concluded the different conditions that influence the process of shrinkage.

The aim of this work was to determine the influences of various drying temperatures and relative humidities on the shrinkage of coffee berries during the drying process. These include changes in area and volume, which directly affect the visual attributes of the fruit. In addition to the above-mentioned parameters, a new correlation between the surface area and weight of coffee berries was investigated.

Material and methods

The present work was conducted in the Laboratory of Physical Properties and Quality Evaluation of Agricultural Products at the National Grain Storage Training Center – CENTREINAR, Federal University of Viçosa, Viçosa, Minas Gerais State, Brazil.

Coffee berries (*Coffea arabica* L.) from Catuaí Vermelho cultivar were purchased from a local farmer. They were manually harvested at the cherry stage with an initial moisture content of 2.27 d.b. Immature, deteriorated or damaged fruits were eliminated, with the purpose of obtaining a more homogeneous material with elevated quality.

Initial moisture contents were achieved by drying the product in an oven at 105 ± 1°C, until a constant mass was obtained. These were performed in triplicate. The reduction of moisture content over drying time was supervised through weight loss until a final moisture content of 0.11 d.b was reached. To monitor weight loss, a digital chronometer and an analytical balance with a precision of 0.01 were utilized.

The drying treatments were placed in a 4 x 4 factorial scheme, with four temperature levels (35, 45, 55 and 65°C) and four levels of relative humidities of drying air (25, 35, 45 and 55%) in a completely randomized design. The treatments conditions were achieved using a temperature-controlled chamber, manufactured by Aminco, model Aminco-Aire 150/300 CFM. Removable perforated trays were placed inside the equipment to allow air to pass through the samples, each of which contained 50 g of product. Airflow was monitored with an anemometer with rotating blades and kept at around 10 m^3 min^{-1} m^{-2}. Temperature and air relative humidity were monitored with a psychrometer installed next to the trays containing the samples.

Product dimensions were measured across the three orthogonal axes of the coffee berries using digital calipers after weighing the samples over a moisture content range of 2.27 to 0.11 d.b., with ten repetitions for each condition, in order to obtain product volumes (V).

Shrinkage

The association between physical processes and time can be described according to the classic kinetic model (Equation 1), in which P represents the physical characteristic being evaluated (in this case, volume); k is the constant rate of change of the characteristic dimension (h^{-1}) that correlates volume and time, dependent on the environment conditions in which the process occurs (LESPINARD et al., 2009); and m is the reaction order degree, which is dimensionless.

$$\frac{dP}{dt} = kP^m \quad (1)$$

For a various number of physical processes, it is difficult to distinguish between zero, first and second order kinetics. This fact can be explained by the small extension of reaction and the small evaluation rate of the studied parameter. Therefore, the m value was considered to be equal to one (first order) to facilitate the calculation procedure, and to linearly analyze the shrinkage phenomenon.

Data was evaluated using a non-linear regression analysis performed via STATISTICA 7.0® software using an exponential model (Equation 2), which was obtained through Equation 1 after mathematical rearrangements.

$$\exp^k = \left(\frac{V_t - V_e}{V_0 - V_e}\right)^{1/t} \quad (2)$$

where:

V_0: initial volume (cm^3);

V_t: volume at a certain time (cm^3);

V_e: equilibrium volume (cm^3);

t: time (h).

The equilibrium volume was obtained from the average of the last three volumes. This was adopted since these values were practically constant, indicating that the equilibrium had been reached.

The best fit model was evaluated via calculation of the mean relative error (MRE) and the standard error estimation (SEE), Equations 3 and 4, respectively. The determination coefficient (R^2) was also utilized.

$$\text{MRE} = \frac{100}{n}\sum_{i=1}^{n}\left(\frac{\left|X_{\text{exp}} - X_{\text{pre}}\right|}{X_{\text{exp}}}\right) \quad (3)$$

$$SEE = \sqrt{\frac{\sum_{i=1}^{n}\left(X_{\text{exp}} - X_{pre}\right)^2}{DF}} \quad (4)$$

where:

n: number of observations, dimensionless;

X_{exp}: experimental values of the equilibrium moisture content (d.b.);

X_{pre}: calculated values of the equilibrium moisture content (d.b.);

DF: degrees of freedom.

For practical purposes, values below 10% of the mean relative error indicate a good fit (MOHAPATRA; RAO, 2005).

Activation Energy

Activation energy is defined as the minimum energy necessary to initiate a chemical reaction. In the drying process, the lower the activation energy is, the higher the water diffusivity in the product will be. This means that lower values of activation energy allow for a higher moisture transfer rate and, consequently, lower drying time and cost.

Arrhenius Equation (Equation 5) is widely used to model and evaluate parameters that modify with temperature, including shrinkage, during the drying process of agricultural products. This equation, in addition with STATISTICA 7.0® software, was employed to obtain the values of activation energy of the cultivar studied in this work. This equation states the relationship between activation energy and the velocity at which the reaction occurs.

$$k = A_0 \exp\left(-\frac{E_a}{RT_{abs}}\right) \quad (5)$$

where:

A_0: pre-exponential factor (h^{-1});

E_a: activation energy ($kJ\ mol^{-1}$);

R: universal gas constant ($8.314\ kJ\ mol^{-1}\ K^{-1}$);

T_{abs}: temperature (K).

Surface area-to-volume ratio (SVR)

SVR was obtained from Equations 6 and 7. It was considered that the coffee berry could be represented by a prolate spheroid shape type (MOHSENIN, 1986), as seen in Figure 1.

$$V = \frac{\pi abc}{6} \tag{6}$$

$$S = \frac{\pi b^2}{2} + \frac{\pi ab}{2e}\sin^{-1} e \tag{7}$$

$$e = \left[1 - \left(\frac{b}{a}\right)^2\right]^{1/2} \tag{8}$$

where:

V: volume (cm^3);
S: surface area (cm^2);
a: longest intercept (mm);
b: longest intercept normal to a (mm);
c: longest intercept normal to a and b (mm);
e: eccentricity (dimensionless).

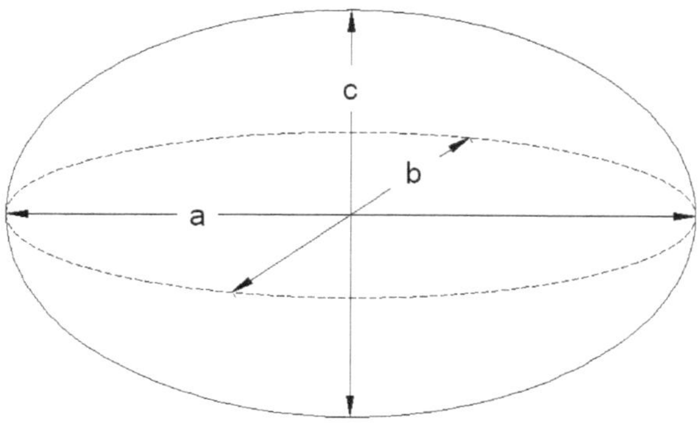

Figure 1. Schematic draw of a coffee berry considered as a prolate spheroid shape type and its characteristics dimensions.

Surface area (S) and weight (W) relationship

Mohsenin (1986) states a relationship between weight and surface area for spheroid products (Equation 9):

$$S = \xi W^d \tag{9}$$

where:

S: surface area, cm^2;
d, ξ: constant, dimensionless;
W: weight (g).

Quantifying the superficial area of agricultural products is a difficult process due to the fact that these foodstuffs do not have a uniform and homogeneous form. As a result, a great amount of time and many replicates are required, resulting in a large range of errors. On the other hand, weight measurement is easy to obtain with a simple weighing scale; therefore this equation provides a simple approach for acquiring values of superficial area by means of product weight. This information is essential for developing and improving drying equipment and helping studies of mass and heat transfer.

Results and discussion

Shrinkage Evaluation

The constant rate values (k) for Catuaí Vermelho cultivar in each T and φ combination are shown in Table 1, in which k values were analyzed varying along with T and φ (Figure 2) in order to assess which factor had the highest influence over k.

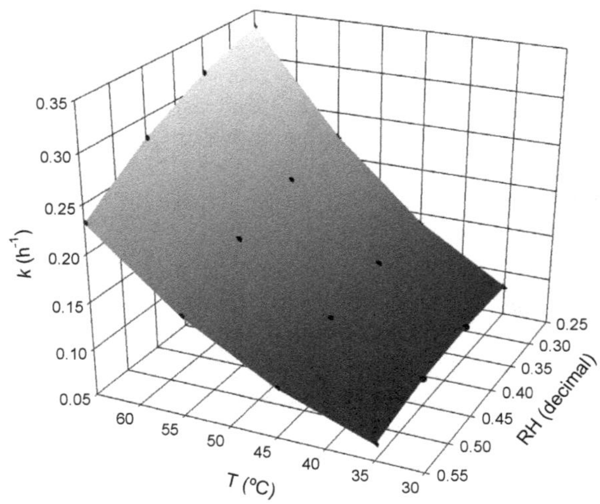

Figure 2. Constant rate of change of characteristic dimension (k) as a function of relative humidity (φ) and temperature (T).

Equation 10 represents the surface shown in Figure 2. The modeling procedure was carried out by means of the software Table Curve 3D® v.4.0, which provided a R^2 of 0.9996.

$$k = -0.0931346875 - 0.0008935T + 0.7045RH + 0.0001187T^2 - 0.50625RH^2 - 0.00973T \cdot RH \tag{10}$$

The figure presented allow us to conclude that k is more dependent on temperature variability, and during the drying process, k values decrease with higher φ values and increase with rises in temperature. These results are supported by other researchers. Krokida et al. (2000) considered that the temperature is the parameter with the most influence over food drying kinetics, and Afonso Júnior et al. (2004) also observed this trend for coffee berries. Ramos et al. (1994) found that temperature increased the rate of cellular shrinkage, with an Arrhenius type behavior.

Figure 3 demonstrates the shrinkage phenomenon of coffee berries at the beginning and the end of the drying process. Ten repetitions were made for each berry continuously during drying, and an average of a, *b* and *c* values were attained in order to obtain the product volume. Volume values ranged from 10.59 x 10^{-7} to 6.43 x 10^{-7} m at the beginning and end of drying, respectively. The total shrinkage was 1.22, 2.23 and 2.18 mm for *a*, *b* and *c* product dimension, respectively.

Figure 3. Total shrinkage representation of coffee berries.

Shrinkage is more relevant at the initial stages of drying. As the drying process occurs, the product tends toward an equilibrium state. As such, it starts to become more difficult for shrinkage to occur, leading to a stability of the product dimension characteristics. In addition, a combination of T and φ that provides a higher drying air potential (high T and low φ) also contributes to a stronger constant rate of change of the characteristic dimension (higher k) during dehydration. However, the product dimensions at equilibrium do not alter with air characteristics, but rather solely with the time to reach the equilibrium, which is diminished.

Activation energy

Activation energies obtained were 39.31, 37.32, 36.28 and 36.22 kJ mol^{-1} for relative humidities of 25, 35, 45 and 55%, respectively. Corrêa et al. (2006) also obtained the activation energy for coffee fruits; however, due to different cultivars and drying conditions used, their value (22.6 kJ mol^{-1}) differs from the present study, denoting the importance of evaluating different agricultural products and cultivars.

Figure 4 illustrates the activation energies found in this study. Examining this figure, it can be noticed that at higher φ (45-55%), the activation energy practically remained constant, marked by a breaking point at 45% φ. Furthermore, E_a increased with a decrease in φ. These features mean that at superior φ importance, it is easier to remove superficial moisture and less energy is required to do so. This trend can be explained by the following: higher φ values leads to lower water vapor pressure difference between the product and the drying air and consequently higher moisture content at the product surface (more water availability). Therefore, it becomes more difficult to remove moisture as the drying process occurs. Therefore, the highest k values are expected for a drying air with the highest potential for decreasing the moisture content of the product (lower φ and higher T), as seen in Table 1.

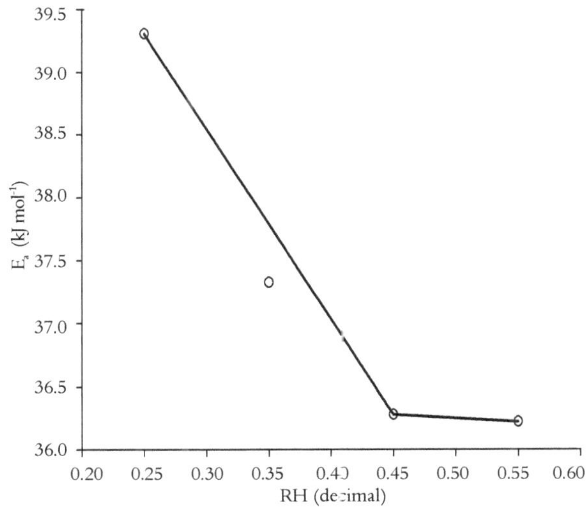

Figure 4. Activation energy (E_a) as a function of relative humidity (φ).

Table 1. Estimated constant rates of change of characteristic dimension (k) for Catuaí Vermelho cultivar.

	Temperature (°C)			
φ (%)	35	45	55	65
25	0.0826	0.1397	0.2231	0.3338
35	0.0843	0.1362	0.2116	0.3126
45	0.0791	0.1236	0.1890	0.2785
55	0.0670	0.1018	0.1550	0.2309

Surface area-to-volume ratio (SVR)

Figure 5 demonstrates that the SVR tends to decrease with an increase in moisture content (d.b.).

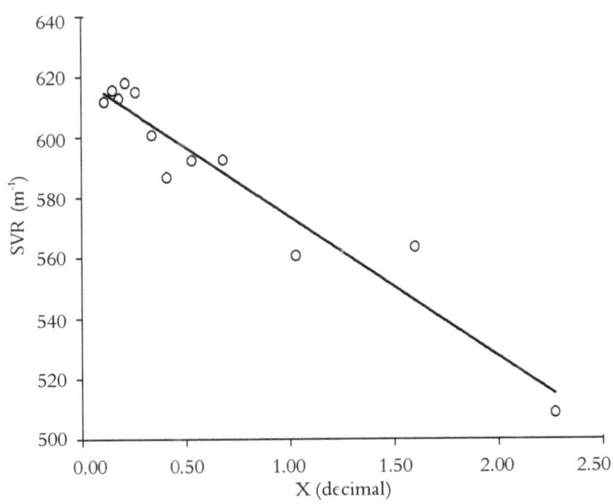

Figure 5. Surface area-to-volume ratio (SVR) as a function of moisture content (X).

As the moisture content increases, both the surface area and volume increase. However, the proportion of the volume increase is higher than that of the surface area, leading to lower SVR values.

Relationship between surface area (S) and weight (W)

The ξ and d parameters were obtained through STATISTICA 7.0® software and provided MRE, SEE and R^2 values of 2.077, 0.1392 and 0.9134, respectively. This indicates that the data fit well to the empirical equation, allowing the use of Equation 11 for predicting surface area using the coffee berry weight, at a moisture content range of 0.11 to 2.27 d.b.

$$S = 3.982756W^{0.278892} \quad (11)$$

Conclusion

The shrinkage phenomenon can be represented through the model here suggested for the Catuaí Vermelho cultivar of coffee berries at the temperatures and relative humidity ranges of 35 to 65°C and 25 to 55%, respectively. SVR had the tendency to decrease as moisture content increased, and a prediction of surface area using the weight of the coffee berry was satisfactory with respect to the empirical equation formulated in the moisture content range studied in this work. Constant rates of change of the characteristic dimension are more dependent on temperature variability than φ.

Acknowledgements

The authors are grateful to CAPES - Coordination of Improvement of Higher Education of Brazil - for financially supporting this research.

References

AFONSO JÚNIOR, P. C.; CORRÊA, P. C.; GONELI, A. L. D.; VILARINHO, M. C.; FRANÇA, A. S. Dinâmica da contração volumétrica do fruto de café durante o processo de secagem. **Revista Brasileira de Armazenamento**, Especial Café, n. 8, p. 29-35, 2004.

AFONSO JÚNIOR, P. C.; CORRÊA, P. C.; PINTO, F. A. C.; SAMPAIO, C. P. Shrinkage evaluation of five different varieties of coffee berries during the drying process. **Biosystems Engineering**, v. 86, n. 4, p. 481-485, 2003.

AGUILERA, B. J. M. Quantification of microstructural changes during first stage air-drying of grape tissue. **Journal of Food Engineering**, v. 62, n. 2, p. 159-164, 2004.

BATISTA, L. M.; ROSA, C. A.; PINTO, L. A. A. Diffusive model with variable effective diffusivity considering shrinkage in thin layer drying of chitosan. **Journal of Food Engineering**, v. 81, n. 1, p. 127-132, 2007.

BENNAMOUN, L.; BELHAMRI, A. Numerical simulation of drying under variable external conditions: application to solar drying of seedless grapes. **Journal of Food Engineering**, v. 76, n. 2, p. 179-187, 2006.

CORRÊA, P. C.; RESENDE, O.; RIBEIRO, D. M. Drying characteristics and kinetics of coffee berry. **Revista Brasileira de Produtos Agroindustriais**, v. 8, n. 1, p. 1-10, 2006.

DISSA, A. O.; DESMORIEUX, H.; BATHIEBO, J.; KOULIDIATI, J. Convective drying characteristics of Amelie mango (*Mangifera Indica* L. cv. 'Amelie') with correction for shrinkage. **Journal of Food Engineering**, v. 88, n. 4, p. 429-437, 2008.

HASHEMI, G.; MOWLAB, D.; KAZEMEINI, M. Moisture diffusivity and shrinkage next term of broad beans during bulk drying in an inert medium fluidized bed dryer assisted by dielectric heating. **Journal of Food Engineering**, v. 92, n. 3, p. 331-338, 2009.

HATAMIPOUR, M. S.; MOWLA, D. Shrinkage of carrots during drying in an inert medium fluidized bed. **Journal of Food Engineering**, v. 55, n. 3, p. 247-252, 2002.

KROKIDA, M. K.; KIRANOUDS, C. T.; MAROULIS, Z. B.; MARINOS-KOURIS, D. Drying related properties of apple. **Drying Technology**, v. 18, n. 6, p. 1251-1267, 2000.

LESPINARD, A. R.; GOÑI, S. M.; SALGADO, P. R.; MASCHERONI, R. H. Experimental determination and modelling of size variation, heat transfer and quality indexes during mushroom blanching. **Journal of Food Engineering**, v. 92, n. 1, p. 8-17, 2009.

MAYOR, L.; SERENO, A. M. Modelling shrinkage during convective drying of food materials: a review. **Journal of Food Engineering**, v. 61, n. 3, p. 373-338, 2004.

McMINN, W. A. M.; MAGEE, T. R. A. Physical characteristics of dehydrated potatoes - Part I. **Journal of Food Engineering**, v. 33, n. 1-2, p. 37-48, 1997.

MOHAPATRA, D.; RAO, P. S. A thin layer drying model of parboiled wheat. **Journal of Food Engineering**, v. 66, n. 4, p. 513-518, 2005.

MOHSENIN, N. N. **Physical properties of plant and animal materials**. New York: Gordon and Breach Publishers, 1986.

RAMOS, I. N.; BRANDÃO, T. R. S.; SILVA, C. L. M. Integrated approach on solar drying, pilot convective drying and microstructural changes. **Journal of Food Engineering**, v. 67, n. 1-2, p. 195-203, 2005.

RAMOS, I. N. A.; SILVA, C. L. M.; SERENO, A. M.; RATTI, C. Shrinkage during drying of foodstuffs. **Journal of Food Engineering**, v. 23, n. 1, p. 91-105, 1994.

RATTI, C. Shrinkage during drying of foodstuffs. **Journal of Food Engineering**, v. 23, n. 1, p. 91-105, 1994.

RESENDE, O.; ARCANJO, R. V.; SIQUEIRA, V. C.; RODRIGUES, S. Modelagem matemática para a secagem de clones de café (*Coffea canephora* Pierre) em terreiro de concreto. **Acta Scientiarum. Agronomy**, v. 31, n. 2, p. 189-196, 2009.

YAN, Z.; SOUSA-GALLAGHER, J. M.; OLIVEIRA, A. R. F. Shrinkage and porosity of banana, pineapple and mango slices during air-drying. **Journal of Food Engineering**, v. 83, n. 3, p. 430-440, 2007.

21

Modified crop model estimation of depleted and potential soybean yield

Thomas Newton Martin[1*], Durval Dourado Neto[2], Pedro Abel Vieira Junior[3], Antonio Roberto Pereira[2], Paulo Augusto Manfron[4] and Pedro Jacob Christoffoleti[2]

*[1]Departamento de Fitotecnia, Centro de Ciências Rurais, Universidade Federal de Santa Maria, Av. Roraima, 1000, 97105-900, Santa Maria, Rio Grande do Sul, Brazil. [2]Departamento de Produção Vegetal, Escola Superior de Agricultura "Luiz de Queiroz", Universidade de São Paulo, Piracicaba, São Paulo, Brazil. [3]Serviço de Negócios RPAa Transferência de Tecnologia, Empresa Brasileira de Pesquisa Agropecuária, Sete Lagoas, Minas Gerais, Brazil. [4]Centro de Educação Norte, Universidade Federal de Santa Maria, Frederico Westphalen, Rio Grande do Sul, Brazil. *Author for correspondence. E-mail: martin.ufsm@gmail.com*

ABSTRACT. Despite the great importance of soybeans in Brazil, there have been few applications of soybean crop modeling on Brazilian conditions. Thus, the objective of this study was to use modified crop models to estimate the depleted and potential soybean crop yield in Brazil. The climatic variable data used in the modified simulation of the soybean crop models were temperature, insolation and rainfall. The data set was taken from 33 counties (28 Sao Paulo state counties, and 5 counties from other states that neighbor São Paulo). Among the models, modifications in the estimation of the leaf area of the soybean crop, which includes corrections for the temperature, shading, senescence, CO_2, and biomass partition were proposed; also, the methods of input for the model's simulation of the climatic variables were reconsidered. The depleted yields were estimated through a water balance, from which the depletion coefficient was estimated. It can be concluded that the adaptation soybean growth crop model might be used to predict the results of the depleted and potential yield of soybeans, and it can also be used to indicate better locations and periods of tillage.

Keywords: leaf area index, water balance, crop model, *Glycine max*.

Introduction

In order to predict and simulate experimental results, the knowledge of soybean crop physiology is fundamental for crop modeler scientists. To construct accurate model predictions for soybean crops, some specific aspects, such as photorespiration and the CO_2 assimilation rate, must be taken into consideration. Knowledge about the successive crop growth stages (phenology), species functionality (physiology) and interactions with the environment (eco-physiology) are necessary for the planning of crop cultivation and for the attainment of high yields and the modeling predictions.

The CROPGRO-Soybean model is a process-oriented model that simulates the carbon, water and nitrogen balance for the soybean crop at the plant and soil level (BOOTE et al., 1998). Its equations express the relationship among plant processes (phenological growth, photosynthesis, respiration, biomass accumulation, photoassimilate partitioning, and soil/plant water balance) and temperature, photoperiod, insolation, and the water and nitrogen stress. The modeling techniques can be used in industrial management to understand better storage formation, commercialization, resources administration, agricultural and political mapping, health areas, sports, research field definitions, and agriculture. Estimating the crop yield for a specific site can be performed by taking into account elements of the weather, such as the air temperature, rainfall and solar radiation, and the resource inputs and soil type.

Carbone et al. (2003) used the CROPGRO-Soybean model v. 35 for a comparison of the simulated and observed yield in 8 research stations and in three maturation groups (V, VI, VII); depending on the site, the crops were analyzed at several ages. The variation between the estimated and observed yield was due to the fact that the model did not take into consideration the influence of weeds, diseases, insects, and wind interference (BATCHELOR et al., 1993). Camargo et al. (1986) used climatic variables to build a crop model that satisfactorily estimated the crop yield; the accuracy of the model was confirmed in the regression by comparing the observed and estimated data (0.76 to 0.87) to calculate the coefficient of determination. More recent studies indicate that the use of an expolinear model is able to estimate and adequately describe the growth and development of the soybean crop conditions in Brazil and Argentina (CONFALONE et al., 2010). Under Brazilian conditions (Paraná State), Berka et al. (2003) have proposed a model for estimating soybean production that is integrated into a Geographic Information System; the model assumes that depletion factor productivity is a function of water stress.

Despite the great importance of soybeans in Brazil, there have been few applications of soybean crop modeling on Brazilian conditions. Also, there have been few studies indicating the parameters for an estimation of the leaf area and the assimilation rate of CO_2. Therefore, by using both modified crop models and parameters generated in Brazil, this study aims to provide estimates of the leaf area, photoassimilate partitioning and the conversion of radiant energy in carbohydrates to estimate the depleted and potential soybean crop yield in Brazil.

Material and methods

The data set of climatic variables (temperature, sunshine and rainfall) for testing and developing the modified soybean crop model were obtained from the "Ciências Exatas" Department of the University of São Paulo – College of Agriculture "Luiz de Queiroz" – Brazil (USP/ESALQ); the Water and Electricity Department of Sao Paulo State - Brazil (DAEE); the "Instituto Agronômico de Campinas" São Paulo State - Brazil (IAC); the Meteorology System of Paraná, Paraná State – Brazil (SIMEPAR); and the National Institute of Meteorology - Brazil (INMET). The data set was from 33 counties (28 Sao Paulo state counties and 5 counties from other states that neighbor São Paulo). Because there is a temporal and spatial variability in the estimates of the sunshine, solar radiation and photosynthetic active radiation (PAR) for the 18 counties of São Paulo State (the same data used in this study), we used all of the data available. Moreover, the variability in sample size for the sunshine, solar radiation and PAR depends on the location and time of year in the state of São Paulo (MARTIN et al., 2008a). Using the data from this study, Martin et al. (2008b) found that the normal distribution occurs in greater frequency and can be used to adequately simulate rainfall.

The Duration of the Vegetative Phase (DVP, d) was estimated, with modifications, according to the model presented by Rodrigues et al. (2001). We used parameters from the soybean cultivar Ocepar 8 (a = 0.05327 day^{-1}, b = -0.00579 h^{-1}, c = 0.001909 D^{-1} °C^{-1} and D = 10 days), which has a long juvenile period, and where flowering varies between 54 and 88 days after emergence. The DVP phase might be calculated as follows:

$$DVP = \frac{1}{a + b.Nm + c.Tm} + D \tag{1}$$

where:

a, b (day h^{-1}) and *c* (°C^{-1}) refer to the empirical parameters of the model, *Nm* is the photoperiod (h day^{-1}) and *Tm* is the average temperature (°C) during the vegetative phase:

$$\mathrm{Nm} = \frac{\sum_{i=1}^{DVP-10} \mathrm{Ni}}{\mathrm{DVP}} \tag{2}$$

$$\mathrm{Tm} = \frac{\sum_{i=1}^{DVP-10} \mathrm{Ti}}{\mathrm{DVP}} \tag{3}$$

In the Duration of the Reproductive Phase (DRP, day), it was assumed that the number of flowering days until the point of physiological maturity was constant and equaled 65 days. Thus, the duration of the cycle (DC, day) can be calculated by:

$$DC = DVP + DRP \tag{4}$$

The following was also considered: the emergence of 300,000 seedling plants per hectare and the effective number of seeds per hectare (Ns = 300,000 seeds ha^{-1}); 16 g mass per 100 seeds (MSR = 0.16 g seed $^{-1}$); the proportion of the embryo in the soybean seed of 15% (Mes = 0.15 g g^{-1} – 0.15 g of embryo per seed gram). Thus, the initial leaf dry biomass ($IDlL_0$, kg ha^{-1}) can be calculated by:

$$\text{ID}\,FSf_0 = \frac{Ns.Mcs.Mes.Mfe}{1000} \tag{5}$$

From the values of the dry biomass of live leaf (DPl, kg ha^{-1}) and the specific leaf area (Sla, m^2 kg^{-1} = m^2 of leaf per kg of dry leaf), which, for the soybean, varies from 15 to 33 m^2 kg^{-1} (DRIESSEN; KONIJN, 1992), the leaf area index can be calculated by (LAI, m^2 m^{-2}) the following:

$$\text{LAI} = \frac{\text{DPl}_i.\text{Sla}_i}{10000} \tag{6}$$

We assumed that, in relation to the relative crop development (Rdi), the temporal variation (daily scale) of the specific leaf area (SLa, m^2 kg^{-1}) can be characterized by the following equations:

$$\text{Sla}_i = \text{Sla}_{max} \ (\text{if Rdi} = 0);$$
$$\text{Sla}_i = \text{Sla}_{mim} - (\text{Sla}_{max} - \text{Sla}_{mim}).\ln(\text{Dr}_i) \tag{7}$$
$$(\text{if Rdi} > 0)$$

$$[7]\ \text{Sla}_i = \text{Sla}_{max} \ (\text{if Sla}_i > \text{Sla}_{max}), \tag{8}$$
in which Sla_{max} and Sla_{min}

These two equations refer to the maximum and minimum limits of the specific leaf area (m^2 kg^{-1}), respectively. The general equation (reduced) corresponding to the photosynthesis can be characterized by the following: $CO_2 + H_2O \rightarrow CH_2O + O_2$. By considering the crop reflecting power (Crp = 0.2 J J^{-1}) and the interception of the photosynthetically active solar radiation (PAR, J m^{-2} s^{-1}), the solar radiation photosynthetically liquid active (PAR_liq, J m^{-2} s^{-1}) can be calculated as:

$$PAR\_liq_i = (1 - \Pr c).PAR_i \tag{9}$$

Thus, the answering curves of the assimilation of CO_2 (ADC, kg ha^{-1} h^{-1}), which correspond to a certain concentration of CO_2 in the atmosphere (CO = 340 vppm), can be described as (GOURDIAAN, 1982):

$$\text{ADC}_i = \text{PLMX}_i.\left[1 - e^{-PLEA.\frac{PAR\_liq_i}{\text{PLMX}_i}}\right]. \left(1 + \beta.\ln\left(\frac{Cx}{Co}\right)\right) \tag{10}$$

where:

Cx refers to the actual concentration of CO_2 in the atmosphere (Cx = 385 vppm), PLEA and β refer to the empirical parameters described by Vries et al. (1983) (PLEA = 0.48 kg CO_2 ha^{-1} h^{-1} (J m^{-2} $s^{-1})^{-1}$ and β = 0.8 for species C_3 and β = 0.4 for species C_4) and PLMX refers to the auxiliary variable (kg ha^{-1} h^{-1}) and is calculated using the following expression:

$$\text{PLMX}_i = 40.\left(\omega_0 + \frac{\omega_1}{\omega_2}.\text{T}_i\right) \tag{11}$$

where:

Ti refers to the average temperature (°C) of the ith day after the emergence, and ω_0, ω_1 and ω_2, depending on the temperature, refers to the empirical parameters that can be obtained in Vries et al. (1983).

The initial efficiency of the use of the absorbed light is characterized by biophysical processes with steady values. The maximum assimilation rate of carbon CO_2 depends on the species, a consideration of the biochemical processes and the ideal physiological conditions. The equation related to the calculation of the assimilation of CO_2 was established for the high temperature conditions (30°C) and the concentration of CO_2 in the atmosphere (340 vppm). Thus, the maximum assimilation of CO_2, under 30°C is 40 kg ha^{-1} h^{-1} (DORNHOFF; SHIBLES, 1970). The correction for the temperature is presented by Hofstra (1972), where the value obtained in the previous function must be multiplied by the value obtained in a linear one. The raw photosynthesis (FBf_i, kg ha^{-1} day^{-1} – kilogram of carbohydrate per hectare of leaf per day) is calculated as follows:

$$FBf_i = \frac{30}{44}.ADC_i.N_i.f\_LAI.f\_Dr \tag{12}$$

where:

f_LAI and f_Dr are factors of correction that refer to the leaf area rate and to the culture relative development, respectively. We began to consider the effect of the auto shading of the leaf area from the value of the leaf area rate factor (f_LAI) and with a leaf area rate (LAI, m^2 m^{-2}) superior to 1 $m^2.m^{-2}$ in accordance with the following equations: $f\_LAI_i = 1$, $LAI_i \leq 1$; $f\_LAI_i = e^{\lambda_0 + \lambda_1 (LAI_{i+} \lambda_2)}$, $f\_LAI_i = 0.28$, in which λ_0, λ_1 and λ_2, refer to the empirical parameters (λ_0 = 0.131 m^2 m^{-2}, λ_1 = -0.14 m^2 m^{-2} and λ_2 = -0.064284 $m^2.m^{-2}$). The photosynthesis rate decreases with the age of the tissues, thus indicating the age of the plant. Therefore the f_Dr factor displays the following functional relationship with the relative culture development (Dr), $f\_Dr_i = 1$, if;

$Rdi \leq Dr\_f_{ini;}$

$$f\_Dr_i = e^{\frac{\ln(f\_Dr_{fim})}{f\_Dr_{ini} - Dr\_f_{ini}}(Dr_i - Dr\_f_{ini})} \tag{13}$$

where:

Dr_f_{ini}, Dr_f_{fim}, f_Rdi_{ni}, and f_Dr_{fim} (Dr_f_{ini} = 0.15, Dr_f_{fim} = 1.00, f_Rdi_{ni} = 1 and f_Dr_{fim} = 0.45) refer to the empirical parameters. The liquid photosynthesis (*FLf*, kg ha^{-1} day^{-1}) per unit of leaf area for a certain day is determined by calculating the difference between the raw photosynthesis (*RP*) from that day and the maintenance respiration of the previous day (MR_{i-1}): $FLf_i = FB_i - RM_{i-1}$. The liquid photosynthesis (*LP*, kg ha^{-1} day^{-1}) per unit of soil area exploited by the culture can be calculated by the following equation: $FL_i = FLf_i.LAI_i$. The formation of the photoassimilate partitioning is determined by fractions allocated to the different organs of the plant. The fraction of allocation will depend on the organ of the considered plant (the root, *f_Ri;* the stem, *f_Hi;* the leaf, *f_Fi;* or the reproductive organs, *f_ORi*) and the relative crop development (*Rdi*) obtained according to the formula of Driessen and Konijn (1992). Also, the fraction of allocation will be determined by the quantity of carbohydrate allocated to the root (CHO_R_i, kg ha^{-1}), the leaf (CHO_F_i, kg ha^{-1}), the stem (CHO_H_i, kg ha^{-1}) and to the reproductive organs (CHO_OR_i, kg ha^{-1}). The efficiency of the carbohydrate conversion of the dry phytomass is determined by the quantity of accumulated carbohydrate multiplied by the efficiency of conversion. Additionally, the calculation of the dry phytomass of the root, stem, leaf and reproductive organs can also be obtained in accordance with Driessen and Konijn (1992).

From the allocated fraction in each of the plant's organs, we can determine the quantity of carbohydrate allocated. That quantity of carbohydrate must be converted into dry biomass. Vries et al. (1983) demonstrated the approximate values of the conversion of carbohydrates into dry biomass for the soybean crop. The efficiency of the conversion varies according to the composition of the dry biomass (carbohydrates, proteins, lipids, lignin and organic acids) formed from the vegetable organ (leaf, stem, root or reproductive organ). However, the primary product of the photosynthesis is the carbohydrate, and its conversion in other organic compounds involves an energetic cost, which is represented by the respiration of synthesis. An approximation is presented by Vries et al. (1983), in which 1.0 kg of photosynthesized carbon results in 404 g of protein, 330 g of lipids, 472 g of lignin, 826 g of structural carbohydrate and 104 g of organic acids. By knowing the percentage of the composition of dry biomass in the soybean plant, we can calculate the efficiency of conversion (Ec) for each organ:

$$Ec = QCE.PCE + QP.PP + QL.PL + QLg.PLg + QAO.PAO \tag{14}$$

where:

QCE refers to the quantity of structural carbohydrate, *PCE* refers to the % composition of structural carbohydrate in the dry biomass, *QP* refers to the quantity of protein, *PP* refers to the percentage composition of protein in the dry biomass, *QL* refers to the quantity of lipids, *Pl* refers to the percentage of the composition. The total dry biomass of the leaf ($FSTf_i$, kg ha^{-1}) and of the root ($FSTr_i$, kg ha^{-1}) accumulated until the ith day after the emergence is given by the following equation:

$$FSTf_i = FSTf_{i-1} + FSf_i - Sf_{i-1}$$
and
$$FSTr_i = FSTr_{i-1} + FSr_i - Sr_{i-1} \tag{15}$$

where:

FSf_i and FSr_i refers to the dry biomass of the leaf and root, respectively, produced in the ith day after the emergence; $FSTf_{i-1}$ and $FSTr_{i-1}$ refer to the total dry biomass of the leaf and root, respectively, accumulated until the ith day after the emergence. Thus,

$$Sf_i = Fsen.FSf_i \quad \text{and} \quad Sr_i = Rsen.FSr_i \tag{16}$$

where:

Fsen and Rsen refer to the leaf and root senescence factors that are the function of the relative development of the culture:

On average, the composition of the botanical seed of the soybean is as follows: 29% of carbohydrate, 37% of protein, 18% of lipids, 6% of liginin and 5% of organiSR.

$Fsen_i = 0$, if Rdi ≤ 0.2;

$$Fsen_i = \frac{f\_Sen\_fim - f\_Sen\_ini}{Dr\_f\_Sen\_fim - Dr\_f\_Sen\_ini}$$

$(Dr\_f\_Sen\_ini + Dr_i)$, if 0.2 < Rdi ≤ 0.65;

$Fsen_i = 1$, if Rdi > 0.65; $Rsen_i = 0$, if Rdi ≤ 0.2;

$$Rsen_i = \frac{r\_Sen\_fim - r\_Sen\_ini}{Dr\_r\_Sen\_fim - Dr\_r\_Sen\_ini}$$

$(Dr\_r\_Sen\_ini + Dr_i)$, if 0.2 < Rdi ≤ 0.65 [17]

where:

$Rsen_i$ = 1 if Rdi > 0.65, and in which Dr_f_Sen_ini, f_Sen_ini, Dr_f_Sen_fim and f_Sen_fim refer to the empirical parameters of the leaf, and Dr_r_Sen_ini, r_Sen_ini, Dr_r_Sen_fim and r_Sen_fim refer to the empirical parameters of the root (Dr_f_Sen_ini = Dr_r_Sen_ini = 0.2, f_Sen_ini = 1. r_Sen_ini = 0, Dr_f_Sen_fim = Dr_r_Sen_fim = 0.65 and f_Sen_fim = r_Sen_fim = 1). Thus, the senescence of the roots was analogically calculated to the senescence of the leaf.

The total dry biomass of the stem ($FSTh_i$, kg ha^{-1}) and the reproductive organ ($FSTor_i$, kg ha^{-1}) accumulated until the ith day after the emergence are given by the following equation:

$$FSTh_i = FSTh_{i-1} + FSh_i \quad \text{and} \quad FSTor_i = FSTor_{i-1} + FSor_i \tag{18}$$

The maintenance respiration can be quantified in each organ of the plant as follows: the root (MRr, kg ha^{-1} day^{-1}), the stem (MRh, kg ha^{-1} day^{-1}), the leaf (MRf, kg ha^{-1} day^{-1}) and the reproductive organs (MRor, kg ha^{-1} day^{-1}). The maintenance respiration of the plant (MR, kg ha^{-1} day^{-1}) is the sum of the maintenance respiration of the different living organs, which depends on the dry biomass and the maintenance respiration coefficient:

$$\begin{aligned} RMr_i &= FSr_i.CRm\_r_i, \\ RMh_i &= FSh_i.CRm\_h_i, \\ RMf_i &= FSf_i.CRm\_f_i, \\ RMor_i &= FSor_i.CRm\_or_i, \\ RM_i &= (RMr_i + RMh_i + RMf_i + RMor_i). \\ & m\_T.m\_Dr \end{aligned} \tag{19}$$

where:

CMR_, CMR_f, CMR_r and CMR_h refer to the maintenance respiration coefficient of the root (CMR_r = 0.01 kg kg^{-1} day^{-1} – kilogram of carbohydrate - CH_2O – per kilogram of dry biomass of the root - FSr – per day), of the stem (CMR_h = 0.015 kg kg^{-1} day^{-1} – kilogram of carbohydrate - CH_2O – per kilogram of dry biomass of the stem - FSf – per day), of the leaf (CMR_f = 0.015 kg kg^{-1} day^{-1} - kilogram of carbohydrate - CH_2O - per kilogram of dry biomass of the leaf - FSf – per day) and the reproductive organs (CMR_or = 0.017 kg kg^{-1} day^{-1} - kilogram of carbohydrate - CH_2O - per kilogram of the dry biomass of the reproductive organ - FSor – per day) respectively. Thus,

$$\begin{aligned} & m\_T_i = e^{0.0607.T_i - 2.1056}; \\ & m\_Dr_i = 1, \text{ if Rdi} \le 0.15; \\ & m\_Dr_i = e^{0.21246579 - 1.41643859.Dr_i} \\ & \text{if } 0.15 < Rd_i \le 1.00 \end{aligned} \tag{20}$$

The dry biomass of the grains (FSg, kg ha^{-1}) is calculated based on the dry biomass of the reproductive organs (FSor, kg ha^{-1}) (flower, pod and botanical seed) and by the following expression: $FSg = 0,8.FSTor_{i=DC}$. In this expression, 0.8 is the proportion of the seeds in relation to the total of the reproductive organs. Thus, the potential yield of grains (*PPg*, kg ha^{-1}) can be calculated by the equation $PPg = \frac{FSg}{1-u}$, in which u refers to the drift of water, and the basis of the mass is determined by the seed (where u = 0.13 g g^{-1} - 13 g of water in 100 grams of wet seed).

The gross daily production of carbohydrates will be depleted in function according to the real quantity of light (sunstroke) and by the real quantity of water available in the soil.

First, the rainy days and the quantity of rain on each day must be simulated. Moreover, a uniform distribution must be simulated. If the generated number is either lower or equal to the probability of the considered day, then the day is considered rainy; otherwise, the rain is equal to zero and the day is not considered rainy. Moreover, the minimum and maximum values that have occurred in the local historical series must be taken into consideration. The quantity of rain calculated for the rainy days is determined by the estimation of two random numbers with a uniform distribution (U_1 and U_2), so that they must be independent. Next, the values V_1 and V_2, $V_1 = 2.(U_1 - 1)$; $V_2 = 2.(U_2 - 1)$ and $VS = V1^2 + V2^2$ must be estimated. If the simulated value (VS) is more than 1, then the previous step must be recalculated.

Next, the accessory value (Vx) for the calculus of the simulated rain (SR) must be determined. This is given by the function:

$$Vx = V1.\sqrt{-2.\log(VS)/VS} \text{ and}$$
$$CS_i = Cm_i + Vx.C_{dp} \quad (21)$$

The process must be re-started if the values of the simulated rain are superior or inferior to the maximum and minimum value of the historical average for the day. The application's normal bivariate simulation, which considers the temperature and photosynthetically active radiation, was tested, and the results were superior to other simulations by Martin et al. (2007), and its applications were used in this study.

The methodology proposed by Thornthwaite (1948) was used to calculate the potential evapotranspiration. The evapotranspiration of the culture, as well as the values of the coefficient of the culture, are presented and defined by Doorenbos and Kassam (1979). The normal hydric balance follows the sequence described by Thornthwaite and Mather (1955). For the soy bean crop, the coefficient of sensibility of the culture to the hydric deficiency was modified by the values described by Doorenbos and Kassam (1979). Thus, the relationship between the real and maximum evapotranspiration can be used to determine the depleted yield (DP, kg ha^{-1}) of the culture, in function to the potential yield (PP, kg ha^{-1}). Therefore,

$$PD = PP\left[1 - Ky_i\left(1 - \left(\frac{ETr_i}{ETc_i}\right)\right)\right] \quad (22)$$

When the Dr ≤ 0.24, then the Ky will be 0.2; when the Dr is between 0.24 and ≤ 0.4, then the Ky will be 0.4; when the Dr is between 0.4 and ≤ 0.8, then the Ky will be 0.6; when the Dr is between 0.48 and ≤ 0.61, then the Ky will be 0.8; when the Dr is between 0.61 and ≤ 0.9, then the Ky will be 0.9; and when the Dr is superior to 0.9, the Ky is equal to 1. The field capacity was 50% and the permanent wilting point was 20%.

Results and discussion

The results of 500 simulations for each location during 36 periods of sowing in Campinas, Pindorama and Cascavel municipalities are shown in Figure 1. The estimates were made for all 33 locations, but, due to the volume of information, we decided to present and discuss the results of three locations. The variation of the potential and depleted yield for the State with simulated sowings in January, May and September are presented in Figure 2. The curves of the potential yield (minimum, maximum and mean) reach the maximum productivity for each situation when the sowing was performed from September to October; during this period, the maximum productivity is reached because the characteristics of the environment are nearer to the comfort zone of the crop.

By the proposed model, the number of days between the germination and flowering is a function of the interaction between the temperature and photoperiod factors (RODRIGUES et al., 2001). The model contemplates the minimum and maximum values for the temperature and photoperiod; these values vary from 19 to 32°C (temperature) and 11 to 14 hours (photoperiod). In this study, there are daily average temperatures less than 19°C, so, in a sense, the model might be deficient, as it will generate contrasting values, such as the productivities generated when the sowing is performed during the period in which the daily average temperature is inferior to that value.

To calculate the number of days for the flowering, it must be taken into account that when the temperature is higher, then the effect of the photoperiod over the number of necessary days for the flowering is lower, which makes the number of days of flowering approximately 35 days (at 32°C). The effect of the genotype that will be used must also be considered, thus making the triple interaction (genotype-temperature-photoperiod) extremely difficult to explain and represent mathematically. By considering this complex triple interrelation, the model reasonably represented the behavior of the potential productivities because it reduced the cycle when the temperature was elevated; however, due to the high availability of solar radiation, the production was superior in comparison to the period with a long photoperiod, in which the number of days of the cycle was higher. Also, the production of the leaf area was low because of the reduced quantity of energy in the environment, thus limiting the source of carbohydrates and, consequently, the yield.

The rate of the leaf area is one of the most important parameters when working on the questions linked to the modeling of agricultural production. Moreover, it is complex because the parameter is a function of the interaction with other factors, which causes many researchers to avoid directly using the rate of the leaf area (WIT, 1978). Many models determine their average calculations during the cycle of the development of the culture. However, in order to make the results more precise and eliminate many empirical questions that aid in solving the problem, the use of mathematical equations that vary the function of the relative development are desired because they determine the daily values of the LAI. Through its auto-regulation systems, the plant, restricts its LAI, which is frequently presented as a parabola or something similar in the explicative texts. Those limitations imposed by the plant were reproduced as factors of the depletion of the relative development and the leaf area index.

The maximum LAI of the plant must be truncated to avoid the presence of a leaf area that consumes more photo assimilated (respiration) than the productive capacity of the plant. Factors which are naturally regulated by the plants and environment, such as the auto-shading and the loss of the capacity of photosynthesis with the aging of the tissues, must be observed in the construction of models. All of those factors must be taken into account, as well as the models proposed by Pereira and Machado (1986). In Piracicaba (São Paulo State), the maximum LAR from the soybean cultures obtained by Câmara and Heiffig (2000) varied from 5 to 8. In the present model, the maximum value of the LAI was limited to 5 to avoid the existence of a major consumption of carbohydrates in relation to its production. The annual variation of the grain yield is also a function of the assimilation of CO_2, which varies according to the energy in the environment. As previously stated, the production is smaller when the sowings are performed in the low temperature and the photoperiod. The model for estimating the assimilation of used CO_2 is interesting because it takes into account not only the solar radiation photosynthetically active, but it also considers the quantity of CO2 in the atmosphere.

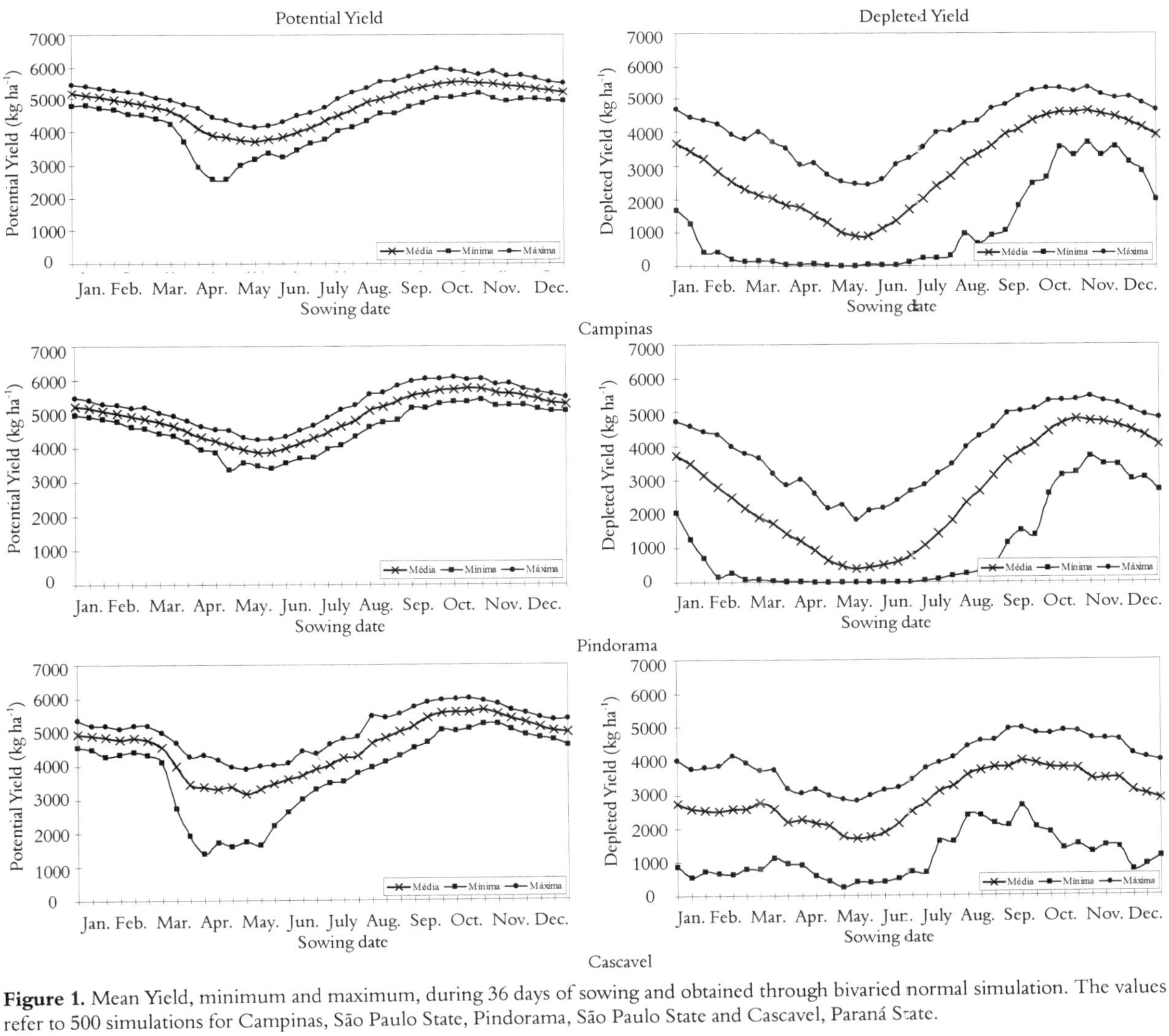

Figure 1. Mean Yield, minimum and maximum, during 36 days of sowing and obtained through bivaried normal simulation. The values refer to 500 simulations for Campinas, São Paulo State, Pindorama, São Paulo State and Cascavel, Paraná State.

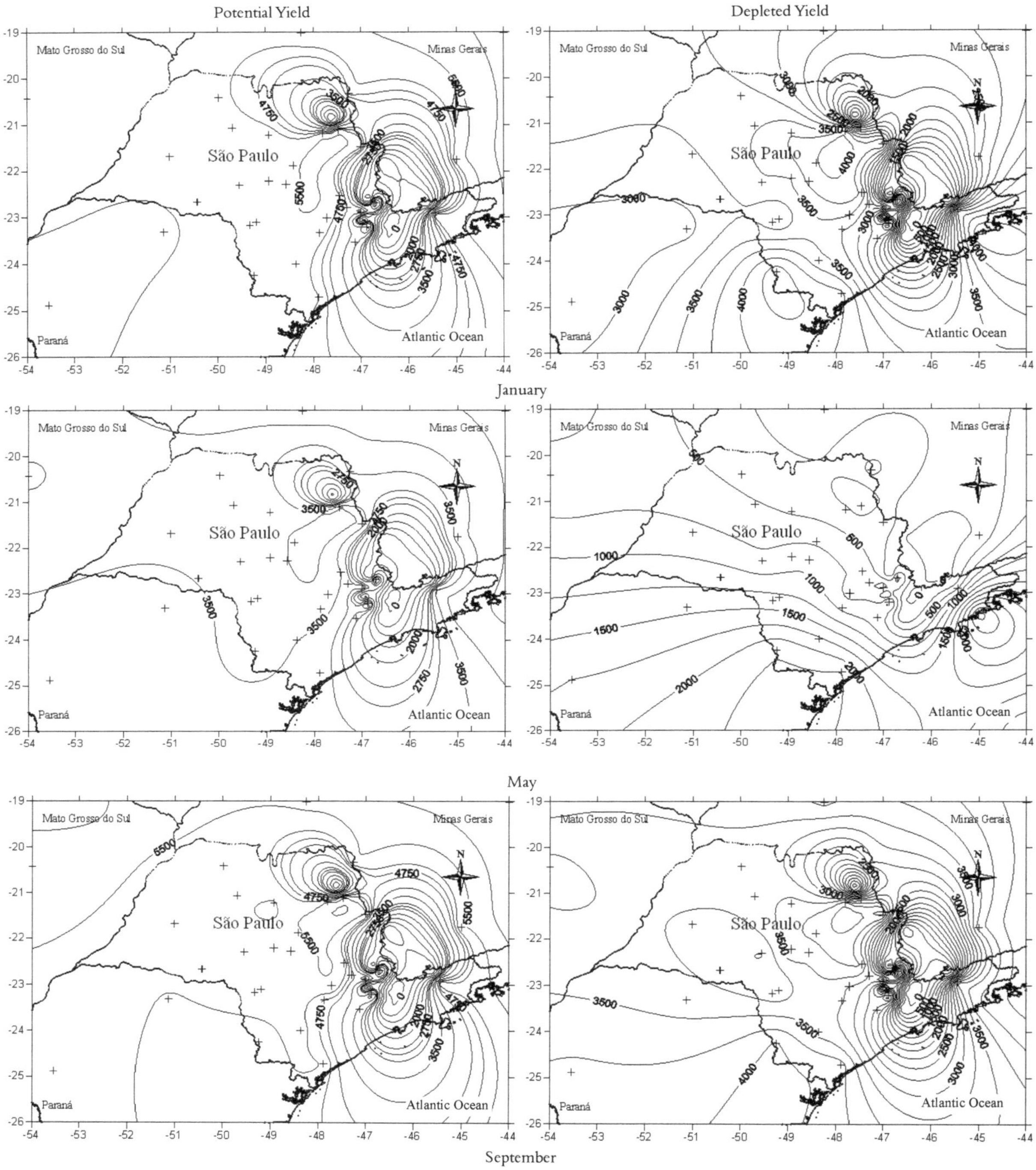

Figure 2. Isolines of the mean potential and depleted yield for the 25 locations. The data were obtained using the Krigagem method of interpolation for January, May and September.

Despite being extremely harmful to the planet, increasing the quantity of CO_2 in the environment would increase the production of soybeans in the region of São Paulo (SIQUEIRA et al., 1994) by approximately 22-32%. The higher rates of CO_2 assimilations occur when the temperature is between 30 and 35°C, and the higher rates occur more often in September (for the present case). When the temperature values differ from those intervals, the factor of reduction must be applied, in which those values and parameters would certainly be different for the national genotypes. The results obtained by the model can be compared to the results obtained by Cooper (2003), in the USA, where the recorded yields varied from 4,000 kg ha^{-1} (1966) to 7,911 kg ha^{-1} (1983). In Australia, the maximum productions varied from 8,004 to 8,604 kg ha^{-1} (1983). Despite those elevated production values, the genetic production of the soybean reaches 18,000 kg ha^{-1} (flowering), before decaying to 4,600 kg ha^{-1} during the period of physiological maturity (VENTIMIGLIA et al., 1999). Other expressive results were obtained by Nishi and Hungria (1996), who obtained productions superior to 4,000 kg ha^{-1} in experimental parcels working with re-inoculation. Other experiments in the field presented elevated values in the soybean production, such as the results from Pires et al. (2000), who estimated an average potential yield of 15,007 kg ha^{-1} in R2 and 10,282 kg ha^{-1} in R5, and, when influenced by the spacing between the rows, estimated the potential yield at 10,962 kg ha^{-1} in 20 cm and 9,602 kg ha^{-1} in 40 cm. In the experiment in Rio Grande do Sul state, the values of the yield varied from 3,929 kg ha^{-1} (OCERPA) to 6,436 kg ha^{-1} (FT-2003) in 96/97 (NAVARRO JÚNIOR; COSTA, 2002). When considering the average productivity obtained in the competition assays of soybeans in Rio Grande do Sul State, Lange and Federizzi (2009) found that the genetic gains ranged from zero to 71.5 kg ha^{-1} during the 19 years of evaluation. According to that study, the average production was 2500 kg ha^{-1} regardless of the maturation cycle, yet the productivity varied according to the regions.

As for the depleted productivity in relation with the precipitation (Figure 1), we verified that the simulated values for the precipitation were similar to the observed values. From the basic statistics (medium, minimum and frequency, data not presented), that type of simulation can be used for the representation of precipitation in a region. We verified that the depletion factor in the function of the sensibility of the culture on the water deficiency (Ky) acted intensely in the depletion of the potential productivities. In terms of the relative development, the major subdivision was in a higher number of the Ky classes, where it proportioned higher reductions on the potential productivities in comparison to the original proposed by Driessen and Konijn (1992). Besides, a representation of reality would be more reliable if the coefficient of the culture was better determined (even on a daily scale) to maximize the present model.

To bring the simulations closer to reality, the daily values of the rain were used in the water balance project. Thus, it becomes possible to verify the existence or absence of water stress conferred on the soybean plant. The development of the plants relies on rain, and it will not develop late in situations where stress is absent. The coefficient of sensibility of the culture to the water stress (Ky) potentiates the effect of the correction factors and the culture senescence, thus obtaining a yield of grains closer to the actual yield. Accordingly, by using the same stress coefficient (Ky), Berka et al. (2003) observed that the agrometeorological model overestimates the values by 10% in the yield when compared to the data obtained by SEAB, (Parana State), which determined the maximum yields of up to 3,070 kg ha^{-1}. Along with the water stress, the reduction of the vegetative (mass formation) and the reproductive period (transference of photoassimilated) must be considered. Thus, the seedlings from the last ten days of January to the first ten days in September were at an extreme risk for low productivity, and there was a significant increase of the frustrations of the crop. Therefore, the seedling from the last ten days of January to the first ten days in September became extremely risky for low productivity and the significant increase of the frustrations of the crop.

Conclusion

The adaptation of the soybean growing model can be used in the prediction of the potential and depleted yield results for the crop, and it can also be used for the indication of better locations and periods of tillage.

Acknowledgements

We would like to thank the Conselho Nacional de Desenvolvimento Científico e Tecnológico (CNPq) for the Ph.D. program scholarship (Forschungszentrum Jülich GmbH, ICG-IV, Jülich, Deutschland). We would also like to give thanks for the data to the Instituto Agronômico de Campinas (IAC) and the Departament of Ciências Exatas, Escola Superior de Agricultura "Luiz de Queiroz", Piracicaba, São Paulo State.

References

BATCHELOR, W. D.; JONES, J. W.; BOOTE, K. J.; PINNSCHMIDT, H. O. Extending the use of crop models to study pest damage. **Transactions of the American Society of Agricultural Engineers**, v. 2, n. 36, p. 551-558, 1993.

BERKA, L. M. S.; RUDORFF, B. F. T.; SHIMABUKURO, Y. E. Soybean yield estimation by an agrometeorological model in a GIS. **Scientia Agricola**, v. 3, n. 60, p. 433-440, 2003.

BOOTE, K. J.; JONES, J. W., PICKERING, N. B. The CROPGRO model for grain legumes. In: TSUJI, G.; HOOGENBOOM, G.; THORNTON, P. K. (Ed.). **Understanding options for agricultural production**. Dordrecht, The Netherlands Kluwer Academic Publ., 1998. p. 99-128.

CÂMARA, G. M. S.; HEIFFIG, L. S. Fisiologia, ambiente e rendimento da cultura de soja. In: **Câmara, G.M.S.** Tecnologia da produção II. Piracicaba: Esalq/USP, 2000. p. 120.

CAMARGO, M. B. P.; BRUNINI, O.; MIRANDA, M. A. C. Modelo agrometeorológico para estimativa da produtividade para a cultura de soja no Estado de São Paulo. **Bragantia**, v. 45, n. 2, p. 279-292, 1986.

CARBONE, G. J.; MEARNS, L. O.; MAVRONATIS, T.; SADLER, E. J.; STOOKSBURRY, D. Evaluating CROPGRO-Soybean perfoMRance for use in climate impact studies. **Agronomy Journal**, v. 95, n. 5, p. 537-544, 2003.

CONFALONE, E. A.; BERNARDES, M. S.; COSTA, L. C.; RIGHI, C. A.; DOURADO NETO, D.; MARTIN, T. N.; MANFRON, P. A.; PEREIRA, C. R. Expolinear model on soybean growth in Argentina and Brazil. **Ciência Rural**, v. 5, n. 40, p. 1009-1016, 2010.

COOPER, R. L. A delayed flowering barrier to higher soybean yields. **Field Crops Research**, v. 20, n. 82, p. 27-35, 2003.

DOORENBOS, J.; KASSAM, A. H. **Yield response to water**. Food and Agriculture Organization of the United Nations. Rome: FAO-Irrigation and Drainage Paper, 1979.

DORNHOFF, G. M.; SHIBLES, R. M. Varietal differences in net photosynthesis of soybean leaves. **Crop Science**, v. 1, n. 10, p. 42-45, 1970.

DRIESSEN, P. M.; KONIJN, N. T. **Land-use systems analysis**. Wageningen: Wageningen Agricultural University, 1992.

GOURDIAAN, J. Potential production process. In: VRIES, F. W. T.; VAN LAAR, H. H. (Ed.). **Simulation of plant growth and crop production**. Wageningen: Simulation Monographs. Pudoc, 1982. p. 8-113.

HOFSTRA, G. Response of soybeans to temperature under high light intensities. **Canadian Journal ofPlant Science**, n. 52, p. 535-543, 1972.

LANGE, C. E.; FEDERIZZI, L. C. Estimation of soybean genetic progress in the south of Brazil using multienvironmental yield trials. **Scientia Agricola**, v. 3, n. 66, p. 309-316, 2009.

MARTIN, T. N.; STORCK, L.; DOURADO NETO, D. Random simulation of photosynthetically active radiation and air temperature through different methods. **Pesquisa Agropecuária Brasileira**, v. 9, n. 42, p. 1211-1219, 2007.

MARTIN, T. N.; DOURADO NETO, D.; STORCK, L.; BURAUEL, P.; SANTOS, E. A. Homogeneous regions and sample size for attributes of the climate in São Paulo State, Brazil. **Ciência Rural**, v. 38, n. 3, p. 690-697, 2008a.

MARTIN, T. N.; DURVAL DOURADO NETO, D.; VIEIRA JUNIOR, P. A.; MANFRON, P. A. Distribution models and spatial-temporal homogeneity for the pluvial precipitation in the State of São Paulo, Brazil. **Revista Ceres**, v. 5, n. 55, p. 476-481, 2008b.

NAVARRO JÚNIOR, H. M.; COSTA, A. C. Relative contribuition of yield componentes for grain production of soybean. **Pesquisa Agropecuária Brasileira**, v. 2, n. 37, p. 67-89, 2002.

NISHI, A. Y. M.; HUNGRIA, M. Efects of soybean [*Glycine max* (L.) Merrill] inoculation in a soil with established population of *Bradyrhizobium* with strains semia 566, 586, 587, 5019, 5079 e 5080. **Pesquisa Agropecuária Brasileira**, v. 5, n. 31, p. 359-368, 1996.

PEREIRA, A. R.; MACHADO, E. C. A dynamic Simulator of the sugarcane crop growth. **Bragantia**, v. 1, n. 45, p. 107-122, 1986.

PIRES, J. L. F.; COSTA, J. A.; THOMAS, A. L.; MAEHLER, A. R. Effect of population and spacing on soybean potential yield during ontogen. **Pesquisa Agropecuária Brasileira**, v. 8, n. 35, p. 1541-1547, 2000.

RODRIGUES, O.; DIDONET, A. D.; LHAMBY, J. C. B.; BERTAGNOLLI, P. F.; LUZ, J. S. Quantitative response of soybean flowering to temperature and photoperio. **Pesquisa Agropecuária Brasileira**, v. 3, n. 36, p. 431-437, 2001.

SIQUEIRA, O. J. F.; FARIAS, J. R. B.; SANS, L. M. A. Potential effects of global climate change for Brazilian agriculture and adaptive strategies for wheat, maize and soybean. **Revista Brasileira de Agrometeorologia**, v. 2, n. 2, p. 115-129, 1994.

THORNTHWAITE, C. W. An approach towards a rational classification of climate. **Geographical Review**, v. 38, n. 1, p. 55-94, 1948.

THORNTHWAITE, C. W.; MATHER, R. J. **The water balance**. New Jersey: Laboratory of Climatology, 1955. (Publication in Climatology).

VENTIMIGLIA, L. A.; COSTA, J. L.; THOMAS, A. L.; PIRES, J. L. F. Soybean yield potential influenced by soil phosphorus content and row spacin. **Pesquisa Agropecuária Brasileira**, v. 2, n. 34, p. 195-199, 1999.

VRIES, P. F. W. T.; LAAR, H. H.; CHARDON, M. C. M. Bioenergetics of growtth of seeds, fruits and storage organs. In: International Rice Research Institute (Ed.). **Symposium on potential productivity of field crops under different environments**. Los Baños: IRRI, 1983. p. 37-59.

WIT, C. T. **Simulation for assimilation, respiration, and transpiration of crops**. Wageningen: A Halsted Press Book; John Wiley, 1978.

22

Effective engineering properties in the design of storage structures of postharvest dry bean grain

Hakan Kibar[1*], Turgut Öztürk[2] and Kadir Ersin Temizel[2]

*[1]Department of Biosystems Engineering, Faculty of Agriculture, Igdir University, Igdir, Turquia. [2]Department of Agricultural Structures and Irrigation, Faculty of Agriculture, Ondokuz Mayis University, Samsun, Turquia. *Author for correspondence. E-mail: hakan.kibar@igdir.edu.tr*

ABSTRACT. Selected engineering properties (physical and mechanical) of dry bean under non-irrigation and drip-irrigation were determined and compared. These properties are necessary for the design of equipments for harvesting, processing, transporting, estimating loads in storage structures for crops and flow problems in silos such as arching, ratholing, irregular flow and segregation. Some engineering characteristics such as: average length, width, thickness, the geometric mean diameter, sphericity, surface area, volume, thousand grain mass (M_{1000}), bulk and true densities, porosity, angle of internal friction, static coefficient of friction, Poisson ratio and pressure ratio were studied. The highest average values for physical properties were recorded at drip-irrigation, the lowest average values at non-irrigation. Differences between both irrigation systems for physical properties were statistically significant, but differences for sphericity of irrigation systems were not significant. The highest average values for mechanical properties (angle of internal friction, static coefficient of friction and pressure ratio) were recorded at drip-irrigation, the lowest average values at non-irrigation. Differences between both irrigation systems for mechanical properties were statistically significant. According to the research results, it is recommended the use of the data obtained from the drip-irrigated plots in the design of handling equipments and storage facilities for dry bean.

Keywords: dry bean, physical properties, mechanical properties, statistical analysis

Introduction

Dry bean is one of the world's major sources of edible legume. Leading producers include India, Brazil, Mexico, China and the United States (FAOSTAT, 2009). In 2008, nearly 20 million tons of dry beans were produced in 28 million ha in the world. Turkey has about 97848 ha of dry bean harvesting area, 154630 tonnes of dry bean production per year (FAOSTAT, 2009).

Grain legumes occupy an important place in human nutrition, especially in the dietary pattern of low-income groups of people in developing countries. Legumes, considered as poor man's meat, are generally good sources of nutrients (THARANATHAN; MAHADEVAMMA, 2003). They are an important and inexpensive source of protein, dietary fiber and starch for a large part of the world's population, mainly in developing countries (PERLA et al., 2003; SHIMELIS; RAKSHIT, 2005). Among the commonly consumed food legumes, dry beans (*Phaseolus vulgaris* L.) occupy an important place in human nutrition in Black Sea and Middle Anatolian regions of Turkey (TOPAK et al., 2009).

Dry bean is sensitive to water stress unlike other common crops (BOUTRAA; SANDERS, 2001). Irrigation is therefore an essential component of dry bean production systems in both arid and semi-arid regions. Net water requirement for a 90-100 d dry bean crop ranges from 350-500 mm depending upon the soil, climate and cultivar (ALLEN et al., 2009).

In the quality classifications and storage operations of dry bean, visual evaluation (dimensions, size, etc.) and analytical evaluation (moisture content, bulk density, angle of internal friction, etc.) related to physical and mechanical properties, can be necessary (BOUMANS, 1985; KORUNIC et al., 1996).

To design the equipment used in seedling, harvesting, transportation, storage and processing of dried dry bean there is a need to know various physical and mechanical properties as a function of moisture content (TAVAKOLI et al., 2009). The physical and mechanical properties of dry bean grains are to be known for the design and improvement of storage strucutres, relevant machines and facilities for harvesting, storing, handling and processing. The dimensions, size and mechanical behaviour of dry bean are important for designing of separating, harvesting, sizing and grinding machines. Bulk density, angle of internal friction, porosity and pressure ratio affects the structures loads (vertical and horizontal loads acting by product on silo wall) and the is important in designing of drying and aeration (natural and mechanical), store systems (silos and warehouses) and transporting structures. The coefficient of friction (static and dynamic) of the grain against the various surfaces is also necessary in designing of conveying, grain flow (mass and hopper) and storage structures (ALTUNTAŞ; YILDIZ, 2007; IXTAINA et al., 2008). These properties have been investigated for locust bean seed (OGUNJIMI et al., 2002), bean (PERLA et al., 2003), Africa oil bean seed (ASOEGWU et al., 2006), faba bean (ALTUNTAŞ; YILDIZ, 2007), barbunia bean (CETIN, 2007), soybean (KIBAR; ÖZTÜRK, 2008) and many others.

The objective of this research was to determine the effect of dry bean at irrigation systems (non-irrigation, drip irrigation) on physical properties (length, width, thickness, sphericity, surface area, volume, bulk density, true density and porosity) and mechanical properties (angle of internal friction, static coefficient of friction, Poisson ratio and pressure ratio).

Material and methods

Experimental design and sample preparation

The field experimental design was a completely randomized block design with three replications. Drip irrigation method was used to irrigate the plots. The variety of dry bean used in the present study was obtained from the crop grown, as a representative of commercial processing, during 2009 in the Bafra lowland zone (41° 35' N, 35° 56' E) of Samsun city, Turkey, which is at an altitude of 20 m. The total average annual rainfall of the lowland is 770 mm (TUMAS, 2009).

The dry beans were sown in May 2009 and harvested in September of the same year. Samples were collected after harvest mature whole raw beans and sun drying them. The dry beans were cleaned manually to remove all foreign matter such as dust, dirt, stones and chaff as well as immature, broken seeds. The initial moisture content of dry bean after harvest was determined by oven drying at 80 ± 5°C for 24h (AOAC, 1984). The dry bean obtained was placed in desiccator and stored at room temperature (23 ± 2°C) before use.

Determination of physical properties

To determine the average size of dry beans, a sample of 100 dry beans was randomly selected. The length (L), width (W) and thickness (T) were measured using a digital caliper reading to ± 0.01 mm accuracy.

According to Mohsenin (1980) and Kibar et al. (2010) the degree of sphericity (%), can be expressed as follows:

$$\phi = \left(\frac{D_g}{L}\right)x100 \tag{1}$$

where:

Geometric mean diameter (D_g) = $(LWT)^{1/3}$.

Surface area (S) in the samples depend on shape of bean and was determined using Equation 2 as described by McCabe et al. (1986), and Jain and Bal (1997).

$$S = \pi D_g^2 \tag{2}$$

Volume (V) of the dry bean was determined using Equation 3 as described by Sreenarayanan et al. (1985), and Jain and Bal (1997).

$$V = \frac{\pi(WT)L^2}{6\left(2L - \sqrt{WT}\right)} \quad (3)$$

The M_{1000} was determined using a digital electronic balance having an accuracy of ± 0.01 g. To evaluate the thousand grain mass, 100 grains were randomly selected from the bulk sample were averaged (TUNDE-AKINTUNDE; AKINTUNDE, 2004; TAVAKOLI et al., 2009).

To determine the bulk density of the experimental samples at non-irrigation and drip-irrigation, the method defined by Mohsenin (1980), and Singh and Goswami (1996) was used. Weight of a bulk density container of 1000 ml volume and 108 mm height was used to determine bulk density. The bulk density container was filled up to 5 cm above the top. The dry beans were then allowed to settle into the container and the bulk density was calculated from the following Equation 4:

$$\rho_b = \frac{G_2 - G_1}{V_b} \quad (4)$$

The liquid displacement method, as described by Singh and Goswami (1996), Öğüt (1998), and Kibar and Öztürk (2009), was used to determine the true density of dry bean samples. In this method, toluene, C_7H_8 (bulk density of toluene at 20°C is 0.87 g cm^{-3}) was used in place of water because it is absorbed to a lesser extent by dry beans and its surface tension (25.52 mN m^{-1} at 20°C) is low. To calculate true density, the air dried weight for samples was first determined. The samples were then submerged in toluene and the displacement volume was determined. In the second stage, the true density of samples was calculated by using Equation 5 as follows:

$$\rho_t = \frac{m_s + m_w}{V_s + V_w} \quad (5)$$

The porosity of dry bean was calculated from bulk and true densities using Equation (6) the relationship given by Mohsenin (1980) and Nelson (2002) as follows:

$$P = \left(\frac{\rho_t - \rho_b}{\rho_t}\right) x100 \quad (6)$$

Determination of mechanical properties

To determine the angle of internal friction of dry bean samples the direct shear method was used according to Uzuner (1996), Zou and Brusewitz (2001), Molenda et al. (2002), Mani et al. (2004), and Kibar and Öztürk (2009). The velocity used during the experiment was 1.16x10^{-5} m s^{-1} and the angle of internal friction of samples was calculated by using Equations 7, 8 and 9.

$$\sigma = \frac{N}{A} * 100 \quad (7)$$

$$\tau = \frac{T_S}{A} * 100 \quad (8)$$

$$\tau = (c + \sigma \, x \tan\varphi) \quad (9)$$

The static coefficients of friction of dry beans were determined according to the method of Kibar and Öztürk (2009). Wood, galvanized steel and concrete (C30) surfaces were used as friction surfaces. During the experiment, the test surface moved at a low velocity (0.024 m s^{-1}). The surfaces were driven by a 12 V, adjustable direct current motor and strength of friction was measured by using a digital dynamometer (Figure 1).

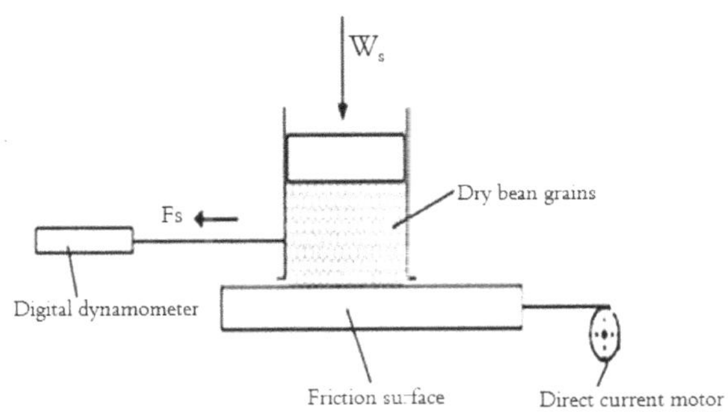

Figure 1. Apparatus to measure the force required to cause two surfaces to slide.

Strength of friction has been taken into consideration as an important parameter to determine static coefficients of friction. Static coefficient of friction was calculated from the constant strength of friction read in the digital dynamometer after movement occurred at the interface. The static coefficients of friction of dry bean samples were calculated by using Equation 10.

$$\mu_S = \frac{F_S}{W_S} \quad (10)$$

The Poisson ratio were calculated with Equation (11) developed by Qu et al. (2001).

$$\nu = \frac{1-\sin\varphi}{2-\sin\varphi} \qquad (11)$$

In this study, the equalities developed by Kézdi (1962), Lohnes (1993) and Eurocode 1 (2003) were used to determine the pressure ratio of dry bean. Accordingly, the three methods were also evaluated in the study. Theoretical equalities relating to this issue are given at Equations 12, 13 and 14. These equalities are:
according to Kézdi (1962):

$$k_K = 1 - \sin\varphi \qquad (12)$$

according to Lohnes (1993):

$$k_L = \frac{(1-\sin\varphi)x(1+\frac{2}{3}x\sin\varphi)}{1+\sin\varphi} \qquad (13)$$

according to Eurocode 1 (2003):

$$k_E = 1.1x(1-\sin\varphi) \qquad (14)$$

Statistical analysis

Some engineering properties of dry bean were determined at non-irrigation and drip-irrigation with at least three replications. Mean, minimum, maximum, standard deviations and t test were calculated by Statistical Package for Social Science (SPSS) 10.0 software. In order to estimate statistical differences between the non-irrigation and drip-irrigation the Student's t-test and Principal Component Analysis (PCA) were applied to investigate the factor structure of irrigation conditions.

Results and discussion

The initial moisture content of dry bean after harvest was determined as a 7.2% (dry basis, db) for drip-irrigated system while for non-irrigated system was 6.1% (db). According to obtained data, moisture content of dry bean is important. Because, the moisture content are highly affected by the physical and mechanical parameters of dry bean (OGUNJIMI et al., 2002; VALVERDE et al., 2006; ALTUNTAŞ; YILDIZ, 2007). Grain moisture content has deep influence on the mechanical properties of grain. Changing the moisture content of grain influences shear stress-strain characteristics, and consequently the determination of strength parameters: the angle of internal friction (φ) and the cohesion (c) (MOLENDA; HORABIK, 2005).

Figure 2 showed that the dimension values of dry bean for both non-irrigation and drip-irrigation conditions. For dry bean at non-irrigation condition, the mean values of the length, width, and thickness varied from 14.92 mm, 7.03 mm and 5.76 mm, respectively. For dry bean at drip-irrigation condition the mean values of the length, width, and thickness varied from 15.40 mm, 7.26 mm and 5.91 mm, respectively. The t test showed that non-irrigation and drip-irrigation length, width, and thickness showed be significant differences ($p < 0.01$).

The plot of scatter diagram dependent on L, W, T in non irrigation and drip-irrigation conditions were presented in Figures 3 and 4. The scatter diagram of non irrigation and drip-irrigation showed that there was a significant level of differences among L, W, T dimensions.

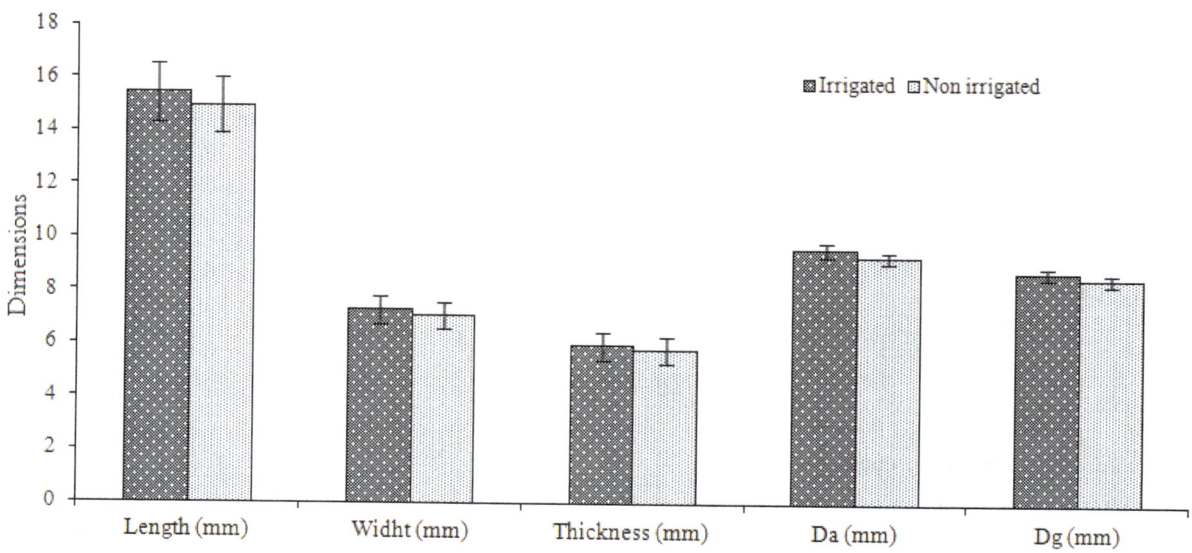

Figure 2. The axial dimensions of dry bean depending upon non-irrigation and drip-irrigation conditions.

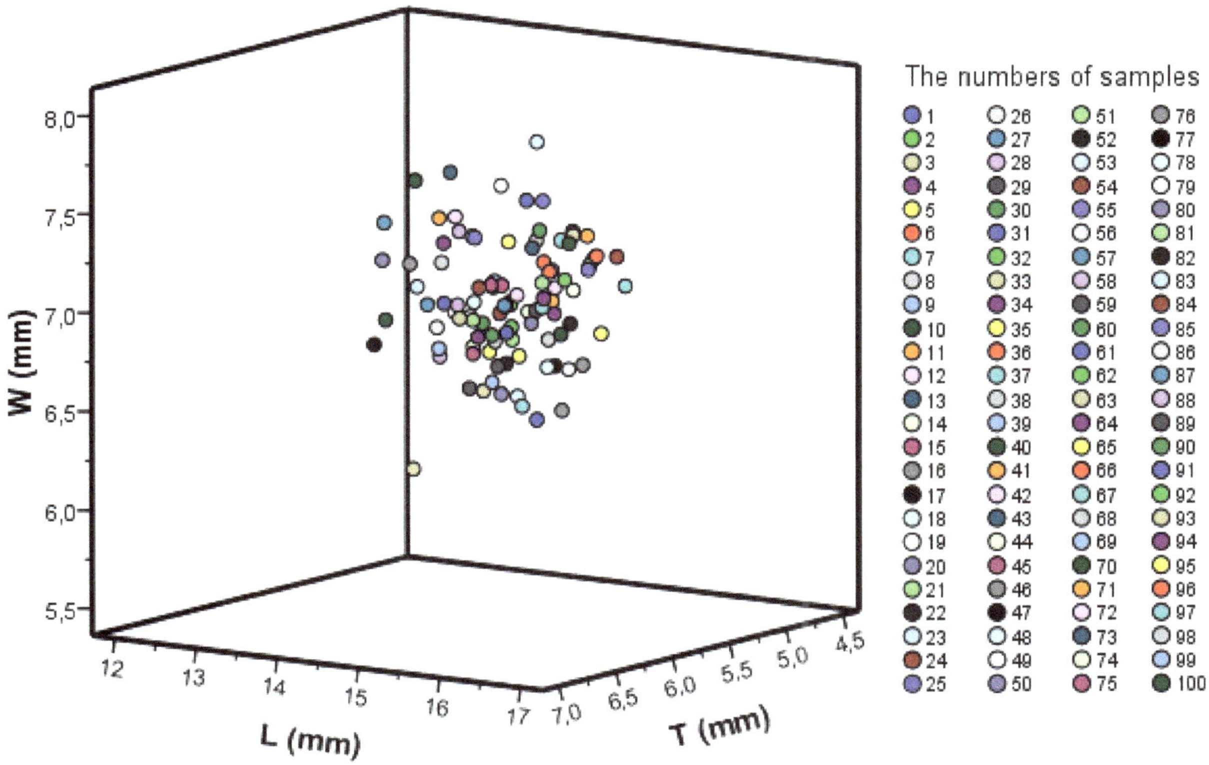

Figure 3. Scatter diagram depending upon L, W, T in non-irrigation conditions.

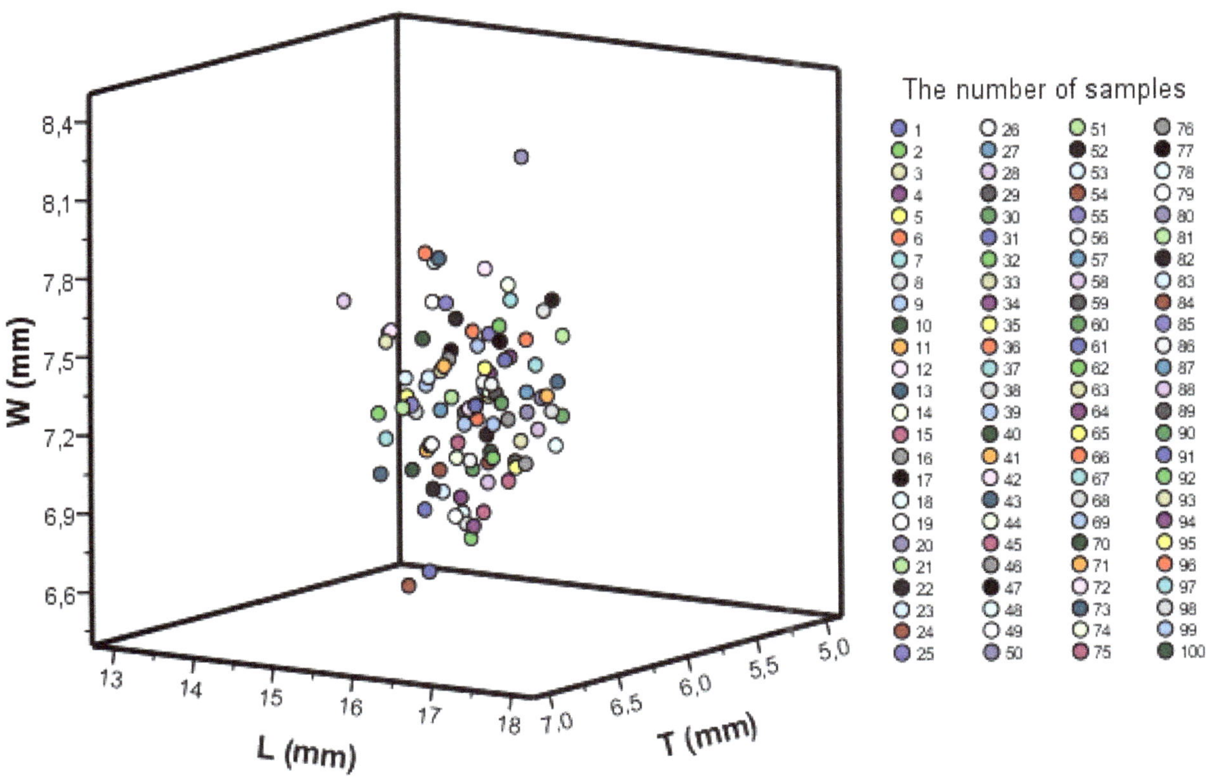

Figure 4. Scatter diagram depending upon L, W, T in drip-irrigation conditions.

Raw materials often occur in sizes that are too large to be used and, therefore, they must be reduced in size. This size-reduction operation can be divided into two major categories depending on whether the material is a solid or a liquid. If it is solid, the operations are called grinding and cutting, if it is liquid, emulsification or atomization. All depend on the reaction to shearing forces within solids and liquids (EARLE; EARLE, 2010). Size reduction is a unit operation widely used in a number of processing industries. Many types of equipment are used in size reduction operations. In a broad sense, size reduction machines may be classified as crushers used mainly for coarse reduction, grinders employed principally in intermediate and fine reduction, ultra-fine grinders utilized in ultra-fine reduction, and cutting machines used for exact reduction (MACCABE et al., 1986). The coefficients of correlation for non-irrigation showed that the dimension ratios of dry bean were not statistically significant. Whereas, it can be seen from the Table 1 dimension ratios of dry bean at drip-irrigation were statistically significant (L/W and L/T $p < 0.05$; L/D_g and L/D_a $p < 0.01$). Regarding this subject, a similar result was reported by Valverde et al. (2006).

Table 1. Ratio of dimensions of dry bean depending upon non-irrigation and drip-irrigation conditions.

Parametres	Non-irrigation					Drip-irrigation				
	Mean	Min.	Max.	SD	Correlation	Mean	Min.	Max.	SD	Correlation
L/W	2.12	1.81	2.38	0.11	0.118^{ns}	2.12	1.80	2.31	0.10	0.228^{*}
L/T	2.59	2.11	2.95	0.12	0.044^{ns}	2.61	2.38	2.89	0.12	0.229^{*}
L/D_g	1.76	1.67	1.87	0.51	0.102^{ns}	1.77	1.69	1.89	0.47	0.267^{**}
L/D_a	1.61	1.50	1.71	0.33	0.105^{ns}	1.63	1.52	1.65	0.29	0.271^{**}

*Significant at 0.05 level, **Significant at 0.01 level, ^{ns}Non-significant; SD: Standard deviation.

As given in Table 2, the sphericity (ϕ) of dry bean for irrigation systems (non-irrigation, drip-irrigation) was found to range from 0.53 to 0.63 and 0.54 to 0.64, respectively.

Table 2. Physical properties of dry bean depending upon non-irrigation and drip-irrigation conditions.

Properties	Non-irrigation				Drip-irrigation			
	Mean	Min.	Max.	SD	Mean	Min.	Max.	SD
ϕ (%)	56	53	63	0.18	56	54	64	0.01
S (mm^2)	223.18	218.55	224.66	4.10	236.99	231.79	240.94	4.70
V (mm^3)	200.62	194.65	204.73	5.29	213.97	199.78	224.72	12.82
M_{1000} (g)	439.8	434.70	450.10	7.84	480.50	466.50	494.20	13.85
ρ_b (kg m^{-3})	804.69	801.12	808.56	3.73	820.00	815.23	818.77	5.49
ρ_t (kg m^{-3})	987.02	980.45	995.41	7.65	1164.44	1000.55	1025.43	12.53
P(%)	18.47	17.53	19.53	1.00	19.00	18.52	19.45	0.46

SD: Standard deviation.

The sphericity values given by drip-irrigation achivied higher values compared with the one by non-irrigation. Differences between the values were not statistically significant at $p > 0.05$ (Table 3). Similar results have been reported by Olajide and Ade-Omowaye (1999). for locust bean seed, Asoegwu et al. (2006) for African oil bean seed and Cetin (2007) for barbunia bean. Sphericity is an important parameter used in fluid flow and heat and mass transfer calculations (SAHIN; SUMNU, 2006).

Table 3. Results of the statistical analysis of physical properties in the dry bean according to the irrigation regimes.

Parametres	Between both irrigation regimes	
	t	Sig.
ϕ	-0.661	0.510^{ns}
S	5.853	0.0005^{***}
V	5.816	0.0002^{***}
M_{1000}	4.100	0.015^{*}
ρ_b	4.533	0.011^{*}
ρ_t	6.518	0.003^{**}
P	-6.601	0.003^{**}

*Significant at 0.05 level ** Significant at 0.01 level *** Significant at 0.001 level ^{ns} Non-significant.

As seen from the Table 2 according to Equation 2, the surface area of dry bean for non-irrigation and drip-irrigation between 218.55-224.66 mm^2 and 231.79-240.94 mm^2 were determined. It was determined that surface area of dry beans harvested from irrigated plots was higher than in the dry beans harvested from non-irrigation plots, because both grain moisture content of dry beans harvested from irrigated plots were higher and has better quality than in the dry beans harvested from non-irrigation plots. Similar trends have been reported by Asoegwu et al. (2006) for African oil bean seed. Differences between both irrigation systems were statistically significant at $p < 0.001$ (Table 3). Surface area is important in all applications where the process is surface dependent. Examples of such applications are mass and heat transfer, flow through packed beds, or fluidization. In food process engineering, combined heat and mass transfer is critical in quality control of many materials where moisture has to be removed to the lowest possible level, but the use of excessive heat may impair sensory attributes (BARBOSA-C´ANOVAS et al., 2005). Abrasiveness and flowability of granular materials is their ability to damage surfaces of equipment with which they are in contact as a result of movement over the surfaces. The degree of abrasiveness and classification of flowability is related to the hardness, shape and size of the grains (MOLENDA; HORABIK, 2005). In this context, the sphericity and surface area of agro-granular materials comes to the fore. The sphericity increases, porosity between the grains decreases and the grains are better compressed. Thus, less energy is consumed for the compression process.

The volume (V) of dry bean under non-irrigation and drip-irrigation ranged from 194.65 to 204.73 mm^3 and 199.78 to 224.72 mm^3 respectively (Table 2). It was determined that volume of dry beans harvested from irrigated plots was higher than in the dry beans harvested from non-irrigation plots, because both moisture content of dry beans harvested from irrigated plots were higher and have higher dimensions (length, width and thickness) than that of the dry beans harvested from non-irrigation plots. Olajide and Ade-Omowaye (1999), also reported the same results for locust bean seed. In this context, differences between both irrigation systems were statistically significant at $p < 0.001$ (Table 3).

The M_{1000} of dry bean under non-irrigation and drip-irrigation were ranged from 434.70 to 450.10 g, and 446.50 to 494.20 g, respectively (Table 2). It was determined that M_{1000} of dry beans harvested from irrigated plots was higher than in the dry beans harvested from non-irrigation plots, because both grain moisture content of dry beans harvested from irrigated plots were higher and have a heavier mass than in the dry beans harvested from non-irrigation plots. A similar trend has been reported for guna seeds (AVIARA et al., 1999). The t test showed that between both irrigation systems significant differences were found, according to $p < 0.05$ (Table 3). This parameter is useful in determining the equivalent diameter which can be used in the theoretical estimation of dry bean volume and in cleaning using aerodynamic forces (TABAK; WOLL, 1998; KHOSHTAGHAZA; MEHDIZADEH, 2006; SOLOMON; ZEWDU, 2009).

The bulk density (ρ_b) of dry bean under non-irrigation and drip-irrigation were ranged from 801.12 to 808.56 kg m^{-3} and 815.23 to 818.77 kg m^{-3} respectively (Table 2). This was due to the fact that an linear increasing in mass owing due to moisture gain in the sample was higher than accompanying volumetric expansion of the bulk. Most of researchers reported linear increasing trends for faba, barbunia bean and African yam beans (ALTUNTAŞ; YILDIZ, 2007; CETIN, 2007; ASOIRO; ANI, 2011). Differences between both irrigation systems were statistically significant at $p < 0.05$ (Table 3). The reported that stress-strain behavior of granular material depends on the bulk density of the sample (MOLENDA; HORABIK, 2005). Dense samples dilate during shear test while loose samples decrease in volume. In dense samples shear stress attains a peak value, and with continuing shear displacement it drops back to a lower ultimate value and remains at that constant level during further shear. In the loose state most granular materials tend to decrease in volume when subjected to shear under constant normal load. Measurement of bulk density is of fundamental use by the industry to adjust storage, processing, packaging and distribution conditions. Chang et al. (1983) reported that distributed filling of silo increases the density of granular material from 5.1 to 9.2% as compared to filling from centrally located spout of conveyor. Stephens and Foster (1976) observed increases in density of the order of 3 to 5% above the bulk density values in condensed filling from spout of a conveyor, and 7% in the case of distributed filling.

For dry bean, depending upon non-irrigation and drip-irrigation conditions the true density (ρ_t) ranges from 980.45 to 995.41 kg m^{-3} and 1000.55 to 1025.43 kg m^{-3}, respectively (Table 2). The increase in true density with increase in moisture content of grain with irrigation might be attributed to the relatively lower true volume as compared to the corresponding mass of dry bean attained due to adsorption of water Pradhan et al. (2009). The same results for grainy products were given by Kibar and Öztürk (2008) and Pradhan et al. (2009). Differences between both irrigation systems were statistically significant at $p < 0.01$ (Table 3).

Porosity (P) was evaluated using mean values of bulk density and true density in Equation 6. The variation of porosity for non-irrigation and drip-irrigation conditions was shown in Table 2. Since the porosity depends on the bulk density as well as densities, the magnitude of variation in porosity depends on these factors only. The reason of low porosity is little of difference between the bulk density and true density of dry beans. In addition, low porosity may be related to the nutritional composition of the product. The same results for grainy products were given by Kibar and Öztürk (2008) and Unal et al. (2009). Differences between both irrigation systems were statistically significant at $p < 0.001$ (Table 3). Porosity is an important physical property characterizing the texture and the quality of dry and intermediate moisture foods. Porosity data is required in modelling and design of various heat and mass transfer processes such as drying, frying, baking, heating, cooling, and extrusion (SAHIN; SUMNU, 2006).

The experimental results for the angle of internal friction (φ) with respect to non-irrigation and drip-irrigation was shown in Table 4.

Table 4. Mechanical properties of dry bean depending upon non-irrigation and drip-irrigation conditions.

Properties	Non-irrigation				Drip-irrigation			
	Mean	Min.	Max.	SD	Mean	Min.	Max.	SD
φ (degree)	21.60	20.70	22.40	0.85	25.60	24.50	27.20	1.40
μ_s								
Wood	0.354	0.346	0.370	0.014	0.430	0.418	0.448	0.016
Galvanized steel	0.303	0.292	0.320	0.015	0.331	0.315	0.342	0.014
Concrete (C30)	0.466	0.456	0.482	0.014	0.588	0.566	0.615	0.025
υ	0.39	0.38	0.39	0.05	0.36	0.35	0.37	0.01
k								
k_K	0.63	0.62	0.65	0.13	0.57	0.54	0.59	0.25
k_L	0.58	0.56	0.58	0.16	0.51	0.49	0.53	0.22
k_E	0.69	0.68	0.71	0.15	0.62	0.60	0.64	0.24

SD: Standard deviation.

The angle of internal friction of dry bean under non-irrigation and drip-irrigation were from 20.70° to 22.40° and 24.50° to 27.20°, respectively. This increasing trend of angle of internal friction with moisture content in drip-irrigation occurs because surface layer of moisture surrounding the particle hold the aggregate of kernel together by the surface tension. Differences between both irrigation systems were statistically significant at $p < 0.01$ (Table 5).

Table 5. Results of the statistical analysis of mechanical properties in the dry bean according to the irrigation regimes.

Parametres	Between both irrigation regimes	
	t	Sig.
φ	4.855	0.008**
μ_s		
Wood	7.200	0.002**
Galvanized steel	2.962	0.041*
Concrete (C30)	8.281	0.00***
υ	-4.000	0.016*
k		
k_K	-4.118	0.015*
k_L	-4.588	0.010**
k_E	-5.277	0.006**

*Significant at 0.05 level **Significant at 0.01 level ***Significant at 0.001 level.

Researches also found in their study that the angle of internal friction increased linearly with increase of moisture content (MOLENDA et al., 1998; MOYA et al., 2002). The angle of internal friction and the angle of friction on the wall material increase with an increase in the moisture content of grain, while the bulk density decreases or increases depending on the pressure level. The angle of friction at the interface of grain and the wall material depends strongly on the roughness of the wall surface. All those three properties of bulk of grain influence the pressure ratio. Consequently, the angle of internal friction is of paramount importance in designing hoper openings, side wall slopes of storage bins and design of wall pressures in storage structures (MOLENDA; HORABIK, 1998).

The static coefficient of friction for dry bean was determined against surfaces of wood, galvanized steel and concrete surfaces. The experimental results related to the static coefficient of friction were given in Table 4. As seen from the Table 4 the value of the static coefficient of friction for non-irrigation conditions was ranged from 0.346 to 0.370 on wood, 0.292 to 0.320 on galvanized steel and 0.456-0.482 on concrete surface. The value of the static coefficient of friction for drip-irrigation conditions was ranged from 0.418 to 0.448 on wood, 0.315 to 0.342 on galvanized steel and 0.566 to 0.615 on concrete surface. The relationships between these coefficients against various surfaces and irrigation systems of dry bean were shown in Table 5. The between both irrigation systems with each of surface were found to be significantly different (Table 5). The coefficient of friction of granular materials of plant origin depends on numerous factors, among which the following are regarded as the most important: moisture content, normal pressure, sliding velocity, surface state and ambient conditions. According to some authors, increase in moisture content caused a decrease in the resulting coefficient of friction, from the presence of liquid water in the contact area (LAWTON, 1980; MOLENDA; HORABIK, 1998). The static coefficient of friction is important for designing of storage bins, hoppers, pneumatic conveying system, screw conveyors, forage harvesters, threshers, etc. (SAHAY; SINGH, 1996; PRADHAN et al., 2008).

The Poisson ratio of dry bean depending upon non-irrigation and drip-irrigation were between 0.38 to 0.39, and 0.35 to 0.37, respectively (Table 4). Poisson ratio of African nutmeg was researched at different moisture content in the range of 8.0-28.7%. In their studies it was found that the Poisson ratio decreased linearly with the increase of moisture content (BURUBAI et al., 2008). Differences between both irrigation systems were statistically significant at $p < 0.05$ (Table 5). Poisson ratio is necessary in design with finite element analysis of wall pressures in storage structures.

The pressure ratio is one of the three most important physical properties of bulk solids, commonly used for calculation of pressure in a silo. Almost all design codes use a Janssen-type pressure distribution to predict silo pressures (WILMS, 1991; MOLENDA; HORABIK, 2005). The variation of pressure ratios for non-irrigation and drip-irrigation conditions was shown in Table 4. Today they are used in calculation of the wall pressures equation proposed by Eurocode 1 (2003). Here the comparison is made between the other methods with Eurocode 1 (2003). The highest value for pressure ratio was determined in k_E (0.71) at non-irrigation and the lowest value in k_L (0.49) at drip-irrigation. Similar results have been reported for some cereal grains (HORABIK; RUSINEK, 2002).

Differences between both irrigation systems with each of the methods (k_K, k_L and k_E) were statistically significant (Table 5). Pressure ratio is necessary in the design of wall pressures (standards such as 30, 55-57) in storage structures. Because postharvest moisture content has higher (7.2%) at products obtained from drip-irrigated plots, moisture content brings about more pressure increases in the design of storage structures. These results were found for different products in the studies related to pressure equations developed by Janssen, Rankine, Reimbert and Eurocode1 (ÖZTÜRK; KIBAR, 2008; ÖZTÜRK et al., 2008). The effect of engineering properties such as bulk density, angle of internal friction, static coefficient of friction, pressure ratio and Poisson ratio in this increase are very important.

At the end of PCA, factor coefficients of identifying qualities were evaluated and the attributes scoring a coefficient value higher than 0.7 in the first three PCA were determined. The plot of the physical and mechanical properties on the first two PCs obtained from analysis of irrigation conditions were presented in Figure 5.

The first two principal components (PCs) accounted for 100% of the total variability among the physical and mechanical properties for drip-irrigation. The first principal component (PC1), which is the most important component, explained 70.162% of the total variability. The second principal component (PC2) had 29.838% of the total variation. PC1 explained 73.249%, while PC2 explained 26.751% of the total variability.

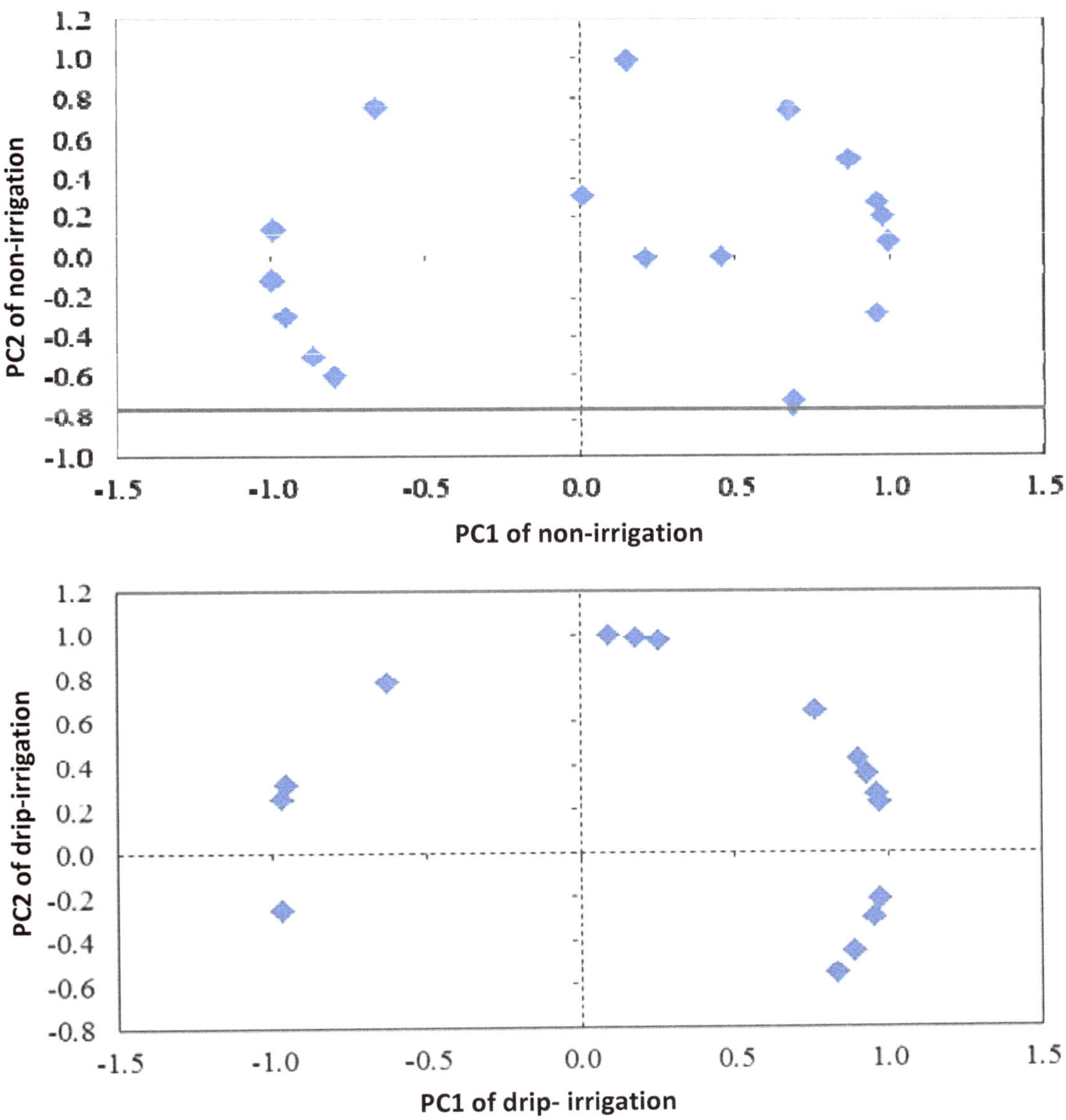

Figure 5. First (PC1) versus second (PC2) principal component according to the method of irrigation.

Conclusion

1. The surface area and the volume of dry beans from irrigated plots are higher than non-irrigation plots.

2. The bulk and true density values in drip-irrigation conditions are higher than non-irrigation conditions.

3. The angle of internal friction is importance in designing hoper openings and wall pressures in storage structures. The significant differences between both irrigation systems were observed.

4. The static coefficient of friction values on three test surfaces in drip-irrigation conditions are higher than non-irrigation conditions.

5. Poisson ratio is necessary in the design with finite element analysis of wall pressures in storage structures. The differences between both irrigation systems were statistically significant.

6. The mean highest value for pressure ratio was determined in k_E (0.68) at non-irrigation and the mean lowest value in k_L (0.48) at drip-irrigation.

Nomenclature

L Length, mm
W Witdh, mm
T Thickness, mm
ϕ Sphericty, %
D_g Geometric average diameter, mm
D_a Arithmetic average diameter, mm
V Volume, mm^3
S Surface area,mm^2
ρ_b Bulk density, kg m^{-3}
G_1 Free weight of bulk density bucket, kg
G_2 Weight of bulk density bucket with rice, kg
V_b Volume of bulk density bucket, m^3
ρ_t True density, kg m^{-3}
m_s Weight of liquid, kg
m_w Weight of air dry sample, kg
V_s Volume of liquid, m^3
V_w Volume of sample, m^3
P Porosity, %
φ Angle of internal friction, degrees
σ Normal stress, kPa
N Load applied on the sample, kg
A Cellular area, cm^2
τ Shear stress, pressure on cutting edge, kPa
T_s Shear force, load on cutting edge, kg
C Coefficient of cohesion
μ_s Static coefficient of friction
F_s Force starting movement at surface interface, kg m^{-2}
W_s Force applied to surface interface, kg m^{-2}
ν Poisson ratio
k_K Pressure ratio according to Kézdi
k_L Pressure ratio according to Lohnes
k_E Pressure ratio according to Eurocode 1

References

ALLEN, R. G; PEREIRA, L. S; RAES, D.; SMITH, M. **Crop evapotranspiration** - guidelines for computing crop water requirements. Rome: FAO , 1998. (FAO Irrigation and Drainage Paper 56).

ALTUNTAŞ, E.; YILDIZ, M. Effect of moisture content on some physical and mechanical properties of faba bean (*Vicia faba* L.) grains. **Journal of Food Engineering**, v. 78, n. 1, p. 174-183, 2007.

AOAC-Association of Official Analytical Chemists. **Official method of analysis**. 14th ed. Washington, D.C.: AOAC International, 1984.

ASOEGWU, S.; OHANYERE, S.; KANU, O.; IWUEKE, C. Physical properties of African oil bean seed (*Pentaclethra macrophylla*). **Agricultural Engineering International: The CIGR Ejournal**, v. VIII, 2006. Available from: <http://www.cigrjournal.org/index.php/Ejounral/article/view/673/667>. Access on: 2012.

ASOIRO, F. U.; ANI, A. O. Determination of some physical properties of African yam beans. **The Pacific Journal of Science and Technology**, v. 12, n. 1, p. 374-380, 2011.

AVIARA, N. A.; GWANDZANG, M. I.; HAQUE, M. A. Physical properties of guna seeds. **Journal of Agricultural Engineering Research**, v. 73, n. 2, p. 105-111, 1999.

BARBOSA-C´ANOVAS, G. V.; ORTEGA-RIVAS, E.; JULIANO, P.; YAN, H. **Food powders**. Physical properties, processing, and functionality. New York: Kluwer Academic/Plenum Publishers, 2005.

BOUMANS, G. Grains, oilseeds, derivatives and substitutes. In: BOUMANS, G. (Ed.). **Grain handling and storage**. Amsterdam: Elsevier Science Publishers, 1985. p. 7-23.

BOUTRAA, T.; SANDERS, F. E. Influence of water stress on grain yield and vegetative growth of two cultivars of bean (*Phaseolus vulgaris* L.). **Journal of Agronomy and Crop Science**, v. 187, n. 4, p. 251-257, 2001.

BRUBAI, W.; AMULA, E.; DAVIES, R. M.; ETEPKE, G. W. W.; DAWORIYE, S. P. Determination of poisson's ratio and modulus of African nutmeg (*Monodora myristicsa*). **International Agrophysics**, v. 22, n. 2, p. 99-102, 2008.

CETIN, M. Physical properties of barbunia bean (*Phaseolus vulgaris* L. cv. 'Barbunia') seed. **Journal of Food Engineering**, v. 80, n. 1, p. 353-358, 2007.

CHANG, C. S.; CONVERSE, H. H.; MARTIN, C. R. Bulk properties of grain as affected by self-propelled rotational type grain spreaders. **Transactions of the ASAE**, v. 26, n. 5, p. 1543-1550, 1983.

EARLE, R. L.; EARLE, M. D. **Unit operations in food processing**: web edition. Palmerston: The New Zealand Institute of Food Science and Technology, 2010.

EUROCODE 1-Basis of design and actions on structures. **Actions in silos and tanks** - part 4. Brussels: European Committee for Standarization, 2003.

FAOSTAT-Food and Agricultural Organization of United Nations Statistical Database. **World crop production statistics**. Rome: Food and Agriculture Organization of The United Nations Publications, 2009.

HORABIK, J.; RUSINEK, R. Pressure ratio of cereal grains determined in uniaxial compression test. **International Agrophysics**, v. 16, n. 1, p. 23-28, 2002.

IXTAINA, V. Y.; NOLASCOA, S. M.; TOMÁS, M. C. Physical properties of chia (*Salvia hispanica* L.) seeds. **Industrial Crops and Products**, v. 28, n. 3, p. 286-293, 2008.

JAIN, R. K.; BAL, S. Properties of pearl millet. **Journal of Agricultural Engineering Research**, v. 66, n. 1, p. 85-91, 1997.

KÉZDI, A. **Earth pressure theories**. Berlin: Springer Verlag, 1962.

KHOSHTAGHAZA, M. H.; MEHDIZADEH, R. Aerodynamic properties of wheat kernel and straw materials. **Agricultural Engineering International: The CIGR Ejournal**, v. 8, p. 1-10, 2006.

KIBAR, H.; ÖZTÜRK, T. Physical and mechanical properties of soybean. **International Agrophysics**, v. 22, n. 3, p. 239-244, 2008.

KIBAR, H.; ÖZTÜRK, T. The effect of moisture content on the physico-mechanical properties of some hazelnut varieties. **Journal of Stored Products Research**, v. 45, n. 1, p. 14-18, 2009.

KIBAR, H.; ÖZTÜRK, T.; ESEN, B. The effect of moisture content on physical and Mechanical properties of rice (*Oryza sativa* L.). **Spanish Journal of Agricultural Research**, v. 8, n. 3, p. 741-749, 2010.

KORUNIC, Z.; FIELDS, P. G.; KOVACS, M. I.; NOLL, J. S.; LUKOW, O. M.; DEMIANYK, C. J.; SHIBLEY, K. J. The effect of diatomaceous earth on grain quality. **Postharvest Biology and Technology**, v. 9, n. 3, p. 373-387, 1996.

LAWTON, P. J. Coefficients of friction between cereal grain and various silo wall materials. **Journal of Agricultural Engineering Research**, v. 25, n. 1, p. 75-86, 1980.

LOHNES, R. A. Lateral stress ratios for particulate materials. **Powder Handling and Processing**, v. 5, n. 4, p. 331-335, 1993.

MANI, S.; TABIL, L. G.; SOKHANSANJ, S. Mechanical properties of corn stover grind. **Transactions of the ASAE**, v. 47, n. 6, p. 1983-1990, 2004.

MCCABE, W. L.; SMITH, J. C.; HARRIOTT, P. **Unit operations of chemical engineering**. New York: McGraw-Hill, 1986.

MOHSENIN, N. N. **Structure, physical characteristics and mechanical properties of plant and animal materials**. New York: Gordon and Breach Science Publishers, 1980.

MOLENDA, M.; HORABIK, J.; ROSS, I. J. Stress and deformation of wheat in direct shear test. **International Agrophysics**, v. 12, n. 2, p. 115-118, 1998.

MOLENDA, M.; MONTROSS, M. D.; HORABIK, J.; ROSS, I. J. Mechanical properties of corn and soybean meal. **Transactions of the ASAE**, v. 45, n. 6, p. 1929-1936, 2002.

MOLENDA, M.; HORABIK, J. Characterization of mechanical properties of particulate solids for storage and handling. In: HORABIK, J.; LASKOWSKI, J. (Ed.). **Mechanical properties of granular agro-materials and food powders for industrial practice, Part I**. Lublin: Insitute of Agrophysucs PAS, 2005. p. 1-145.

MOYA, M.; AYUGA, F.; GUAITA, M.; AGUADO, P. Mechanical properties of granular agricultural materials. **Transaction American Society of Agricultural Engineers**, v. 45, n. 5, p. 1569-1577, 2002.

NELSON, S. O. Dimensional and density data for seeds of cereal grain and other crops. **Transaction American Society of Agricultural Engineers**, v. 45, n. 1, p. 165-170, 2002.

OGUNJIMI, L. A. O.; AVIARA, N. A.; AREGBESOLAI, O. A. Some engineering properties of locust bean seed. **Journal of Food Engineering**, v. 55, n. 2, p. 95-99, 2002.

OLAJIDE, J. O.; ADE-OMOWAYE, B. I. O. Some Physical Properties of Locust Bean Seed. **Journal of Agricultural Engineering Research**, v. 74, n. 2, p. 213-215, 1999.

ÖĞÜT, H. Some physical properties of white lupin. **Journal of Agricultural Engineering Research**, v. 69, n. 3, p. 273-277, 1998.

ÖZTÜRK, T.; KIBAR, H. The design parameters of storage load in the cylindrical storage structures for hazelnut. **Journal of Faculty of Agriculture, University of Ondokuz Mayis**, v. 23, n. 2, p. 98-103, 2008.

ÖZTÜRK, T.; ESEN, B.; KIBAR, H. The design loads depend on the grain moisture content in a cylindrical storage structure for corn. **Journal of Faculty of Agriculture, University of Ondokuz Mayis**, v. 23, n. 2, p. 110-115, 2008.

PERLA, O;. LUIS, A. B.; SONIA, G. S.; MARIA, P. B.; JUSCELINO, T.; OCTAVIO, P. L. Effect of processing and storage time on in vitro digestibility and resistant starch content of two bean (*Phaseolus vulgaris* L.) varieties. **Journal of the Science of Food and Agriculture**, v. 83, n. 12, p. 1283-1288, 2003.

PRADHAN, R. C.; NAIK, S. N.; BHATNAGAR, N.; SWAIN, S. K. Moisture-dependent physical prope rties of Karanja (*Pongamia pinnata*) kernel. **Industrial Crops and Products**, v. 28, n. 2, p. 155-161, 2008.

PRADHAN, R. C.; NAIK, S. N.; BHATNAGAR, N.; VIJAY, V. K. Moisture-dependent physical properties of jatropha fruit. **Industrial Crops and Products**, v. 29, n. 3, p. 341-347, 2009.

QU, Q.; NEGI, S. C.; JOFRIET, J. C. Storage aof cohesive material in silos-Part 2: Parametic study. **Powder Handling Processing**, 2001.

SAHAY, K. M.; SINGH, K. K. **Unit operation of agricultural processing**. New Delhi: Vikas Publishing House Pvt. Ltd., 1996.

SAHIN, S.; SUMNU, S. G. **Physical properties of foods**. New York: Springer Science, 2006.

SHIMELIS, E. A.; RAKSHIT, S. K. Proximate composition and physico-chemical properties of improved dry bean (*Phaseolus vulgaris* L.) varieties grown in Ethiopia. **LWT-Food Science and Technology**, v. 38, n. 4, p. 331-338, 2005.

SINGH, K. K.; GOSWAMI, T. K. Physical properties of cumin seed. **Journal of Agricultural Engineering Research**, v. 64, n. 2, p. 93-98. 1996.

SREENARAYANAN, V. V.; SUBRAMANIAN, V.; VISVANATHAN, R. Physical and thermal properties of soybean. **Proceedings of Indian Society of Agricultural Engineers**, v. 3, n. 2, p. 161-169, 1985.

SOLOMON, W. K.; ZEWDU, A. D. Moisture-dependent physical properties of niger (*Guizotia abyssinica* Cass.) seed. **Industrial Crops and Products**, v. 29, n. 1, p. 165-170, 2009.

STEPHENS, L. E.; FOSTER, G. H. Grain bulk properties as affected by mechanical grain spreaders. **Transactions of the ASAE**, v. 19, n. 2, p. 354-358, 1976.

TABAK, S.; WOLL, D. Aerodynamic properties of cottonseeds. **Journal of Agricultural Engineering Research**, v. 70, n. 3, p. 257-265, 1998.

TAVAKOLI, H.; RAJABIPOUR, A.; MOHTASEBI, S. S. Moisture-dependent some engineering properties of soybean grains. **Agricultural Engineering International: The CIGR E-journal**, v. XI, 2009. Available from: <http://www.cigrjournal.org/index.php/Ejounral/article/viewFile/1110/1152>. Access on: 2012.

THARANATHAN, R. N.; MAHADEVAMMA, S. A review: grain legumes a boon to human nutrition. **Trends in Food Science and Technology**, v. 14, n. 2, p. 507-518, 2003.

TOPAK, T.; ACAR, B.; SÜHERI, S. Drip and sprinkler irrigation of dry bean (*Phaseolus vulgaris* L.) in the Konya Basin, Turkey. **The Philippine Agricultural Scientist**, v. 92, n. 2, p. 186-192, 2009.

TUMAS-Turkey Meteorological Archive System. **Climate statistics**. Ankara: Turkish State Meteorological Service Publications, 2009.

TUNDE-AKINTUNDE, T. Y.; AKINTUNDE, B. O. Some physical properties of sesame seed. **Biosystems Engineering**, v. 88, n. 1, p. 127-129, 2004.

UNAL, H.; SINCIK, M.; IZLI, N. Comparison of some engineering properties of rapeseed cultivars. **Industrial Crops and Products**, v. 30, n. 1, p. 131-136, 2009.

UZUNER, B. A. **Soil mechanics**. Ankara: Technique Press, 1996.

VALVERDE, M.; MADRID, R.; GARCÍA, A. L. Effect of the irrigation regime, type of fertilization, and culture year on the physical properties of almond (cv. Guara). **Journal of Food Engineering**, v. 76, n. 4, p. 584-593, 2006.

WILMS, H. Criteria for evaluation of silo design codes. **Bulk Solids Handling**, v. 11, n. 1, p. 55-61, 1991.

ZOU, Y.; BRUSEWITZ, G. H. Angle of internal friction and cohesion of consolidated ground marigold petals. **Transactions of the ASAE**, v. 44, n. 5, p. 1255-1259, 2001.

23

Gamma-ray computed tomography to evaluate changes in the structure of a clayey soil due to agricultural traffic

Luiz Fernando Pires

Departamento de Física, Setor de Ciências Exatas e Naturais, Universidade Estadual de Ponta Grossa, Av. Carlos Cavalcanti, 4748, 84030-900, Ponta Grossa, Paraná, Brazil. E-mail: luizfpires@gmail.com

ABSTRACT. A first generation gamma-ray computed tomography (CT) equipped with a ^{137}Cs source was used to investigate changes in soil structure due to agricultural machinery traffic. Core samples with volume of 55 cm^3 were collected at the surface (0-10 cm) of a Brazilian clayey soil. Significant alterations in soil structure occurred after two passages of tractor wheels, with important decreases in soil porosity. CT data allowed a continuous analysis of soil bulk density and also soil porosity distribution for 1.25 and 1.20 mm layers, which cannot be provided by traditional methods commonly used in soil physics.

Keywords: gamma-ray attenuation, soil bulk density, soil porosity, image analysis, soil physics.

Introduction

Gamma and X-ray computed tomography (CT) is an imaging technique based on the computation of a large number of transmission measurements of gamma or X-photons. These photons are used to reconstruct an image of a cross-section distribution of a physical property of a scanned material from projections taken from a number of different directions (KAK; SLANEY, 1988). CT produces high-quality images of porous materials, such as soil, being an interesting alternative to reliably evaluate changes in soil physical properties due to natural or artificial processes (BEUTLER et al., 2005).

The first studies using CT were carried out in the medical science field, and lately with the success of the technique, it began to be used in other areas of knowledge. In soil science, CT has been used in a broad range of studies allowing measurements of soil porosity (ϕ), bulk density (ρ_b), and pore size distribution with a good resolution, without interfering with the physical integrity of the sample (BALOGUN; CRUVINEL, 2003; GANTZER; ANDERSON, 2002; PIERRET et al., 2002).

The environmental impact of agricultural machinery traffic on soils presents as main consequences changes in water infiltration and retention due to soil structure modification from compaction (HORN et al., 2003; RADFORD et al., 2000). Soil compaction results in increases in ρ_b and reductions in ϕ, which occurs when the soil particles are rearranged (forced close together) after the agricultural machinery traffic. Therefore, modifications caused by compaction in ρ_b, which is represented by the mass of dry soil by its bulk volume, has important consequences in the physical quality of a soil such as changes in its pore size distribution, root system distribution, water and nutrient transport, and heat transfer.

Important parameters that have been used to quantify soil structural damages are its porosity and bulk density, and non-destructive inspection techniques, *e.g.* CT, allow measurements of these physical properties point by point on a millimetric or micrometric scale. As soil quality is directly related to good structure, it becomes very important to characterize the impact of agricultural machinery traffic on its pore system. However, literature concerning the effect of tractor traffic on the structure of tropical soils is scarce.

The objective of this study was to investigate the sensibility of the gamma-ray CT technique as a method for assessing the structural deformations to soils compacted by agricultural machinery traffic and to characterize the soil bulk density and porosity modifications with millimetric resolution.

Material and methods

Experimental area and sampling

The study area is located in the Jaguariúna region in São Paulo State (SE Brazil - 22°25' S; 46°35' W; 580 m a.s.l.), which presents average annual rainfall, relative humidity, and air temperature of 1,335 mm per year, 76%, and 21.7°C, respectively. The dry season covers June to August, July being the driest month. During spring-summer, October to March, very high intensity rainfall events are common.

Samples of a clayey soil characterized as Rhodic Ferralsol (450 g kg^{-1} clay, 25.5 g kg^{-1} organic matter, 1.05 g cm^{-3} dry bulk density, 2.65 g cm^{-3} particle density, 5.8 pH, 73.5 mmol dm^{-3} CTC), according to FAO classification (FAO, 1998), were collected in a corn (*Zea mays*) field cultivated every year and in a native mixed forest. Eighteen samples were collected from the soil surface layer (0-10 cm) with steel cylinders (h = 3.0 cm, D = 4.8 cm, V = 55 cm^3). Details regarding sampling procedures can be found in Pires et al. (2006).

The vehicle utilized for soil preparation was a tractor with rubber tracks (MF 275F) of 2,763 kg of mass with ballast and 56 kW engine power (Figure 1). During management, the experimental area was submitted to none and two-wheel tractor passages on the same track. Twelve samples were collected at the corn field in a randomized block design in areas with none (0P) and two passages of tractor wheels (2P) and six others in a native forest with no traffic and soil management.

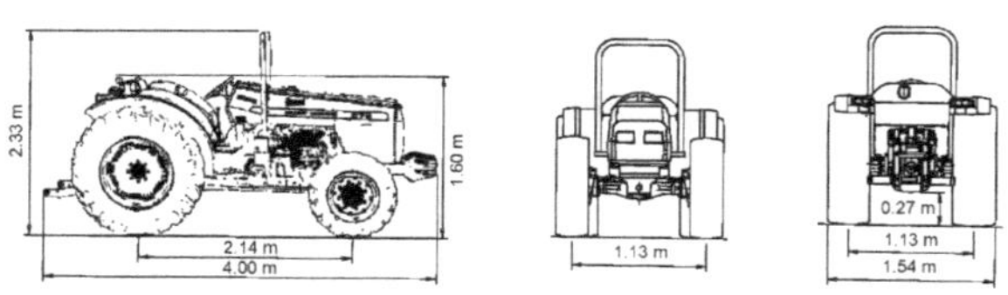

Figure 1. Drawing of the tractor (MF 275F) used for soil management.

CT scanner

The CT scanner used was a first-generation system with a fixed source-detector arrangement and translation/rotational movements of samples (Figure 2). Gamma-rays (661.6 keV) were emitted by a 7.4 GBq ^{137}Cs radioactive source. Circular lead collimators (1 mm) were adjusted between the source and the detector. A crystal scintillator detector of NaI(Tl) (7.62 x 7.62 cm) coupled to a photomultiplier tube was used to detect the gamma-monoenergetic photons. Samples were rotated over 180° at 2.25° steps; the linear movement steps were 0.14 cm. The voxel size was 1.14 x 1.14 x 1.00 mm^3 (as calculated from the ratio of the diameter of the soil sample to the number of pixels of the reconstruction matrix and the diameter of the gamma-ray beam). The acquired data were stored on a PC and the reconstruction algorithm Microvis (2000), developed by Embrapa Agricultural Instrumentation (CNPDIA – São Carlos, Brazil), was used to obtain the CT images. The calibration of the CT scanner was obtained through the correlation between linear attenuation coefficients μ (cm^{-1}) of different homogeneous materials (water, nylon, acrylic, alcohol, aluminum and brass) and their respective tomographic units (TUs) obtained by the image reconstruction program.

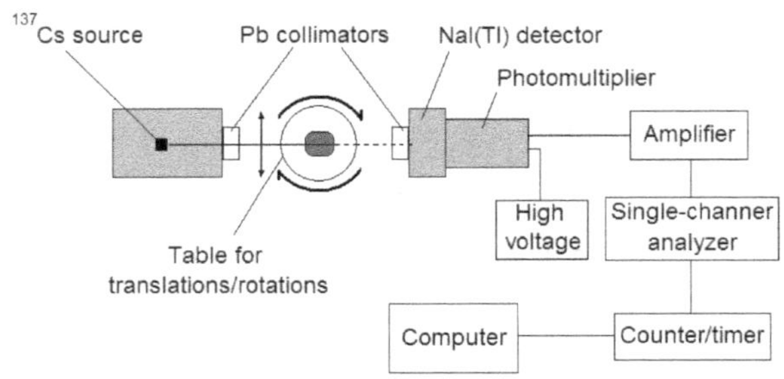

Figure 2. Schematic diagram of the first generation gamma-ray computed tomography (CT) scanner.

Image processing and data analysis

CT analyses were performed on all soil samples of the forest region and submitted to none and two-wheel tractor passages. The tomographic images of the soil samples were taken in the vertical planes that included the axes of the cylinders. 2-D maps of TUs allowed a continuous analysis of the soil bulk density variation with depth and its porosity with width.

The ρ_b (g cm^{-3}) by CT was calculated by substituting α (inclination) of the calibration curve and mass attenuation coefficients μ_m ($cm^2\,g^{-1}$) of water (w) and soil (s), respectively, in the Beer-Lambert equation. The following equation was derived to obtain ρ_b:

$$\rho_b = \frac{\left(TU/\alpha\right) - \mu_{mw}.\theta_r.\rho_w}{\mu_{ms}} \tag{1}$$

where: θ_r ($cm^3\ cm^{-3}$) is the residual soil water content and ρ_w ($g\ cm^{-3}$) is the water density. Samples had been air-dried in order to minimize the effect of water inside the soil on the ρ_b measurements by CT.

The measurement of the soil sample porosity by image (ϕ_i) was made using the following equation:

$$\phi_i\ (\%) = \left(1 - \frac{\rho_b}{\rho_p}\right).100 \qquad (2)$$

where: ρ_p ($g\ cm^{-3}$) represents the soil particle density. Details about the method used to measure ρ_p can be found in Flint and Flint (2002).

The increment ratio of density (Γ_n) was utilized as a compaction criterion (MARSILI et al., 1998) and it was calculated according to:

$$\Gamma_n = \frac{(\lambda_n - \lambda_0)}{\lambda_0} = \left(\frac{\lambda_n}{\lambda_0}\right) - 1 \qquad (3)$$

where: λ_0 is the initial ρ_b measured for the forest region and λ_n is the density after none and two-wheel tractor passages.

SAS software (SAS, 1996) was employed for data processing. An analysis of variance (ANOVA) and Duncan test ($p < 0.05$) were performed to measure statistical differences and to discriminate means.

Results and discussion

A good linear correlation between the tomographic units and the linear attenuation coefficients of different materials (Figure 3A) was obtained, which is essential to obtain quality CT images (CRESTANA et al., 1985). Soil and water mass attenuation coefficients, for the 661.6 keV photons, were 0.0836 ± 0.0007 and 0.0823 ± 0.0008 $cm^2\ g^{-1}$, respectively. TU increases linearly with dry ρ_b (Figure 3B), showing a linear dependence for different soil compaction states and bulk density (VAZ et al., 1989).

From the 2-D images taken through the gamma-ray CT, it is possible to observe the changes in soil structure induced by the machinery traffic (Figure 4). Figure 4A presents the soil sample not submitted to passages of tractor wheels, showing the existence of several areas of lower density (white areas) and compacted regions (black areas) next to the edges, which represent the photon attenuation by the steel cylinder. However, a very compacted portion can be observed near the lower sample surface.

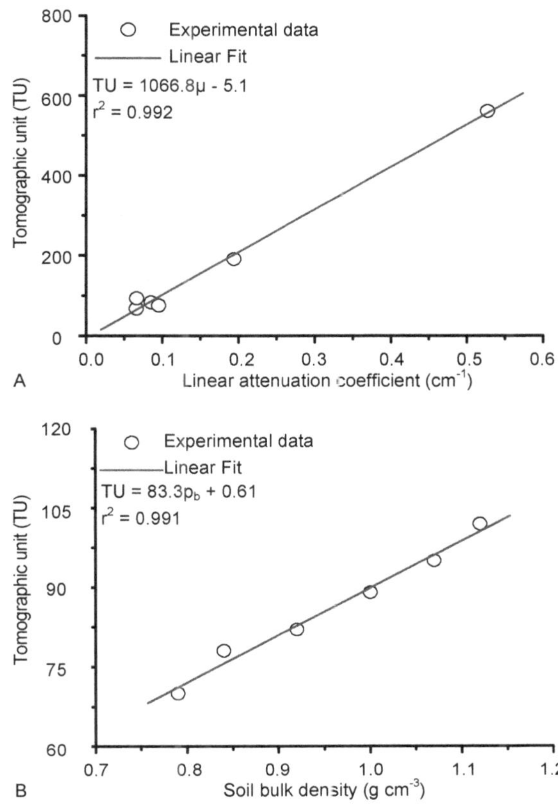

Figure 3. (A) Calibration curve between the tomographic units (TUs) and linear attenuation coefficients (μ) of different materials for ^{137}Cs gamma-photons and (B) TUs versus soil bulk density (ρ_b) of samples with different compaction states.

In Figure 4B, one can see a great increase in regions of higher ρ_b inside the soil sample, which represents a direct result of the two-wheel (2P) tractor passages. As these effects appear to cover almost the entire sample, it is likely that this tropical clayey soil is very susceptible to great changes in its soil pore system after agricultural machinery traffic. According to Pagliai et al. (2003), important damages can be induced in soil structure after passages of rubber tracked tractors, due mainly to the reduction of elongated pores following the same trend of total porosity.

However, it is not possible to analyze changes in the shape of soil pores through images of millimetric resolution, as obtained in this study. This type of analysis can only be made by using CT systems with micrometric resolution (VAN GEET et al., 2003).

Soil bulk density variations before (0P) and after (2P) tractor wheel traffic are shown in Figure 5. It is possible to observe up to about 10 mm in depth a more dense area for the sample 0P than 2P, which can be explained by the presence of surface crusting for 0P (PIRES et al., 2007). Samples not submitted to tractor wheel traffic were collected exactly at the soil surface and it was not necessary to trim off the top surface.

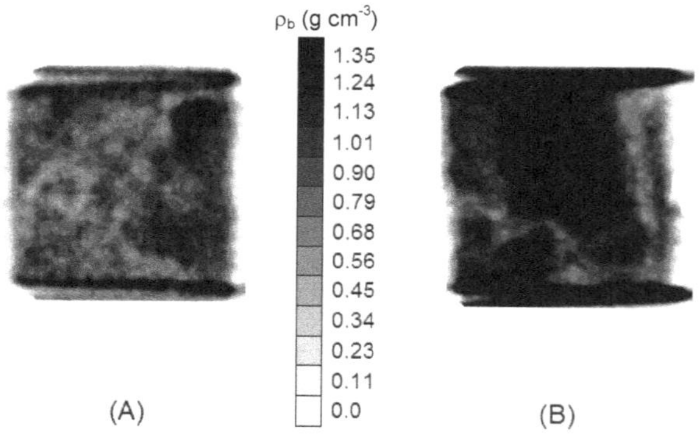

Figure 4. 2-D tomographic images of core samples used for evaluating the effect of passages of tractor wheels on the structure of a Brazilian clayey soil. (A) Sample not submitted to passages of tractor wheels, and (B) sample submitted to two-wheel tractor passages. The scale represents the soil bulk density (ρ_b) distribution.

However, those submitted to machinery traffic were collected in a 3 to 7 cm deep layer, being the excessive soil carefully trimmed off and top and bottom surfaces of samples made flat. Below 10 mm in depth, a significant increase in ρ_b can be observed after 2P when compared to 0P (Table 1). The ρ_b increment ratio for 0P and 2P in relation to samples collected in the native forest shows a difference of 14% between these two treatments. The increment ratio of density (Γ_n) parameters obtained in this study are higher than those measured by Marsili et al. (1998) also comparing 0P and 2P (Γ_n=0.13).

Table 1. Mean values of soil bulk density (ρ_b), standard deviations ($\sigma\rho_b$), coefficient of variation (CV), and its increment ratio (Γ_n) for soil submitted to 0 and 2 passages of tractor wheels.

Treatments	ρ_b (g cm^{-3})	$\sigma\rho_b$ (g cm^{-3})	CV (%)	Γ_n
0 passes (0P)	1.00 b	0.06	6.3	0.08
2 passes (2P)	1.13 a	0.07	6.2	0.22
Forest (F)	0.92 b	0.03	2.9	-

Standard deviation represents the scatter of six ρ_b values for each treatment; Letters represent subsets of mean ρ_b values that are statistically identical according to the Duncan test ($p < 0.05$).

Through the analysis of ρ_b profiles and Γ_n, it is possible to obtain a better comprehension of the soil structure damages from traffic and to measure density values for layers of 1.25 mm, which cannot be provided by traditional techniques of analysis in a non-destructive way.

The increase in ρ_b after passages of tractor wheels can be explained by macropore decreases (MARSILI et al., 1998; PAGLIAI et al., 2003), once soil volume is reduced during soil surface load application by heavy vehicle traffic (Figure 6). This compaction effect can be worsened after several passages of tractor wheels (HORN et al., 2003; PAGLIAI et al., 1998). Richards et al. (2001) observed that the structure of compacted ploughed layers is massive with very few visible pores, whereas the uncompacted ploughed layers has fragmented structure with distinct aggregates and clods. A normal distribution of ϕ_i evaluated by image analysis can be observed before tractor wheel traffic, with the highest concentration of porosity values ranging from 62 to 68%. After two-wheel tractor passages, the ϕ_i distribution changes greatly with a significant reduction in the number of high porosity values found in the matrix of TU, which can be probably explained by the decrease in the amount of large pores inside the samples. The highest concentration of ϕ_i values for 2P ranged from 44 to 50%. Horn et al. (2003) showed that the pore size distribution after machinery traffic is changed to smaller diameters with a reduction of inter-aggregate pores, while single aggregates remain stable.

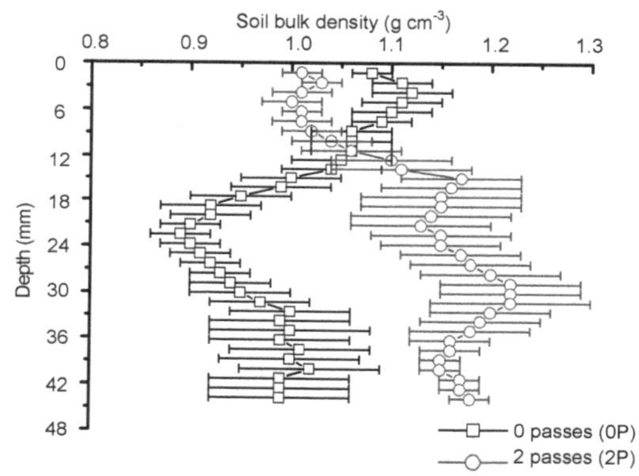

Figure 5. Soil bulk density (ρ_b) variations along depth for core samples not submitted and submitted to two-wheel tractor passages. Errors bars are 95% confidence intervals.

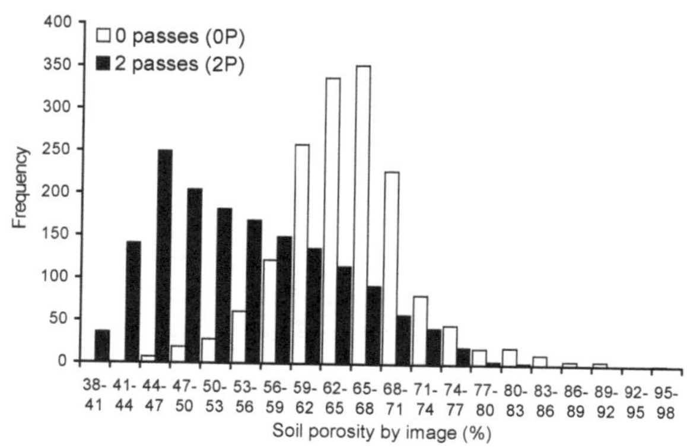

Figure 6. Changes in the soil porosity by image (ϕ_i) for core samples not submitted and submitted to two-wheel tractor passages. The frequency represents the number of points added in the matrix of soil bulk density (ρ_b) used to reconstruct the 2-D tomographic images.

Table 2 presents data of soil porosity by image variation between samples before (0P) and after (2P) tractor's wheel traffic. These ϕ_i were obtained by using CT data (Equation 2) and represent values of TU for one selected line of the matrix of TU data. The results show that ϕ_i suffers important changes (decreases) after machinery traffic confirming the qualitative analysis of soil structure damages observed in Figure 4. These damages are mainly related to decreases in soil macropores, which in clayey soils affect the hydraulic conductivity and root development that are closely associated with the macropores configuration (PAGLIAI et al., 2003; PERRET et al., 1999).

Table 2. Mean values of soil porosity by image (ϕ_i) obtained by the computed tomography (CT) method and the relative error (%RE) between values of ϕ_i for samples submitted to 0 and 2 passages of tractor wheels.

Layer*	1	2	3	4	5	6	7	8	9	10
0 passes (0P)	61.4	61.0	60.2	64.9	64.3	61.9	66.4	65.7	59.9	55.9
2 passes (2P)	50.2	55.0	53.2	42.0	45.6	48.2	44.3	50.4	63.2	60.7
%RE **	18.2	9.8	11.7	35.2	29.1	22.1	33.3	23.2	5.5	8.6

*Each layer (4.8 mm) represents mean values of soil porosity by image (ϕ_i) obtained by the sum of 4 layers of 1.2 mm each; **The analysis of ϕ_i variation was not performed at the same samples. The relative error was obtained by using the following equation: $\%RE = \left|(\phi_{i0P} - \phi_{i2P})/\phi_{i0P}\right|.100$

Conclusion

Computed tomography (CT) data analysis showed that machinery traffic has negatively affected the structural state of a Brazilian clayey soil caused by an increase of soil bulk density with depth for samples collected at the soil surface. Using CT, it was possible to measure average densities for thin soil layers allowing a better characterization of soil structure modification by machinery traffic. It was also possible by using CT data to obtain a soil porosity distribution by image before and after passages of tractor wheels. The data show that the total soil porosity by image is greatly affected after two passages of tractor wheels by a reduction of the percentage of macropores.

Acknowledgements

The author is thankful to Dr. Osny O. S. Bacchi and Dr. Klaus Reichardt of the Laboratory of Soil Physics of the Center for Nuclear Energy in Agriculture for the discussions regarding the gamma computed tomography method and to the Conselho Nacional de Desenvolvimento Científico e Tecnológico (CNPq) for PQ grant.

References

BALOGUN, F. A.; CRUVINEL, P. E. Compton scattering tomography in soil compaction study. **Nuclear Instruments and Methods in Physics Research A**, v. 505, n. 1-2, p. 502-507, 2003.

BEUTLER, A. N.; CENTURION, J. F.; FREDDI, O. S.; ANDRIOLI, I. Efeito da compactação do solo na estabilidade de agregados e no conteúdo gravimétrico de água. **Acta Scientiarum. Agronomy**, v. 27, n. 2, p. 193-198, 2005.

CRESTANA, S.; MASCARENHAS, S.; POZZI-MUCELLI, R. S. Static and dynamic three-dimensional studies of water in soil using computed tomographic scanning. **Soil Science**, v. 140, n. 5, p. 326-332, 1985.

FAO-Food and Agriculture Organization. **World reference base for soil resources**. Rome, 1998.

FLINT, A. L.; FLINT, L. E. The solid phase: Particle density. In: DANE, J. H.; TOPP, G. C. (Ed.). **Methods of soil analysis**: part 4. Physical methods. Madison: Soil Science Society of America Book Series, 2002. p. 229-240.

GANTZER, C. J.; ANDERSON, S. H. Computed tomographic measurement of macroporosity in chisel-disk and no-tillage seedbeds. **Soil and Tillage Research**, v. 64, n. 1-2, p. 101-111, 2002.

HORN, R.; WAY, T.; ROSTEK, J. Effect of repeated tractor wheeling on stress/strain properties and consequences on physical properties in structured arable soils. **Soil and Tillage Research**, v. 73, n. 1-2, p. 101-106, 2003.

KAK, A. C.; SLANEY, M. **Principles of computerized tomographic imaging**. New York: IEEE Press, 1988.

MARSILI, A.; SERVADIO, P.; PAGLIAI, M.; VIGNOZZI, N. Changes of some physical properties of a clay soil following passage of rubber- and metal-tracked tractors. **Soil and Tillage Research**, v. 49, n. 3, p. 185-199, 1998.

MICROVIS. **Programa de reconstrução e visualização de imagens tomográficas**. São Carlos: Embrapa-CNPDIA, 2000. (Guia do usuário).

PAGLIAI, M.; MARSILI, A.; SERVADIO, P.; VIGNOZZI, N.; PELLEGRINI, S. Changes in some physical properties of a clay soil in central Italy following the passage of rubber tracked and wheeled tractors of medium power. **Soil and Tillage Research**, v. 73, n. 1-2, p. 119-129, 2003.

PAGLIAI, M.; ROUSSEVA, S.; VIGNOZZI, N.; PIOVANELLI, C.; PELLEGRINI, S.; MICLAUS, N. Tillage impact on soil quality. I. Soil porosity and related physical properties. **Italian Journal of Agronomy**, v. 2, n. 1, p. 11-20, 1998.

PERRET, J.; PRASHER, S. O.; KANTZAS, A.; LANGFORD, C. Three-dimensional quantification of macropore networks in undisturbed soil cores. **Soil Science Society of America Journal**, v. 63, n. 6, p. 1530-1543, 1999.

PIERRET, A.; CAPOWIEZ, Y.; BELZUNCES, L.; MORAN, C. J. 3D reconstruction and quantification of macropores using X-ray computed tomography and image analysis. **Geoderma**, v. 106, n. 3-4, p. 247-271, 2002.

PIRES, L. F.; BACCHI, O. O. S.; DIAS, N. M. P. Gamma-ray beam attenuation to assess the influence of soil texture on structure deformation. **Nukleonika**, v. 51, n. 2, p. 125-129, 2006.

PIRES, L. F.; BACCHI, O. O. S.; REICHARDT, K.; DIAS, N. M. P. Gamma-ray computed tomography as a tool to evaluate porosity changes along depth for surface crusted soils. **Nukleonika**, v. 52, n. 3, p. 125-131, 2007.

RADFORD, B. J.; BRIDGE, B. J.; DAVIS, R. J. Changes in the properties of a Vertisol and responses of wheat after compaction with harvester traffic. **Soil and Tillage Research**, v. 54, n. 3-4, p. 155-170, 2000.

RICHARDS, G.; COUSIN, I.; SILLON, J. F.; BRUAND, A.; GUÉRIF, J. Effect of compaction on the porosity of a silty soil: influence on unsaturated hydraulic properties. **European Journal of Soil Science**, v. 52, n. 1, p. 49-58, 2001.

SAS-Statistical Analisys System. **System for information**. Cary: SAS Institute Inc., 1996.

VAN GEET, M.; LAGROU, D.; SWENNEN, R. Porosity measurements of sedimentary rocks by means of microfocus X-ray computed tomography (μCT). In: MEES, F.; SWENNEN, R.; VAN GEET, M.; JACOBS, P. (Ed.). **Applications of X-ray computed tomography in the geosciences**. London: Geological Society, 2003. p. 51-60.

VAZ, C. M. P.; CRESTANA, S.; MASCARENHAS, S.; CRUVINEL, P. E.; REICHARDT, K.; STOLF, R. Computed tomography miniscanner for studying tillage induced soil compaction. **Soil Technology**, v. 2, n. 3, p. 313-321, 1989.

24

Moisture content effect in the relationship between apparent electrical conductivity and soil attributes

Marcelo Marques Costa[1*], Daniel Marçal de Queiroz[1], Francisco de Assis de Carvalho Pinto[1], Elton Fialho dos Reis[2] and Nerilson Terra Santos[3]

*[1]Departamento de Engenharia Agrícola, Universidade Federal de Viçosa, Av. P. H. Rolfs, s/n, 36570-000, Viçosa, Minas Gerais, Brazil. [2]Departamento de Engenharia Agrícola, Universidade Estadual de Goiás, Anápolis, Goiás, Brazil. [3]Departamento de Estatística, Universidade Federal de Viçosa, Viçosa, Minas Gerais, Brazil. *Author for correspondence. E-mail: marcelo.costa@ufv.br*

ABSTRACT. To map the spatial variability of a field to define the variable rate application, an intensive sampling of the soil-plant system is necessary. The apparent soil electrical conductivity (ECa) has been used for soil mapping because it correlates well with soil attributes, allows for dense sampling and can be obtained at low cost. However, ECa is influenced by soil moisture content, and the variability of this attribute can reduce the reliability of the ECa maps to explain the physical and chemical soil attributes. The objective of this study was to identify conditions that maximize the correlations between the ECa and the soil attributes. The results show that the mean soil moisture content of soil sampled on different dates was correlated with the mean of the ECa. The ideal time for measuring ECa occurred when the mean moisture content of the soil was higher. In this condition, the coefficient of variation for the soil moisture content was lower, there was no correlation between ECa and soil moisture content, and ECa was more correlated with other soil attributes evaluated in this work.

Keywords: precision agriculture, temporal variability, soil moisture content.

Introduction

Precision farming involves the management of crops in consideration of spatial variability. This spatial variability is generally represented by maps of variables associated with the soil-plant system. Generally, to obtain these maps, a large number of samples are necessary to accurately represent the variability. The cost of soil analysis limits denser sampling, which may result in less accurate maps. Therefore, researchers have been investigating indirect methods that have lower costs and present high correlation with variables of agronomic interest. For this purpose, the apparent soil electrical conductivity (ECa) has been widely accepted in precision agriculture. ECa is a geophysical method used to characterize the physical and chemical soil variability, and it has been used to delimit management zones (HOSSAIN et al., 2010; LÜCK et al., 2009).

In non-saline soils, ECa results from the interaction between attributes that determine their ability to conduct electricity. These attributes are generally correlated with the crop yield. Thus, soil ECa represents a tool with great potential for precision agriculture (CORWIN et al., 2006, SHANE et al., 2008). However, according to LÜCK et al. (2009), the spatial variability of the soil moisture content makes the measurement of ECa unreliable for predicting the variability of soil physical and chemical attributes.

Ekwue and Bartholomew (2010) observed in their experiments a strong positive correlation between the ECa and the soil moisture content for three soil types evaluated in laboratory and field conditions. In the midwest of the United States, Brevik et al. (2006) monitored the ECa and volumetric moisture content of soil from June 1999 to October 2000 and concluded that ECa is strongly influenced by soil moisture content. The same authors stated that the soil moisture content should be considered in the investigation of ECa and recorded in the reports generated. This approach suggests that the spatial variability of the soil moisture content at different sampling leads to a variability in the ECa maps that is not due to soil properties, but to the variation of moisture water, which can result in a low repeatability of the measurements. Thus, the comparison of maps of ECa collected under different conditions of moisture content may not provide reliable information about the variability of soil physical and chemical attributes.

The soil moisture content varies in time and space, and this variability may be the main cause of the ECa variability. Therefore, if at ECa data acquisition the soil moisture content presents a large spatial variability, the obtained ECa values may have a low correlation with the soil properties. According to Lück et al. (2009), the soil moisture content affects both the level and the range of the variation of ECa, and to generate reliable maps, the authors suggest the use of a system of multiple sensors that can adjust the ECa measured with different levels of soil moisture.

Rhoades et al. (1989) suggested that ECa should be measured when the soil has a moisture content close to its field capacity. In this situation, the ECa value is approximate to the soil electrical conductivity of the saturation extract. The authors state that in this condition, the soil moisture content tends to be more spatially homogeneous. However, it is difficult to know whether the soil moisture content is sufficiently close to the field capacity at the time of sampling. Another problem is that corn and soybean planting in Brazil starts as soon as there is enough water content in the soil for seed germination, which is especially true in regions where the rainy season allows two crops per year and farmers need to start planting just after the ideal soil moisture content is reached. After planting, the ECa measurement equipment, which is pulled by a tractor, has restricted access due to the mechanical damage caused to plants. During crop development, the ECa can be measured with portable devices by using scouting techniques.

The hypothesis of this work is that the soil moisture content is a determining factor for ECa; however, when the mean soil moisture content is high, its variability in the field is smaller. Thus, the ECa values, measured when the soil moisture content is higher, should have a low correlation with the soil moisture content and a higher correlation with the soil attributes that define the total amount of available charges in the soil. The objective of this study is to identify the conditions in which an ECa sample is more correlated with soil physical and chemical attributes.

Material and methods

The study was performed in a 75 ha field at a farm located in Goianapolis, Goias State, midwest region of Brazil. The study area can be viewed at the coordinates 16° 27' 48" S, 48° 59' 38" W. According to the Köppen classification, the climate is tropical and is designated as Cwa. The altitude is 1017 m above sea level, and the mean annual rainfall is 1400 mm. The soil in the study area is classified as 'Latossolo Vermelho Amarelo' according to the Brazilian soil classification system (EMBRAPA, 2006). In the 2009/2010 crop year, the field was planted with corn, and in the 2010/2011 crop year, the field was planted with soybeans.

Data acquisition

The limits of the field were obtained using a Magellan Promark III DGPS. The differential correction was post-processed using a database of the Brazilian Institute of Geography and Statistics (IBGE), located in Brasilia. With the aid of GIS, a regular grid of points was created within the field for ECa measurement as well as for collecting samples for soil moisture content determination and for measuring soil physical and chemical attributes, as shown in Figure 1. The mesh were consisted of two points per hectare, resulting in a total of 150 points for collecting samples for determination of soil physical and chemical attribute, at the onset of the rainy season . Later, the ECa data and moisture content soil were acquired on five different days: November 11th, 2010; November 23rd, 2010; December 4th, 2010; December 13th, 2010 and January 26th, 2011. For each sampling date, the Eca was measured in the same 150 points which was collected soil samples. The moisture content of the soil was measured at the five dates together with the ECa. However, the moisture content soil was measured at only 30 points of 150 points, due to the difficulty of the operation. These 30 points were selected as shown in Figure 1 and the moisture content soil was measured in the same 30 points for five sampling dates.

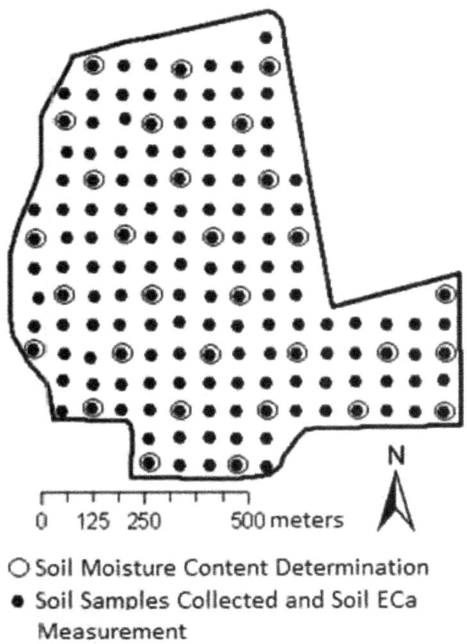

Figure 1. Location of the points for ECa measurement, for collecting samples for moisture content and for physical and chemical soil attribute determination.

A Landviser LandMapper ERM 02 handheld device was used to measure the soil apparent electrical conductivity. This device uses the principle of electrical resistivity (ER), which is measured by four-electrode contact with the soil. The Wenner array configuration of the electrodes, in which the electrodes are arranged in lines and equally spaced, was used as described by Corwin and Lesch (2003). The space between each pair of electrodes was set to 0.20 m; therefore, the ECa measurements correspond to the apparent electrical conductivity of the soil for a depth of 0 to 0.20 m. For each point, five measurements were performed randomly at a distance of up to 5 m from the point. During the measurement of Eca, soil samples were collected in the layer from 0 to 0.20 m and packed in sealed container with lid, for soil moisture determination. The gravimetric moisture of soil was determined at dry basis by the standard method of greenhouse. In this method, the mass of water is obtained from the difference between the mass of soil moist and the mass of soil dried at oven, a temperature of 105°C for 48 hours. The moisture content of the soil is represented by the percentage weight of water relative to the mass of dry soil sample.

The soil samples for chemical and physical analysis were collected using an auger soil sampler for the depth from 0 to 0.20 m. Each sample was composed of 10 sub-samples collected at a distance of up to 5 m from the point. The chemical analysis consisted of the determination of the following: pH, potential acidity (H + Al), Ca^{2+}, Mg^{2+}, K, P, Al^{3+}, remaining P, organic matter content, sum of bases (SB), cation exchange capacity at pH 7 (CEC), base saturation (V) and the levels of Zn, Fe, Mn and Cu. The physical analysis consisted of the determination of clay, silt and sand content.

Data analysis

First, data related to the soil attributes were analyzed using summary statistics, not considering the spatial distribution. The following parameters were calculated: mean, standard deviation, maximum and minimum values and coefficient of variation (CV). This analysis allowed a general evaluation of the soil fertility, characterizing it according to the mean values of its physical, chemical and physico-chemical attributes.

After performing the summary statistics, a Pearson correlation analysis was used to test the hypothesis of this work. For this purpose, was used the SAS (2008) statistical software. The correlation between the mean values of ECa and the mean values of soil moisture content using the data of five sampling dates was found. To check whether the soil moisture content showed a greater variability in the field when the samples were collected under conditions of lower moisture content, a correlation analysis between the mean values of the soil moisture content and their coefficients of variation, including all five dates of collection, was performed.

For each date, a file with 30 data points of soil moisture content and ECa was generated. Therefore, it was possible to perform a correlation analysis between the ECa and the soil moisture content for each sampling date. It was then possible to test the hypothesis that a weaker correlation between the ECa and the soil moisture content occurs when the mean soil moisture content is higher. The hypothesis that ECa is more correlated with soil attributes when it is less correlated with soil moisture content was verified by the correlation between the soil properties and ECa on the different dates.

Results and discussion

The summary statistics of the soil attributes show typical characteristics of soils of the Brazilian cerrado region, with low values of cation exchange capacity at pH 7 (CEC) and low values of organic matter (Table 1). The high value of base saturation (V) is due to application of lime as a corrective for soil acidity.

Table 1. Summary statistics of soil attributes.

Variable	Unit	Mean Value	Standard Deviation	Minimum Value	Maximum Value	Coefficient of Variation (%)
pH [1]		6.74	0.30	5.80	7.30	4.41
P [2]	mg dm^{-3}	6.70	3.63	1.70	18.40	54.12
K [3]	mg dm^{-3}	52.20	14.12	24.00	108.00	27.04
Ca^{2+} [4]	cmolc dm^{-3}	3.24	0.44	1.90	4.20	13.70
Mg^{2+} [5]	cmolc dm^{-3}	0.83	0.12	0.60	1.20	15.12
Al^{3+} [6]	cmolc dm^{-3}	0,00	0,00	0,00	0,00	0,00
H + Al [7]	cmolc dm^{-3}	1.70	0.90	0.33	5.61	52.61
SB [8]	cmolc dm^{-3}	4.20	0.53	2.63	5.24	12.56
CEC [9]	cmolc dm^{-3}	5.90	0.84	4.17	9.89	14.32
V [10]	%	72.13	11.05	37.00	93.00	15.32
OM [11]	dag kg^{-1}	3.04	0.30	2.50	4.30	9.73
P_rem [12]	mg L^{-1}	17.45	3.37	9.50	27.40	19.30
Zn [13]	mg dm^{-3}	3.73	2.21	1.50	19.90	59.28
Cu [14]	mg dm^{-3}	1.33	0.60	0.80	6.70	45.02
Fe [15]	mg dm^{-3}	21.97	5.58	11.00	41.10	25.41
Mn [16]	mg dm^{-3}	26.89	9.05	13.70	66.70	33.67
B [17]	mg dm^{-3}	0.26	0.09	0.10	0.50	34.36
S [18]	mg dm^{-3}	11.38	2.39	6.60	25.60	21.04
Clay content	%	32.97	3.28	26.00	43.00	9.95
Silt content	%	10.64	2.98	6.00	20.00	28.00
Sand content	%	56.38	4.40	45.00	65.00	7.81
ECe1:5 [19]	μS cm^{-1}	72.45	11.54	49.42	106.30	15.93

[1]pH, active acidity in water; [2]P, phosphorous; [3]K, potassium; [4]Ca^{2+}, calcium; [5]Mg^{2+}, magnesium; [6]Al^{3+}, exchangeable acidity; [7]H + Al, potential acidity; [8]SB, exchangeable sum of bases; [9]CEC, cation exchangeable capacity at pH 7; [10]V, basis saturation; [11]OM, organic matter; [12]P-rem, remaining phosphorous; [13]Zn, Zinc; [14]Cu, copper; [15]Fe, iron; [16]Mn, manganese; [17]B, boron; [18]S, sulfur; [19]ECe1:5, electrical conductivity of soil extract at 1:5.

Consequently, the soil presents an mean pH close to neutral. Under these conditions of low acidity, Al is fully neutralized and has zero value. High pH values reduce the availability of some micronutrients, such as Fe, Cu, Mn and Zn. Foliar fertilization is required to meet the plants' need for these micronutrients. It can also be observed that clay and sand contents present low variability in the field, expressed by the low coefficient of variation.

To characterize the conditions of moisture content during the ECa measurements, soil samples were collected for soil moisture content determination. The mean values of ECa and soil moisture content, as well as the coefficient of variation for the moisture content, are presented in Table 2.

Table 2. Mean values of soil moisture content and apparent soil electrical conductivity (ECa) measured on different dates.

Day of Measurement	Mean ECa Value (ms m^{-1})	Soil Moisture Content	
		Mean Value (%)	Coefficient of Variation (%)
November 11th, 2010	4.8	20.5	11.8
November 23rd, 2010	5.9	22.0	9.6
December 4th, 2010	4.6	20.1	12.1
December 13th, 2010	4.2	20.2	12.3
January 26th, 2011	2.6	18.0	17.2

The correlation coefficient between the mean ECa values and the mean soil moisture content was 0.989, and the probability of correlation to be equal to zero was less than 0.001. This result shows that in the study area, higher ECa values were obtained when the mean soil moisture content was higher and that there is a strong correlation between the two attributes. Ekwue and Bartholomew (2010) obtained similar results when evaluating the relationship between the ECa and moisture content in field and laboratory conditions for three soil types.

The correlation coefficient between the mean values of soil moisture content and their coefficient of variation was -0.985. The negative correlation indicates that when the moisture content was higher, its variability was lower in the field. Thus, the condition that resulted in a more homogenous soil moisture content occurred on November 23rd, 2010, which was the day with the highest mean moisture content value. Scherpinski et al. (2010), when analyzing the spatial variability of soil hydraulic conductivity and soil water infiltration, observed a moderate spatial dependence with a range of 96.22 and 206.32 m, respectively. Starting from an initial condition in which the soil moisture content is equivalent to the field capacity, the spatial variability of physical-hydric attributes indicates that as the soil looses water, the variability of the moisture content becomes higher.

Some regions lose water faster than others, causing a greater variability in soil moisture as the soil dries. The correlation coefficients between the soil moisture content and the ECa values are presented in Table 3. The table shows the change in the coefficient of correlation between these attributes and associations with the variability of soil moisture content.

The farmer may are interested in investigating the spatial variability of soil water content. In this situation it is recommended the sampling of the ECa when the soil water content is lower. Some regions may dry faster than others and this would increase the influence of water content on the values of ECa. Thus, the ECa sampling conducted on January 26th, 2011, would be more appropriate because it showed a higher correlation with the soil water content. The highest correlation coefficient between ECa and the soil moisture content was observed on this date (Table 3). This observation is justified because the soil attributes that have the greatest effect on the response of rainfed crops are those that control the soil's ability to retain water and, therefore, its availability; these attributes include soil texture and organic matter, topography and type of vegetation cover (GUASTAFERRO et al., 2010; MORAL et al., 2010).

Table 3. Correlation coefficients between apparent soil electrical conductivity and soil moisture content for each day of measurement.

Date	r	p –value
November 11th, 2010	0.34	0.146
November 23rd, 2010	0.10	0.671
December 4th, 2010	0.43	0.060
December 13th, 2010	0.56	0.010
January 26th, 2011	0.69	0.001

r – correlation coefficient; p-value – probability of accepting H_0 ($r = 0$).

The results presented in Table 3 indicate that the lowest coefficient of correlation between soil moisture content and soil ECa was obtained on November 23rd, 2010, which had the highest mean soil moisture content (Table 2). This result implies that the ECa sampled on that day presented a higher correlation with soil properties because the variability of the moisture content did not significantly influence the variability of the ECa. The coefficients of correlation between the ECa and the physical and chemical attributes of the soil were performed using the 150 data points, and the results are presented in Table 4.

On November 23rd, 2010, the day with the highest mean soil moisture content, there was no significant correlation between soil moisture content and ECa values. Of the 23 soil attributes evaluated, the ECa was significantly correlated with 18 of them. For the other four dates, the highest number of significant correlations between ECa and the soil attributes evaluated was 14. The highest correlation coefficient (r) was 0.53, and this value was observed on November 23rd, 2010, for the correlation between the ECa and the soil pH and on December 13th, 2010, between the ECa and Mg. The second highest value was 0.52, also observed on November 23rd, 2010, for the correlation between the ECa and base saturation. ECa was significantly correlated with the soil electrical conductivity of saturation extract and the total cation exchange capacity only on November 23rd, 2010, which is in agreement with the results presented by Rhoades et al. (1989).

Table 4. Correlation coefficients between apparent soil electrical conductivity and soil attributes for the five days of measurements.

Variable	D1[1] (M=20,5%)	D2[2] (M=22%)	D3[3] (M=20,1%)	D4[4] (M=20.2%)	D5[5] (M=18%)
pH[6]	0.29^{**}	0.53^{**}	0.39^{**}	0.41^{**}	0.45^{**}
P[7]	-0.05^{ns}	-0.06^{ns}	-0.19^{*}	-0.15^{ns}	-0.09^{ns}
K[8]	-0.03^{ns}	-0.08^{ns}	0.03^{ns}	0.02^{ns}	0.07^{ns}
Ca^{2+}[9]	0.46^{**}	0.44^{**}	0.35^{**}	0.39^{**}	0.44^{**}
Mg^{2+}[10]	0.49^{**}	0.43^{**}	0.42^{**}	0.53^{**}	0.50^{**}
H+Al[11]	-0.14^{ns}	-0.45^{**}	-0.32^{**}	-0.30^{**}	-0.35^{**}
CEC[12]	0.16^{ns}	-0.19^{*}	-0.09^{ns}	-0.03^{ns}	-0.07^{ns}
V[13]	0.21^{**}	0.52^{**}	0.34^{**}	0.37^{**}	0.41^{**}
OM[14]	0.20^{**}	0.04^{ns}	0.08^{ns}	0.15^{ns}	0.11^{ns}
P_rem[1]	-0.22^{**}	-0.32^{**}	-0.38^{**}	-0.34^{**}	-0.35^{**}
Zn[16]	0.23^{**}	0.23^{**}	0.13^{ns}	0.24^{**}	0.27^{**}
Cu[17]	-0.09^{ns}	0.05^{ns}	0.06^{ns}	0.00^{ns}	0.02^{ns}
Fe[18]	-0.19^{*}	-0.39^{**}	-0.16^{*}	-0.17^{*}	-0.22^{**}
Mn[19]	0.05^{ns}	0.17^{*}	0.22^{**}	0.21^{**}	0.13^{ns}
SB[20]	0.50^{**}	0.47^{**}	0.40^{**}	0.46^{**}	0.49^{**}
B[21]	0.12^{ns}	0.11^{ns}	0.11^{ns}	0.1^{ns}	0.16^{ns}
S[22]	-0.1^{ns}	-0.17^{*}	-0.13^{ns}	-0.14^{ns}	-0.11^{ns}
Clay content	0.18^{*}	0.17^{*}	0.25^{**}	0.24^{**}	0.26^{**}
Silt content	0.11^{ns}	0.27^{**}	0.19^{**}	0.24^{**}	0.21^{**}
Sand content	-0.21^{*}	-0.30^{**}	-0.32^{**}	-0.34^{**}	-0.34^{**}
ECe1:5[23]	-0.14^{ns}	0.29^{**}	0.04^{ns}	0.08^{ns}	0.11^{ns}

[1]D1, measurements obtained on November 11th, 2010, M represents the mean soil moisture content; [2]D2, November 23rd, 2010; [3]D3, December 4th, 2010; [4]D4, December 13th, 2010; [5]D5, January 26th, 2011; [6]pH, active acidity in water; [7]P, Phosphorous; [8]K, Potassium; [9]Ca, Calcium; [10]Mg, Magnesium; [11]H + Al, potential acidity; [12]CEC, cation exchangeable capacity at pH 7; [13]V, base saturation; [14]OM, Organic Matter; [15]P-rem, remaining phosphorous; [16]Zn, Zinc; [17]Cu, Copper; [18]Fe, Iron; [19]Mn, Manganese; [20]SB, exchangeable sum of bases; [21]B, Boron; [22]S, Sulfur; [23]ECe1:5, electrical conductivity of soil extract at 1:5. *Significant at a 5% significance level. **Significant at a 1% significance level. ^{ns}not significant.

The results of this research reinforce the need to measure the soil moisture content when measuring ECa because discriminating the effects of variation in soil moisture content measurements on ECa is not possible with current technologies (LÜCK et al., 2009). Under conditions of higher soil moisture content, a lower correlation coefficient was observed between the soil moisture content and ECa, which indicates that under these conditions, the variability of the ECa is less influenced by the soil moisture content. Rhoades et al. (1989) argued that a greater homogeneity of the soil moisture content occurs when it is close to field capacity and that the ECa measured in these conditions approximates the soil electrical conductivity of the saturation extract, which was also observed in this work. However, the soil matric potential, which defines the moisture content at field capacity, is influenced by particle-sized structures, which means that the soil moisture content at field capacity can vary if the soil texture is variable. Bernardo (1987) reported that water tension at field capacity vary depending on the soil texture, and this can providing variation from 8% to 30% on soil moisture content at field capacity. Thus, in areas where the soil texture has a high variability, the soil water content in its field capacity tends to follow the variability of texture. Thus, it is recommended the area be divided into sub-regions where soil texture is more homogeneous.

When the variability of the soil texture in the field is not high, its correlation with ECa should be lower or even zero, and the soil moisture content at field capacity should be more homogeneous. The low variability of the soil moisture content in these conditions implies a lower participation of the moisture content in the ECa variability, and the map generated in this condition should be more related to the soil chemical properties. In field conditions, it is difficult to know if the moisture content at the time of sampling is high enough. To evaluate the reliability of the ECa map, it is recommended that soil samples be taken for soil moisture determination at the time of ECa measurements at representative points of the area that coincide with the ECa measurements. Subsequently, a correlation analysis between the ECa and soil moisture content should be performed. Strong correlations indicate that the generated map was influenced by soil moisture content, and the map can be different if obtained at different conditions of soil moisture content, because the soil can dry unevenly. This unevenness depends not only on the soil texture but also on the vegetation cover and soil organic matter, among other factors.

In this work, ECa measured with the highest mean soil moisture content was more correlated with the base saturation. The base saturation represents an attribute of great agronomic value as a basis of calculation for recommending corrective measures to neutralize the acidity and the exchangeable Al in the soil. A high correlation between the ECa and base saturation is a good argument for using ECa maps in the delimitation of management zones.

Conclusion

Less variability of the soil moisture content was observed when its mean value was higher.

The correlation between apparent soil electrical conductivity and soil moisture content was lowest when the soil had higher moisture. In this scenario the correlation was higher with soil properties.

The correlation analysis between the apparent soil electrical conductivity and soil moisture content is a good criterion for evaluating whether the measured apparent soil electrical conductivity values represent mainly the variability of the soil moisture content.

Acknowledgements

The authors wish to thank the Coordenação de Aperfeiçoamento de Pessoal de Nível Superior (CAPES), the Fundação de Amparo à Pesquisa do Estado de Minas Gerais, Brazil (FAPEMIG), and the Conselho Nacional de Desenvolvimento Científico e Tecnológico (CNPq) of Brazil for financially supporting this research.

References

BERNARDO, S. **Manual de irrigação**. 4. Ed. Viçosa: UFV, 1987.

BREVIK, E. C.; FENTON, T. E.; LAZARI, A. Soil electrical conductivity as a function of soil water content and implications for soil mapping. **Precision Agriculture**, v. 7, n. 6, p. 393-404, 2006.

CORWIN, D. L.; LESCH, S. M. Application of soil electrical conductivity to precision agriculture: Theory, principles, and guidelines. **Agronomy Journal**, v. 95, n. 3, p. 455-471, 2003.

CORWIN, D. L.; LESCH, S. M.; OSTER, J. D.; KAFFKA, S. R. Monitoring management-induced spatio-temporal changes in soil quality through soil sampling directed by apparent electrical conductivity. **Geoderma**, v. 131, n. 3-4, p. 369-387, 2006.

EKWUE, E. I.; BARTHOLOMEW, J. Electrical conductivity of some soils in Trinidad as affected by density, water and peat content. **Biosystems Engineering**, v. 108, n. 2, p. 95-103, 2010.

EMBRAPA-Empresa Brasileira de Pesquisa Agropecuária. **Sistema brasileiro de classificação de solo**. Rio de Janeiro: Embrapa, 2006.

GUASTAFERRO, F.; CASTRIGNANÒ, A.; BENEDETTO, D.; SOLLITTO, D.; TROCCOLI, A.; CAFARELLI, B. A comparison of different algorithms for the delineation of management zones. **Precision Agriculture**, v. 11, n. 6, p. 600-620, 2010.

HOSSAIN, M. B.; LAMB, D. W.; LOCKWOOD, P. V.; FRAZIER, P. EM38 for volumetric soil water content estimation in the root-zone of deep vertosol soils. **Computers and Electronics in Agriculture**, v. 74, n. 1, p. 100-109, 2010.

LÜCK, E.; GEBBERS, R.; RUEHLMANN, J.; SPANGENBERG, U. Electrical conductivity mapping for precision farming. **Near Surface Geophysics**, v. 7, n. 1, p. 15-25, 2009.

MORAL, F. J.; TERRÓN, J. M.; SILVA, J. R. M. D. Delineation of management zones using mobile measurements of soil apparent electrical conductivity and multivariate geostatistical techniques. **Soil and Tillage Research**, v. 106, n. 2, p. 335-343, 2010.

RHOADES, J. D.; MANTEGHI, N. A.; SHOUSE, P. J.; ALVES, W. J. Soil electrical conductivity and soil salinity: new formulations and calibrations. **Soil Science Society of America Journal**, v. 53, n. 2, p. 433-439, 1989.

SAS-Statistical Analysis System. **User's guide**. Version 9.2. Cary: SAS Institute Inc., 2008.

SCHERPINSKI, C.; URIBE-OPAZO, M. A.; BOAS, M. A. V.; SAMPAIO, S. C.; JOHANN, J. A. Variabilidade espacial da condutividade hidráulica e da infiltração da água no solo. **Acta Scientiarum. Agronomy**, v. 32, n. 1, p. 7-13, 2010.

SHANER, D. L.; KHOSLA, R.; BRODAHL, M. K.; BUCHLEITER, G. W.; FARAHANI, H. J. How well does zone sampling based on soil electrical conductivity maps represent soil variability?. **Agronomy Journal**, v. 100, n. 5, p. 1472-1480, 2008.

PERMISSIONS

All chapters in this book were first published in ASA, by Editora da Universidade Estadual de Maringá; hereby published with permission under the Creative Commons Attribution License or equivalent. Every chapter published in this book has been scrutinized by our experts. Their significance has been extensively debated. The topics covered herein carry significant findings which will fuel the growth of the discipline. They may even be implemented as practical applications or may be referred to as a beginning point for another development.

The contributors of this book come from diverse backgrounds, making this book a truly international effort. This book will bring forth new frontiers with its revolutionizing research information and detailed analysis of the nascent developments around the world.

We would like to thank all the contributing authors for lending their expertise to make the book truly unique. They have played a crucial role in the development of this book. Without their invaluable contributions this book wouldn't have been possible. They have made vital efforts to compile up to date information on the varied aspects of this subject to make this book a valuable addition to the collection of many professionals and students.

This book was conceptualized with the vision of imparting up-to-date information and advanced data in this field. To ensure the same, a matchless editorial board was set up. Every individual on the board went through rigorous rounds of assessment to prove their worth. After which they invested a large part of their time researching and compiling the most relevant data for our readers.

The editorial board has been involved in producing this book since its inception. They have spent rigorous hours researching and exploring the diverse topics which have resulted in the successful publishing of this book. They have passed on their knowledge of decades through this book. To expedite this challenging task, the publisher supported the team at every step. A small team of assistant editors was also appointed to further simplify the editing procedure and attain best results for the readers.

Apart from the editorial board, the designing team has also invested a significant amount of their time in understanding the subject and creating the most relevant covers. They scrutinized every image to scout for the most suitable representation of the subject and create an appropriate cover for the book.

The publishing team has been an ardent support to the editorial, designing and production team. Their endless efforts to recruit the best for this project, has resulted in the accomplishment of this book. They are a veteran in the field of academics and their pool of knowledge is as vast as their experience in printing. Their expertise and guidance has proved useful at every step. Their uncompromising quality standards have made this book an exceptional effort. Their encouragement from time to time has been an inspiration for everyone.

The publisher and the editorial board hope that this book will prove to be a valuable piece of knowledge for researchers, students, practitioners and scholars across the globe.

LIST OF CONTRIBUTORS

Robson Gonçalves Trentin
Departamento de Engenharia Mecânica, Universidade Tecnológica Federal do Paraná, Via do Conhecimento, km 01, 85501-970, Pato Branco, Paraná, Brasil

Alcir José Modolo, Thiago de Oliveira Vargas and José Ricardo da Rocha Campos
Departamento de Agronomia, Universidade Tecnológica Federal do Paraná, Pato Branco, Paraná, Brasil

Paulo Fernando Adami
Departamento de Agronomia, Universidade Tecnológica Federal do Paraná, Dois Vizinhos, Paraná, Brasil

Murilo Mesquita Baesso
Faculdade de Zootecnia e Engenharia de Alimentos, Universidade de São Paulo, Pirassununga, São Paulo, Brasil

Paulo Cesar Corrêa, Fernando Mendes Botelho, Gabriel Henrique Horta Oliveira, André Luis Duarte Goneli, Osvaldo Resende and Sílvia de Carvalho Campos
Departamento de Engenharia Agrícola, Centro de Ciências Agrárias, Universidade Federal de Viçosa, Cx. Postal 270, 36571-000, Viçosa, Minas Gerais, Brazil

Victor Brunini Moreto and Glauco de Souza Rolim
Faculdade de Ciências Agrárias e Veterinárias, Universidade Estadual "Julio de Mesquita Filho", Via de Acesso Prof.Paulo Donato Castellane s/n.°, 14884-900, Jaboticabal, São Paulo, Brazil

Fernanda Machado Baptestini and Paulo Cesar Corrêa
Universidade Federal de Viçosa, Departamento de Engenharia Agrícola e Ambiental, Avenida Peter Henry Rolfs, s/n, Campus Universitário, Viçosa, Minas Gerais, Brazil

Gabriel Henrique Horta de Oliveira
Instituto Federal de Educação, Ciência e Tecnologia Sudeste de Minas Gerais, Campus de Manhuaçu, Manhuaçu, Minas Gerais, Brazil

Paulo Roberto Cecon
Universidade Federal de Viçosa, Departamento de Estatística, Viçosa, Minas Gerais, Brazil

Nilda de Fátima Ferreira Soares
Universidade Federal de Viçosa, Departamento de Tecnologia de Alimentos.Viçosa, Minas Gerais, Brazil

Leonardo Amaral Monteiro
Programa de Pós-graduação em Engenharia de Sistemas Agrícolas, Escola Superior de Agricultura "Luiz de Queiroz", Universidade de São Paulo, Av. Pádua Dias, 11, 13418-900, Piracicaba, São Paulo, Brazil

Paulo Cesar Sentelhas
Departamento de Engenharia de Biossistemas, Escola Superior de Agricultura "Luiz de Queiroz", Universidade de São Paulo, Piracicaba, São Paulo, Brazil

Paulo Carteri Coradi
Universidade Federal de Santa Maria, Campus Cachoeira do Sul, Rua Ernesto Barros, 1345, Santo Antônio, 96506-322, Cachoeira do Sul, Rio Grande do Sul, Brazil

Ângelo Francisco Calegare Lemes
Universidade Federal de Mato Grosso do Sul, Campus Chapadão do Sul,Chapadão do Sul, Mato Grosso do Sul, Brazil

Narendra Veranagouda Ganganagowdar
Departamento de Engenharia da Computação, Instituto de Tecnologia de Manipal, Universidade de Manipal, Manipal, Karnataka, India-576104

Hareesha Katiganere Siddaramappa
Departamento de Computação Aplicada, Instituto de Tecnologia de Manipal, Universidade de Manipal, Manipal, Karnataka, India 576104

Zenil Ricardo Cunha Rodrigues de Oliveira, Domingos Sarvio Magalhães Valente, Francisco de Assis de Carvalho Pinto and Nara Silveira Velloso
Departamento de Engenharia Agrícola, Universidade Federal de Viçosa, Av. P.H. Rolfs, s/n, Campus Universitário, 36570-900, Viçosa, Minas Gerais, Brazil

Fábio Lúcio Santos
Departamento de Engenharia, Universidade Federal de Lavras, Lavras, Minas Gerais, Brazil

Kelly Aparecida de Sousa, Osvaldo Resende, Thaís Adriana de Souza Smaniotto and Daniel Emanuel Cabral de Oliveira
Instituto Federal de Educação, Ciência e Tecnologia Goiano, Câmpus Rio Verde, Rodovia Sul Goiana, Km 01, Zona Rural, 75901-970, Rio Verde, Goiás, Brazil

André Luis Duarte Goneli
Departamento de Engenharia Agrícola, Faculdade de Ciências Agrárias, Universidade Federal da Grande Dourados, Dourados, Mato Grosso do Sul, Brazil

Valdiney Cambuy Siqueira
Programa de Pós-graduação em Engenharia Agrícola, Departamento de Engenharia, Universidade Federal de Lavras, Cx. postal 3037, 37200-000, Lavras, Minas Gerais, Brazil

Osvaldo Resende and Tarcísio Honório Chaves
Instituto Federal de Educação, Ciência e Tecnologia Goiano, Rio Verde, Goiás, Brazil

Josiane Rosa Silva de Oliveira, Fernando Falco Pruski and José Márcio Alves da Silva
Departamento de Engenharia Agrícola, Universidade Federal de Viçosa, Av. P.H. Rolfs, s/n, 36570-000, Viçosa, Minas Gerais, Brazil

Danilo Paulúcio da Silva
Departamento de Estudos Básicos e Instrumentais, Universidade Estadual do Sudoeste da Bahia, Rod. BR-415, Km 3, s/n, 45700-000, Itapetinga, Bahia, Brazil

Daiane de Cinque Mariano
Departamento de Agronomia, Universidade do Estado de Mato Grosso, Tangará da Serra, Mato Grosso, Brazil

Rogério Teixeira de Faria
Departamento de Engenharia Rural, Faculdade de Ciências Agrárias e Veterinárias, Universidade Estadual Paulista, Jaboticabal, São Paulo, Brazil

Paulo Sérgio Lourenço de Freitas and Bruno Patias Lena
Departamento de Agronomia, Universidade Estadual de Maringá, Av. Colombo, 5790, 87020-900, Maringá, Paraná, Brazil

Andre Luiz Johann
Departamento de Engenharia de Biossistemas, Escola Superior de Agricultura "Luiz de Queiroz", Universidade Estadual Paulista, Piracicaba, São Paulo, Brazil

Evandro Chaves de Oliveira
Instituto Federal do Espírito Santo, BR-259, Km 70, Cx. Postal 256, 29709-910, Campus Itapina, Colatina, Espírito Santo, Brazil

José Maria Nogueira da Costa and Flávio Barbosa Justino
Departamento de Engenharia Agrícola, Centro de Ciências Agrárias, Universidade Federal de Viçosa, Viçosa, Minas Gerais, Brazil

Trazilbo José de Paula Júnior
Empresa de Pesquisa Agropecuária de Minas Gerais, Zona da Mata, Viçosa, Minas Gerais, Brazil

Williams Pinto Marques Ferreira
Empresa Brasileira de Pesquisa Agropecuária, Sete Lagoas, Minas Gerais, Brazil

Leonardo de Oliveira Neves
Instituto Federal Catarinense, Rio do Sul, Santa Catarina, Brazil

Daniel Emanuel Cabral de Oliveira
Instituto Federal de Educação, Ciência e Tecnologia Goiano, Campus Iporá, Avenida Oeste, 350, Parque União, 76200-000, Iporá, Goiás, Brazil

Osvaldo Resende and Lílian Moreira Costa
Instituto Federal de Educação, Ciência e Tecnologia Goiano, Rio Verde, Goiás, Brazil

Hellismar Wakson da Silva
Universidade Federal de Lavras, Lavras, Minas Gerais, Brazil

Welliam Chaves Monteiro da Silva and Aristides Ribeiro
Departamento de Engenharia Agrícola, Universidade Federal de Viçosa, Av. P. H. Rolfs, s/n, 36570-000, Campus Universitário, Viçosa, Minas Gerais, Brazil

Júlio César Lima Neves and Nairam Félix de Barros
Departamento de Solos, Universidade Federal de Viçosa, Viçosa, Minas Gerais, Brazil

Fernando Palha Leite
Celulose Nipo Brasileira S.A., Ipatinga, Minas Gerais, Brazil

Eloy Lemos de Mello
Universidade Estadual do Oeste do Paraná, Rua Universitária 2069, 85819-110, Jardim Universitário, Cascavel, Paraná, Brazil

Fernando Falco Pruski, Lucas Mucida Costa, Gleiph Ghiotto Menezes and Nívia Carla Rodrigues
Universidade Federal de Viçosa, Viçosa, Minas Gerais, Brazil

Ildegardis Bertol
Universidade do Estado de Santa Catarina, Lages, Santa Catarina, Brazil

Rivanildo Dallacort, Paulo Sergio Lourenço de Freitas, Rogério Teixeira Faria, Antonio Carlos Andrade Goncalves, Roberto Rezende and Rachel Muylaert Locks Guimarães
Departamento de Agronomia, Centro de Ciências Agrárias, Universidade Estadual de Maringá, Av. Colombo, 5790, 87020-900, Maringá, Paraná, Brazil

Mariana de Fátima Sato, Dayana Carla Rigoni, Alessandro Nogueira and Gilvan Wosiacki
Departamento de Engenharia de Alimentos, Setor de Ciências Agrárias e de Tecnologia, Universidade Estadual de Ponta Grossa, Av. General Carlos Cavalcanti, 4748, 84030-900, Ponta Grossa, Paraná, Brazil

Maria Helene Giovanetti Canteri
Coordenação de Alimentos, Universidade Tecnológica Federal do Paraná, Ponta Grossa, Paraná, Brazil

Carmen Lúcia de Oliveira Petkowicz
Centro Politécnico, Universidade Federal do Paraná, Curitiba, Paraná, Brazil

Adilson Pacheco de Souza
Instituto de Ciências Agrárias e Ambientais, Universidade Federal de Mato Grosso, Campus de Sinop, Av. Alexandre Ferronatto 1200, Distrito Industrial, 78557-267, Sinop, Mato Grosso, Brazil

Clóvis Manoel Carvalho Ramos
Departamento de Engenharia Rural, Faculdade de Ciências Agronômicas, Universidade Estadual Paulista "Júlio de Mesquita Filho", Botucatu, São Paulo, Brazil

Adriano Dawison de Lima and Helenice de Oliveira Florentino
Departamento de Bioestatística, Instituto de Biociências, Universidade Estadual Paulista "Júlio de Mesquita Filho", Botucatu, São Paulo, Brazil

João Francisco Escobedo
Departamento de Recursos Naturais e Ambientais, Faculdade de Ciências Agronômicas, Universidade Estadual Paulista "Júlio de Mesquita Filho", Botucatu, São Paulo, Brazil

Gabriel Henrique Horta Oliveira
Instituto Federal de Brasília, DF-480, Lote 1, Setor de Múltiplas Atividades, 72405-980, Brasília, Distrito Federal, Brazil

Paulo Cesar Corrêa
Departamento de Engenharia Agrícola, Centro de Ciências Agrárias, Universidade Federal de Viçosa, Viçosa, Minas Gerais, Brazil

Fernando Mendes Botelho
Universidade Federal de Mato Grosso, Sinop, Mato Grosso, Brazil

André Luis Duarte Goneli
Faculdade de Ciências Agrárias, Universidade Federal da Grande Dourados, Dourados, Mato Grosso do Sul, Brazil

Paulo Cesar Afonso Júnior
Empresa Brasileira de Pesquisa Agropecuária, Centro Nacional de Pesquisa do Café, Brasília, Distrito Federal, Brazil

Sílvia Carvalho Campos
Empresa Brasileira de Pesquisa Agropecuária, Embrapa Agrossilvipastoril, Sinop, Mato Grosso, Brazil

Thomas Newton Martin
Departamento de Fitotecnia, Centro de Ciências Rurais, Universidade Federal de Santa Maria, Av. Roraima, 1000, 97105-900, Santa Maria, Rio Grande do Sul, Brazil

Durval Dourado Neto, Antonio Roberto Pereira and Pedro Jacob Christoffoleti
Departamento de Produção Vegetal, Escola Superior de Agricultura "Luiz de Queiroz", Universidade de São Paulo, Piracicaba, São Paulo, Brazil

Pedro Abel Vieira Junior
Serviço de Negócios RPAa Transferência de Tecnologia, Empresa Brasileira de Pesquisa Agropecuária, Sete Lagoas, Minas Gerais, Brazil

Paulo Augusto Manfron
Centro de Educação Norte, Universidade Federal de Santa Maria, Frederico Westphalen, Rio Grande do Sul, Brazil

Hakan Kibar
Department of Biosystems Engineering, Faculty of Agriculture, Igdir University, Igdir, Turquia

Turgut Öztürk and Kadir Ersin Temizel
Department of Agricultural Structures and Irrigation, Faculty of Agriculture, Ondokuz Mayis University, Samsun, Turquia

Luiz Fernando Pires
Departamento de Física, Setor de Ciências Exatas e Naturais, Universidade Estadual de Ponta Grossa, Av. Carlos Cavalcanti, 4748, 84030-900, Ponta Grossa, Paraná, Brazil

Marcelo Marques Costa, Daniel Marçal de Queiroz and Francisco de Assis de Carvalho Pinto
Departamento de Engenharia Agrícola, Universidade Federal de Viçosa, Av. P. H. Rolfs, s/n, 36570-000, Viçosa, Minas Gerais, Brazil

Elton Fialho dos Reis
Departamento de Engenharia Agrícola, Universidade Estadual de Goiás, Anápolis, Goiás, Brazil

Nerilson Terra Santos
Departamento de Estatística, Universidade Federal de Viçosa, Viçosa, Minas Gerais, Brazil

Index

www.ingramcontent.com/pod-product-compliance
Lightning Source LLC
LaVergne TN
LVHW080651210926
848485LV00003B/363
* 9 7 8 1 6 4 7 4 0 0 5 1 4 *